3rd International Conference on Sustainable Cities (ICSC 2018)

AA006670

IOP Conference Series: Earth and Environmental Science
Volume 177

Moscow, Russia
18 May 2018

ISBN: 978-1-5108-6862-5
ISSN: 1755-1307

Printed from e-media with permission by:

Curran Associates, Inc.
57 Morehouse Lane
Red Hook, NY 12571

Some format issues inherent in the e-media version may also appear in this print version.

Copyright© (2018) by the Institute of Physics
All rights reserved. The material featured in this book is subject to IOP copyright protection, unless otherwise indicated.

Printed by Curran Associates, Inc. (2018)

For permission requests, please contact the Institute of Physics at the address below.

Institute of Physics
Dirac House, Temple Back
Bristol BS1 6BE UK

Phone: 44 1 17 929 7481
Fax: 44 1 17 920 0979

techtracking@iop.org

Additional copies of this publication are available from:

Curran Associates, Inc.
57 Morehouse Lane
Red Hook, NY 12571 USA
Phone: 845-758-0400
Fax: 845-758-2633
Email: curran@proceedings.com
Web: www.proceedings.com

Table of contents

Volume 177

International Conference on Sustainable Cities

18 May 2018, Moscow, Russian Federation

Accepted papers received: 19 July 2018
Published online: 10 August 2018

Preface

Papers

Strategic guidelines for sustainable development of the urban environment to improve the quality of life and restore nature

Poster presentation

IOP Conf. Series: Earth and Environmental Science **177** (2018) 011001 doi:10.1088/1755-1315/177/1/011001

Preface

The 2018 3rd International Conference on Sustainable (ICSC 2018) cities was successfully held on May 18, 2018 in Moscow, Russia. The conference is devoted to new technologies that enable sustainable cities development and promote a stable, comfortable, modern space for life.

There were 3 major sections of the scientific track of the conference:

- Strategic frontiers in sustainable city development

- Transportation for sustainable cities

- Green technologies for sustainable business

- New energy solutions for sustainable development of modern cities

This volume of IOP Conference Series: Earth and Environmental Science (EES) is a compilation of the accepted papers in ICSC 2018 and represents contributions that were presented in the conference. This year we had experienced significant increase in the number of submitted for the review quality papers and only 41 of them where accepted for the oral or poster presentation.

On behalf of the organization committee I would like to thank all of the conference sponsors, partners and volunteers who made conference possible.

Looking forward to meeting you during ICSC 2018.

On behalf of organizing and program committees of the ICSC 2017

Igor Balk, ICSC 2017 co-chair.

Content from this work may be used under the terms of the Creative Commons Attribution 3.0 licence. Any further distribution of this work must maintain attribution to the author(s) and the title of the work, journal citation and DOI.
Published under licence by IOP Publishing Ltd

IOP Conf. Series: Earth and Environmental Science **177** (2018) 011002 doi:10.1088/1755-1315/177/1/011002

Peer review statement

All papers published in this volume of *IOP Conference Series: Earth and Environmental Science* have been peer reviewed through processes administered by the proceedings Editors. Reviews were conducted by expert referees to the professional and scientific standards expected of a proceedings journal published by IOP Publishing.

Content from this work may be used under the terms of the Creative Commons Attribution 3.0 licence. Any further distribution of this work must maintain attribution to the author(s) and the title of the work, journal citation and DOI.
Published under licence by IOP Publishing Ltd

Transforming Yekaterinburg into a Safe, Resilient-Smart and Sustainable City

S A Timashev[1, 2, 4], V N Alekhin[1, 2], L V Poluyan[1, 2], I Fontanals[3] and A Gheorghe[4]

[1] Science & Engineering Center "Reliability and Safety of Large Systems and Machines", Ural Branch, RAS, 54-A Studencheskaya str., Ekaterinburg, 620049, Russia
[2] Ural Federal University named after the first President of Russia B. N. Yeltsin, 19 Mira str., Ekaterinburg, 620002, Russia
[3] OptiCits, 42 carrer Sant Joan de La Salle, Sant Gervasi - la Bonanova, Barcelona, 08022, Spain
[4] Old Dominion University, Norfolk, VA, 23529, USA

E-mail: timashevs@gmail.com

Abstract. The initiative (since 2014) project described in this paper is a product of a joint innovative research and implementation effort of the Civil Engineering and Architecture Institute, Ural Federal University, the Science and Engineering Centre "Reliability and Safety of Large Systems and Machines", Ural Branch Russian Academy of Sciences (both Yekaterinburg), Start-up OptiCits, Barcelona, Spain and the Old Dominion University, Norfolk, VA, USA. The project is based on using the MAICS convergent technology [1] to create a versatile multi-purpose tool for optimizing the science and art of risk based governance of resilience-smart and sustainable city infrastructure and communities operating in usual and extreme conditions. The tool being developed is tailored to the needs of the City of Yekaterinburg–the capital of the Urals Region and allegedly the third most important and vibrant city of Russia. It is also being offered to the Yekaterinburg City Administration as an every-day decision-support work-tool and addendum to the Strategic Program "Yekaterinburg 2030 – a Safe City"[2] during preparation of the city for winning and conducting the World Expo-2025. Authors believe that the findings of this research would also be useful to the Sverdlovsk Oblast cities of every size and type of communities that inhabit them, including, first and foremost, Nizhny Tagil, Kamensk Uralsky, Serov, Pervouralsk, Revda, Verkhnyaya Pyshma, multiple mono-cities et al. The project also incorporates block-chain technology, smart contracts and digital currency as an effective tool for implementing the project.

1. Introduction

This project is an integrative multi-sector research as it addresses both technological and social dimensions of a resilience-smart and sustainable city using the proposed in [1] MAICS convergent technology (Digital Computing Mechanics and Design, Artificial Intelligence (AI), Information Theory (IT), Cognitive and Social Sciences). The description of how the dimensions are integrated together is the core problem of the whole research. It is tethered to the realities of the City of Yekaterinburg and addresses following strategic domains: energy, urban civil infrastructures' resilience and planning, disaster (including draught and flooding due to climate change) mitigation and

Content from this work may be used under the terms of the Creative Commons Attribution 3.0 licence. Any further distribution of this work must maintain attribution to the author(s) and the title of the work, journal citation and DOI.
Published under licence by IOP Publishing Ltd

response, education, environmental footprint, life expectancy, and vital social services (including telecommunications, and transportation). Research is conducted by intersecting and cross-pollinating previously unrelated bodies of research from different branches of sciences and engineering.

The project uses following four novel generalized optimization criteria: 1) structural and system resilience (SRes); 2) regional (local) life expectancy in good health (RLE) [3, 4]; 3) modified life quality index (MLQI); and 4) two types of regional entropy (RE): [of creation REo (organization) and of chaos REc (deterioration)].

It is postulated that these four unifying optimization criteria cover all major areas of the wellbeing of contemporary and future communities of smart and sustainable cities. Moreover, these integral criteria permit convoluting the multidimensional, multidisciplinary and multifaceted heterogeneous problem of regional resilience, strategic preparedness and risk into a (quasi) single dimension problem, thereby greatly simplifying the problem of creating smart and connected communities that inhabit sustainable cities of any size (large, medium and small).

The foundation of this research is the following working hypothesis: The system of systems (SoS) of interdependent critical infrastructures (ICIs) of the built urban environment can be considered as an intermediary between the natural environment and the resource demands of a modern society and everyone who is a member of this society. Urban ICI networks are also the principal source of socio-technological hazards of the city. A point failure anywhere in the ICIs can rapidly propagate through the city with broad impacts on the citizens and the environment [5, 6]. Hence, the resilience of the urban ICIs could be used as one of the generalized independent variables of the sustainable city problem, as it allows quantitatively connecting the tangible with the intangibles. At the same time, it reveals the boundaries of this quantifiable connectedness, beyond which the relationships start to be too fuzzy [5] and uncertain. Hence, it stands for reason that management of complex urban risk may be boiled down (as a first approximation) to management of economic, environmental and physical risk for the whole urban system of the ICIs.

2. City DNA

Major cities, such as Yekaterinburg, are growing into megacities and are currently subjected to a multitude of pressures, out of which the following four are of greatest concern: 1) rapid evolution of digital technology (IoT, IIoT) and globalization; 2) fast socioeconomic changes; 3) obvious climate change; 4) the growing needs of demanding citizens, due to a combination of the above elements. All these pressures create new threats that could be effectively mitigated by the novel concept of urban resilience.

Modern ICIs have following indispensable components: Risk-based diagnostic subsystems; Monitoring and/or control sub-system(s); Risk based integrity maintenance subsystems; Assets safety and security/defense subsystems. The total risk of operating ICIs is carried by its full group of possible scenarios. All these specifics should be consistently accounted for during the design, operation and risk assessment of urban potentially dangerous objects (PDOs) and ICIs [7].

3. City Resilience

The concept of resilience emerged currently as a central theme of industrial and urban development (there are more than 120 definitions of resilience, most of them are qualitative) [8–12]. For instance, the European Commission defines resilience as «*the ability of ... a community, a country or a region to withstand, adapt, and quickly recover from stresses and shocks such as drought, violence, conflict or natural disaster*». As an intrinsic property of any socio-technological system, it can serve as the basis and tool for solving the most urgent issues of modern civilization, including strategic investments by leading development institutions and humanitarian communities.

Despite the importance of critical infrastructures and systems and expected growth of future climatic hazards, relatively few studies have addressed these issues and no methodology for the analysis of such an impact has ever reached a consensus. In the theory of infrastructures resilience is defined as a multi-attribute measure that describes the ability of a system of interdependent CIs to withstand a disaster shock, and its ability to recover within an acceptable envelope of time and cost. As of now, it seems (to our knowledge) that there is no quantitative definition of resilience and strategic preparedness to which a majority would subscribe.

The "Yekaterinburg-Resilient-Smart City" research covers: 1) critical infrastructure as a net; 2) urban interdependent system of critical infrastructures. The regional and infrastructure resilience/preparedness are defined in quantitative, probabilistic terms, as most of the multiple parameters on which resilience / preparedness are dependent, are either ill-defined, fuzzy or random variables (RVs), or random functions/fields (RFs). The main components of urban resilience probability vector include the physical and spatial resilience, as well as environmental (disaster), economic resilience. industrial resilience, organizational (functional), and social (human) resilience. The randomness of the parameters which describe resilience quantitatively could be of different nature (epistemic, aleatory, fuzzy, vague, uncertain, indefinite, etc.). Each specific resiliencies can be parsed into partial resiliencies as related to different aspects of the considered type of resilience. The physical and spatial resilience of a system of critical infrastructures is defined through their reliability and operational risk [13].

The conditionality of the probabilities is due to the time of analysis, structural, financial, social and other restrictions for which the resilience is assessed. This multidisciplinary and multifaceted approach is used is applied to regional (municipal) critical and strategic infrastructures of different nature. It is shown how to use the concept of quantitative resilience in design, operation and mitigation of the consequences of an industrial (leak, rupture of oil/gas, water or sewage pipelines or a plant explosion/fire) or a natural disaster (e.g., incident related to global warming), and how to assess and manage regional risk and quantify the interdependence between different types of infrastructures. The super-urgent problem is formulated on how to connect the physical and spatial (core) resiliencies with the functional, organizational, economic and social resiliencies.

Urban resilience is the driver and, simultaneously, a precious quality of sustainable urban development. Considering a city as a system of systems (SoS), resilience recognizes all of them as dynamic and complex systems that must continuously adapt to various challenges of stochastic, probabilistic, uncertain, or vague character in an integrated and holistic manner. Each part of these systems has an inherent reliance on all the other parts.

In general, factors that influence city resilience include: the range and severity of hazards; the risk to life, limb, health and property; the vulnerability and exposure of human, social, and environmental systems to different types of hazards, and the degree of (strategic) preparedness of the physical, social and the governance systems to any natural, urban or industrial shocks and stressors and their consequences during an incident, accident, malicious act or catastrophe. Our resilience concept adopts a multiple-hazards approach, considering resilience against all types of plausible hazards, and refers not only to reducing risks and damages from disasters [i.e. loss of lives, limbs, health, elements of Nature (flora & fauna) and assets], but also to the quantified ability to quickly recover to the pre-event state after a physical or social disaster or catastrophe at minimal cost.

The essence and components of urban resilience consists of working to: 1) prevent any potential threat; 2) withstand any impact caused; 3) react to the crises derived from the impact; 4) recover the city's functionalities; 5) learn from the experience. All this is achieved when the city (in our case, Yekaterinburg) becomes smart. In its own right, sustainable cities of the 21-st century have to be first resilient, smart and safe in order to then become sustainable.

This concept is expanded to investigate the existing (but yet not discovered or fully understood and quantitatively described) connections [e.g., (semi) quantitative dependencies/correlations] between physical resilience of a system of systems of ICIs and the psychological/societal resilience of people who live inside and extensively use this system of ICIs in the context of different types of communities– from megacities, to big, medium size and small towns, to villages/ settlements and tribal areas. As one of the results of this research the urgent problem is formulated: how to assess the connections between the physical and spatial (core) resiliencies and the functional, organizational and social resiliencies, which largely depend on the type of the society.

The described above multidisciplinary and multifaceted approach is applied to the analysis of critical and strategic infrastructures of different nature of the city in consideration. A methodology is being developed to study how to assess and manage: 1) reliability/safety levels of interdependent infrastructures taking into account the human factor, including errors and mishaps of human beings who operate or use the ICIs [13]; 2) physical regional resilience by managing reliability/safety of the systems of ICIs, and 3) the vector of societal resilience by managing the probabilistic vector of the physical resilience of the ICIs or with purely social tools (which are being defined during the research).

All the above is applied to: urban critical civil infrastructures (energy, transportation (including oil/gas, water and sewage pipelines [14, 15]) and urban planning, disaster mitigation and response, education and learning, health and wellness, including healthcare, social resilience, safety, and social services.

This research is also implementing the life quality index (LQI) and the willingness to pay concept (WTP) for designing, constructing and operating components of urban infrastructure, managing not only the economic loss mitigation but also, simultaneously, the risk to life safety and health. This promotes and expedites reaching the National 80-years-life-expectancy goal by 2030.

Results of the conducted research are useful to the decision makers (DMs) at the regional, municipal and plant level, considering short-term and long-term priorities associated with sustainable growth of the interdependent critical infrastructures (ICIs) under their jurisdiction using the above criteria. The DMs will also be able to monitor how their decisions influence the quality of life and the level of happiness of their constituents, as related to the decisions they make in disaster and ordinary times, including optimal distribution of their budgets. The envisioned methodology, as developed, may also serve as a useful tool for managing risk of potentially dangerous objects (PDOs), interdependent critical infrastructures (ICI) of different nature, and their systems, according to the above criteria.

4. Urban resilience system architecture

The architecture of the urban resilience system (URS) mimics the long time existing on the market different monitoring and maintenance optimization systems designed to enhance performance of critical industrial infrastructures. The difference is in that the urban infrastructure, in its entirety, is a very specific complex system of interdependent systems (SoIS), being widely spread over the whole territory of a municipality, is 1) intensely used by the city community and 2) elements of its transportation infrastructures (cars, trams, buses, metro carriages) continuously move. The URS is designed to provide, in the first place, raw and processed data about how this SoIS functions and degrades in time.

To implement the resilience methodology for creating a smart and sustainable city with a well connected community it is necessary to have in place an urban resilience subsystem (URS). A typical URS is comprised by: sensors/CCTV, distributed over the whole territory of the city; geolocation subsystem; information subsystem; security subsystem, and the situation room SR, which serves as the ultimate place where the decision makers formulate, simulate and calibrate their actions in response to different incidents, emergency situations, and catastrophes. This URS identifies the weakest spots in the urban SoIS and reacts faster and more efficiently during and after an impact or crisis. Administrations of some cities (i.e., Barcelona, Bristol, Lisbon, Norfolk, et al.) that have the smart cities as a goal for their status, established the position of a Chief Resilience Officer (CRO), whose responsibility is dealing with urban crises and systemic stress.

The problem of urban resilience management consists of: 1) Assessing the full possible damage and all its components; 2) Designing means and methods for reduction the potential consequences of an initial failure in the system of ICIs as related to the physical and the social component of the smart community. This problem can be solved only through interdisciplinary approach, and by convoluting the heterogeneous parameters, which define the operation of the CI, into few integral parameters, which should be simple to understand and use.

The main conceptual problem of assessing, monitoring, and managing resilience/risk of smart and connected communities embedded into ICIs is defined by following three factors: the dimension of the problem is huge (could be tens of thousands of interdependent parameters); the problem is multi-disciplinary, and the parameters involved when solving the problem are from different sciences and branches of engineering, and currently are, as a rule, hard, if not impossible to convolute; the ICI risk cannot be adequately described without explicitly accounting for the Human Factor (HF). Hence, before attempting to solve the problem in consideration, it is necessary to introduce some unified measures of safety/risk, which account for the human factor in socially meaningful terms.

The last problem that crowns the full solution of urban community risk management is designing and implementing risk mitigation control means using the LQI and the WTP concepts. There are two approaches to solve this problem. The direct problem is posed as follows: With given means for improving CIs safety S_{giv} choose such a set of measures that maximizes reduction of incident A probability $Q_i = P / A$. The inverse problem is formulated in following terms: With minimal expenditures EX choose such a set of measures, implementation of which lowers the incident probability $Q_i = P / A$ down to an acceptable (preassigned) level $P_{acc}(A)$ [16].

Technological dimensions that are explored in this research: (1) analysis of the epistemic and aleatory digital data flow from different urban critical infrastructures [13], its integration and management; (2) constructing and using new algorithms and frameworks for digging and correct interpreting large volumes of diverse and complex deterministic, probabilistic, fuzzy and uncertain data related to urban infrastructures and community; (3) innovative engineering management and systems engineering approaches for integrating cyber [10], physical, and social parameters and security, safety and wellbeing concerns in a large-scale system-of-systems context with multiple stakeholders.

Social dimensions that are being researched, include: (1) qualitative and quantitative understanding of institutional changes and social behavior and responses to different types of intelligent technological change within different types of communities; (2) short- and long-term responses of communities to natural catastrophes, industrial and malicious disasters; (3) methods for measuring and predicting community challenges and opportunities.

5. Integration of interdependent technological and social dimensions

The described research considers and demonstrates following cases of integrating the interdependent technological and social dimensions in time: (1) IT and AI based innovative concepts and methods of supporting communities in their quest to improve quality of life; (2) novel methodologies for seamless integration of flexible urban infrastructure systems into the fabric of sustainable cities; (3) prediction, analysis, and mitigation of physical and/or institutional challenges to sustainable cities from new technologies, forms of data, and infrastructures; (4) algorithms and representations for enabling design of human-centered infrastructures; (5) new technologies and practices to improve community-level decision making under uncertainty associated with highly complex systems and infrastructures; (6) the dual role of emerging technologies in shaping innovative and harmonious human-technology-environment partnerships leading to career longevity via life-long learning of new skills of the 21-st century in communities of practice.

Following the above working hypothesis, the research aims at creating a meaningful and objective picture of how the physical infrastructure system supports and at the same time provides (generates) most of the risks to the modern communities that live in such surroundings, which are supposed to be resilience-smart, sustainable and safe cities. There are two competing concepts of such cities [16, 17]. The first concept is used to equip the urban infrastructures and services for optimal managing of the city as the path to create a resilient city. In the second concept the optimal management of a city goes through already resilient network of services and infrastructures equipped with smart technology to create a smart and sustainable city. In the latter case the concept of a resilient-smart and sustainable city is formulated and builds up around optimizing implementation of following five key ideas:

- the win-win exchanging/sharing of goods and services between citizens and communities, using the common heritage or private property;
- the minimal environmental consumption and energy efficiency (minimal environmental footprint of the city), by recomposing the mix of energy consumption and the self-production of renewables;
- the free and fluid communication among social stakeholders (citizens, communities, companies, and institutions) using new technologies;
- city-wide integration of new information and communication technologies, robotics and intelligent systems that maximize delivering the needed information to end users ubiquitously and with transparency;
- the reliable and safe network operation, which is the basis of resilience, to: 1) achieve maximum security of supply of goods and services with the right energy and environmental consumption; 2) make good use of the available infrastructure and 3) provide the necessary social communication that will enable the city to adapt and recover functionalities in case of an impact.

Implementation of these ideas may include changes in the design and management of: 1) infrastructures, with emphasis on the redundancies and interconnections; 2) interdependent services, focusing on the ways they could support each other in case of an incident; 3) behavior of citizens in ordinary and, especially, in critical situations (the fundamental strategic element for improving urban resilience and sustainability). To meaningfully solve the problem of smart and sustainable cities, the second approach is used.

6. Governance and decision support software

The hotspot in implementing the smart resilience concept in every-day management of cities is in creating and using specially tailored software for managing the interdependent urban critical infrastructures. In the described research the well established software product, originally developed by OptiCits, HAZUR, is used [18]. OptiCits has accumulated considerable expertise, using HAZUR, in providing practical support to such smart resilient cities as Barcelona and its sattelites La Garrotxa, Sent Cugat de Valles, Terrassa, Tremp, (all Spain), Lisbon (Portugal) and Bristol (UK) [19, 20, 21,22]. In this research the enhanced version of HAZUR, based on joint findings of the described above endeavour, is used. It permits: 1) processing the data from diagnostics and monitoring sensors for studying, analyzing and managing the city resilience, 2) optimizing the compatibility, coordination, operation of different services (agents) based on gathered data, and 3) running simulations to illustrate how the city community and its ICIs will react in the case of an impact [23]. HAZUR supports city leaders and executive officers, public administration staff, city operators and professionals and advisors willing to work with all city stakeholders.

7. Stakeholders of sustainable cities

The stakeholders of urban resilience, safety and sustainability are four major different groups of organizations: 1) multilateral bodies (key to the urban safety and resilience market), as they provide financing for UR improvement projects in emerging and developed cities, management and its benefits to citizens, economic operators, and decision-makers, support of policies for improving safety, resilience and sustainability; 2) research centers, which play a fundamental and vital role in this regard; 3) businesses that catalyze UR development; 4) governments of all caliber cities that are the end users of the UR product. It goes without saying that residents of the communities that live in cities that want to be smart and sustainable are the ultimate beneficiaries of urban sustainability, safety and resilience and, hence, should be the most active component of the same.

For this reason one of the main problems of sustainable cities is the formation of a motivated citizenry interested in making their home safe, smart and sustainable. Analysis gives that for achieving this goal it is necessary to preserve and expand the size of the middle class of each smart, sustainable and connected community. The main existential threat to the middle class are the fast-changing intelligent technologies, that historically soon will make many types of middle class jobs obsolete, among them, just to name a few: car and truck drivers, conveyor belt and construction workers, welders, translators, low-level accountants and bookkeepers, librarians, sales persons. Some forecasters even mention medical diagnosticians, surgeons of average traumas. It is known that computers already can prove some original theorems (provided they are well formulated by humans).

Hence, the main existential problem of every resident of a 21-st century smart and sustainable city/community is to figure out how to be permanently useful in a fast-changing world and thereby make a decent living allowing following her/his aspirations. It is also very important that smart sustainable cities/communities as entities provide a tangible input to peace and happiness to its members.

Community engagement in this research is achieved by creating a virtual community occupying a virtual City of Yekaterinburg, which will have a Chief Resilient Officer (CRO) and a Situation Room that also serves as a living lab. The above stakeholders of the smart community conduct pilot projects which include reacting to some representative disaster scenarios. During this process they identify key issues, plan and implement projects, make decisions, and evaluate outcomes, support data collection and/or interpretation within the community context. They may also be instrumental in providing data, facilities, resources, and expertise to the project. The scenarios are executed via 3D computer simulation [23, 24].

8. Conclusion

1. Initial results of an interdisciplinary project on developing a methodology of urban risk management via risk governance of ICIs systems are presented.

2. Results of the research may be useful to the municipal level decision makers (DMs), who make decisions related to optimal distribution of their budgets, taking into account sustainable growth of entities under their jurisdiction. They will also be able to monitor how their decisions influence the quality of life / level of happiness of their constituents as related to the decisions they make in the disaster and ordinary times.

3. In order to implement the resilience methodology to create a smart sustainable city it is necessary to build up for it an urban resilience subsystem URS, its architecture outlined in this paper, and create in its frame work a Resilience Office that is the core of dealing with urban crises and systemic stress. This URS would identify the weakest spots in the urban System of Systems and react faster and more efficiently during and after an impact or crisis.

4. Results of this research provide additional impacts in: (1) novel methods of optimization of the type and number of means for mitigating system risk; (2) more effective engineering engagement of societal problems from a new and novel perspective based on inclusion of previously disparate streams of knowledge; (3) path-finding methods for predicting, analyzing and mitigating physical challenges to sustainable cities and its communities; (4) considerable improvements of decision support methods and decision making under uncertainties; (5) advances in modern ways of learning [25].

References

[1] Timashev S A 2016 Infranetics-Off-Shoot of Infranomics – a New Convergent Science of the 21-st Century *Nauka Urala* **5-6**

[2] 2018 *Strategic Program "Yekaterinburg 2030 – a Safe City"* (Yekaterinburg: City Administration)

[3] Tatarkin A I, Timashev S A, Kozlova O A et al 2015 Assessing the Impact of Factors on the Average Life Expectancy of the Population in the Municipality (on the Example of Yekaterinburg) *International Journal of Applied and Fundamental Research* **9-2** 128–32

[4] Timashev S A 2014 Average Life Expectancy as a Criterion for Regional Risk Management *Journal of Risk Analysis and Crisis Response* (Paris: Atlantis Press) **4(1)** 10–19

[5] Oboskalov V, Timashev S, Bushinskaya A and Gusev S 2016 Stochastic Network Reduction Technique for Calculations of Electrical Power System Structural Reliability *2nd Int.l Symp. on Stochastic Models in Reliability Engineering, Life Science, and Operations Management, SMRLO 2016* 144–49

[6] Robert B, Morabito L and Quenneville O 2007 The preventive approach to risks related to interdependent infrastructures. *Int. J. of Emergency Management* **4(2)** 166–82
[7] *ISO 31000:2009*. Risk management – Principles and guidelines
[8] Gheorghe A V, Vamanu D V, Katina P F and Pulfer R 2018 *Critical Infrastructures, Key Resources, and Key Assets* (Switzerland: Springer International Publishing)
[9] Holling CS 1973 Resilience and stability of ecological systems *Annual Review of Ecology and Systematics* **4(1)** 1–23
[10] Hollnagel E, Woods D and Leveson N. (Eds.) 2009 *Resilience engineering: Concepts and precepts* (New York: CRC Press. International Organization for Standardization)
[11] Jackson S and Ferris T L J 2013 Resilience principles for engineered systems *Systems Engineering* **16(2)** 152–64
[12] Timashev S A and Krimgold F 2014 Resilience and strategic preparedness of interdependent critical infrastructures. *Proc. of the 11th All Russia Science and Technology Conference Safety of Critical Infrastructures and Territories Yekaterinburg/Abzakovo, UB, RAS P 11 ALL RUSS SCI TE* 43–6
[13] Timashev S A and Bushinskaya A V 2017 Assessment of Probability of Failure of Building Structures in Uncertainty Conditions *Proc. of the Inter. Conf. on Construction, Architecture and Technosphere Safety (ICCATS 2017)* **262**
[14] Timashev S A and Bushinskaya A V 2016 Diagnostics and Reliability of Pipeline Systems *Springer Verlag*
[15] Timashev S A 2015 Predictive maintenance – key component of providing integrity and safety of pipelines *Proc.of 8th Annual Global Pipe Tech Summit* (Budapest, Hungary)
[16] Timashev S A 2017 Regional Resilience as a Functional of Infrastructures Operational Risk *Proc. of the 26th Annual Conference/Meeting of the Society for Risk Analysis – Europe (SRA-E)* (Lisbon, Portugal)
[17] Timashev S A 2017 Resilient Urban Infrastructures – Basics of Smart Sustainable Cities *IOP Conference Series: Materials Science and Engineering (ICCATS 2017)* **262**
[18] 2017 *HAZUR* Software (Barcelona)
[19] Evans B, Chen A S, Prior A, Djordjevic S, Savic D A, Butler D, Goodey P, Stevens J R and Colclough G 2017 Mapping Urban infrastructure interdependencies and fuzzy risks *7-th Int. Conf. on Building Resilience ICBR* (Bangkok, Thailand) 816–23
[20] Velasco M, Russo B, Maigrat P, Fontanals I and Vendreli E 2018 Holistic resilience assessment to cope with climate input in urban areas-implementation in Barcelona *SmartMobility World Congress*
[21] Fontanals L, Tricas J, Canalias F and Fontanals I 2014 Territorial resilience, service management vector. Case Study of the Garrotxa Business Studies **1 144**
[22] Fontanals L, Tricas J, Sempere J and Fontanals I 2012 Improving the resilience of cities: Industrial knowledge applied to the Management of the City *CONAMA 2012* (Spain)
[23] Gheorghe A and Timashev S A 2015 Designing resiliency of city infrastructures comprised of present and next generation systems using gaming technology *Int. Conf. «Economic & Technical Aspects of Civil Engineering Critical Infrastructures» SAFETY2015* (Yekaterinburg) 5–6
[24] Guryev E S, Poluyan L V and Timashev S A 2014 Construction of Dynamic Risk Maps for Large Metropolitan Areas *J. of Risk Analysis and Crisis Response* (Paris: Atlantis Press) **4-2** 72–6
[25] Timashev S A, Alekhin V N, Poluyan L V and Guryev E S 2015. Innovative Master's program "Safety of civil engineering critical infrastructures and territories" *WEEF2015 «World Engineering Education Forum»* (Italy)

IOP Conf. Series: Earth and Environmental Science **177** (2018) 012002 doi:10.1088/1755-1315/177/1/012002

The influence of the industrial complex on the atmospheric air in the Sverdlovsk region

A V Korobeinikova, M N Ivantsova, I S Selezneva and M A Bezmaternykh

Ural Federal University, 28 Mira str., Yekaterinburg, 620002, Russia

E-mail: mivantsova@mail.ru

Abstract. The work is dedicated to studying the level dynamics of polluting substances emissions into the atmosphere of Yekaterinburg city and the Sverdlovsk region in the period of 2012–2016. In the process of our work we found out an insignificant reduction of main indicators values such as SI, HR, and IPA to estimate the atmospheric air quality. Nevertheless, the level of the air pollution in Yekaterinburg city remained high in the stated period. We also suggested some methods to reduce enterprises emissions in the article, which will lead to improving the air quality. In the result of our work it was established that the main pollutants are suspended particles, nitrogen dioxides, phenol, and benzene. Moreover, the connection between the atmospheric air pollution degree and the Sverdlovsk region population's health was determined.
Key words: atmospheric air, environment, pollutants, TLV, industrial complex, population morbidity

1. Introduction

At present, the Earth's biosphere is constantly exposed to increasing anthropogenic impact. Chemical pollution of the environment is the largest impact and is due to the release of non-natural substances into the atmosphere, hydrosphere and lithosphere. Thus, gaseous and aerosol pollutants of industrial origin enter the atmosphere. The content of carbon dioxide continues to grow, which contributes to the increase in the average annual temperature on our planet. The consequences of human impact on the environment are the focus of environmentalists around the world, as they are the cause of such global problems as the "greenhouse effect", the destruction of the ozone layer, acid rain [1–8].

Industrial regions are the main sources of air pollution. The greatest contribution is made by North America, East Asia and Europe. They account for more than a half of all pollutants which are emitted into the atmosphere. In large cities, there is an environmental problem associated with an increased threshold limit value (TLV) of harmful substances in the air. In Russia these are Moscow, St. Petersburg, Samara, Chelyabinsk, Yekaterinburg and others.

Pollution of the atmosphere also affects the human body, especially the respiratory system. The natural resistance to diseases decreases due to harmful secondary health effects [9–15].

Thus, the current problem of the global scale is the organization of control over atmospheric pollution on the one hand and the limitation of hazardous emissions by industrial enterprises and motor vehicles, on the other.

Content from this work may be used under the terms of the Creative Commons Attribution 3.0 licence. Any further distribution of this work must maintain attribution to the author(s) and the title of the work, journal citation and DOI.
Published under licence by IOP Publishing Ltd

2. Methods

Data acquisition on environmental pollution is carried out during the monitoring of pollution on the basis of the state observation network. In our work we have analyzed the data of state reports on air pollution in the Sverdlovsk region cities in 2012–2016.

Three main indicators are used to assess air quality:

1. SI, dimensionless – the standard index, the largest impurity concentration measured during the considered period of time, which is divided by the corresponding value of the TLV, from measurements at the post for one impurity, or at all posts after one impurity, or at all posts for all admixtures.
2. HR, % – the highest repeatability of exceeding the TLV from measurements at the post for one impurity, or at all posts for one impurity, or at all posts for all impurities.
3. IPA, dimensionless – a complex index of air pollution in five priority substances, which determines the state of atmospheric pollution in the city (it is defined as the sum of single pollution indices of the five priority pollutants reduced to the harmfulness of sulfur dioxide).

Assessment of the level of air pollution is carried out in four categories: low, heightened, high and very high. The level of atmospheric air pollution in the city is determined by the maximum value of one of the three criteria: SI, HR, and IPA.

We have analyzed the changes in the level of air pollution by the main pollutants such as suspended particles, nitrogen dioxides, phenol, and benzene. On the base of the obtained data we have plotted corresponding diagrams, graphs and schemes, which let us present the found correlations visually.

3. Results and discussion

The Sverdlovsk region is one of the largest industrial centers of Russia. Therefore, human economic activity has a very significant impact on the environment including the atmosphere. State supervision in order to monitor human activity and prevent abuse in the sphere of nature management in the Sverdlovsk region is carried out [16–18]. This supervision is represented by:

- The Department of the Federal Supervisory Natural Resources Management Service in the Urals Federal District;
- Ministry of Natural Resources and Environment of the Sverdlovsk Region;
- The Ural Department of the Federal Environmental, Industrial and Nuclear Supervision Service of Russia;
- The prosecutor's office, etc.

Adherence to standards and laws in the field of the environment provides many benefits, including savings, for example, no costs to pay fines for environmental pollution; reduction in the number of payments to employees on sick leave due to reduced morbidity. The other benefit is an increase in the company's income, namely an increase in the number of potential customers of the company's products and a positive image of the enterprise, which cares about its personnel and consumers. The most important result is an improvement of the overall environmental situation on the territory near the enterprise and therefore an increase in the number of people who move to this area for permanent residence.

Criminal, administrative, civil, disciplinary and special sanctions depending on the degree of public danger are provided for offense of the environmental legislation [19–23].

At present time, the main stationary sources of pollution on the territory of the Sverdlovsk region are located in such cities as Yekaterinburg, Nizhny Tagil, Kamensk-Uralskiy, Pervouralsk, etc. The quality of atmospheric air in the region is determined by the emissions of pollutants from stationary mobile sources (motor transport) (figure 1):

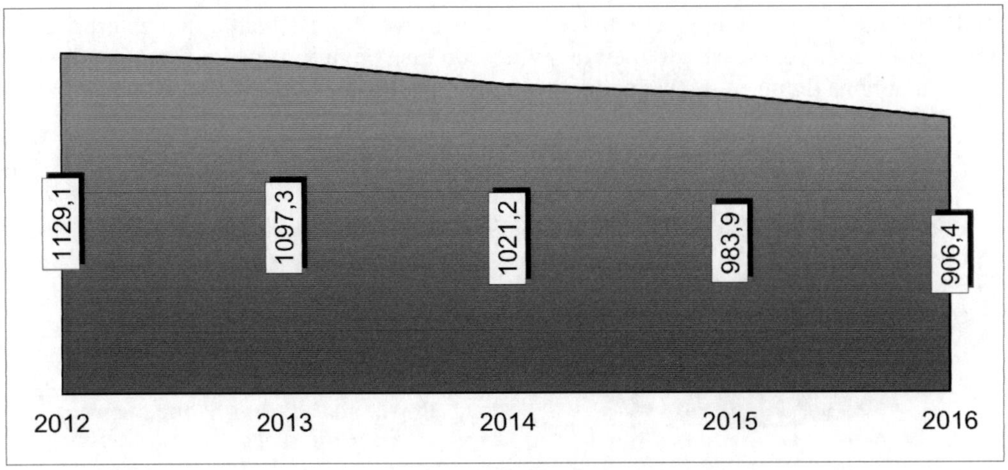

Figure 1. Emissions of pollutants from stationary sources, thousand tons [17]

In 2016 (in compare to 2015) emissions of pollutants into atmospheric air from stationary sources were decreased by 77.5 thousand tons (7.9 %), including solids – 33.9 thousand tons (20.4 %); emissions of liquid and gas-different substances – 43.6 thousand tons (5.3 %), including reduction of emissions of sulfur dioxide – 28.9 thousand tons (10.9 %); nitrogen oxides – 10.4 thousand tons (7.1 %); volatile organic compounds – 1.2 thousand tons (10.8 %); hydrocarbons – 3 thousand tons (2.4 %), and carbon oxide – 0.6 thousand tons (0.2 %).

The decrease in the total amount of air emissions was mainly due to a decrease in electricity generation, fuel consumption, decrease in production at a number of enterprises, reduction in the volume of repair work on the linear parts of the main gas pipelines, and the implementation of environmental measures. Emissions of pollutants into the air in the Sverdlovsk region were decreased by 222.7 thousand tons (19.7 %) in compare to 2012.

3.1. Enterprises and their contribution to air pollution

The diagram (figure 2) presents a list of stationary sources that contributed mainly to air pollution on the territory of the Sverdlovsk region in 2016 (81.4 % of the total emission in the region) [17].

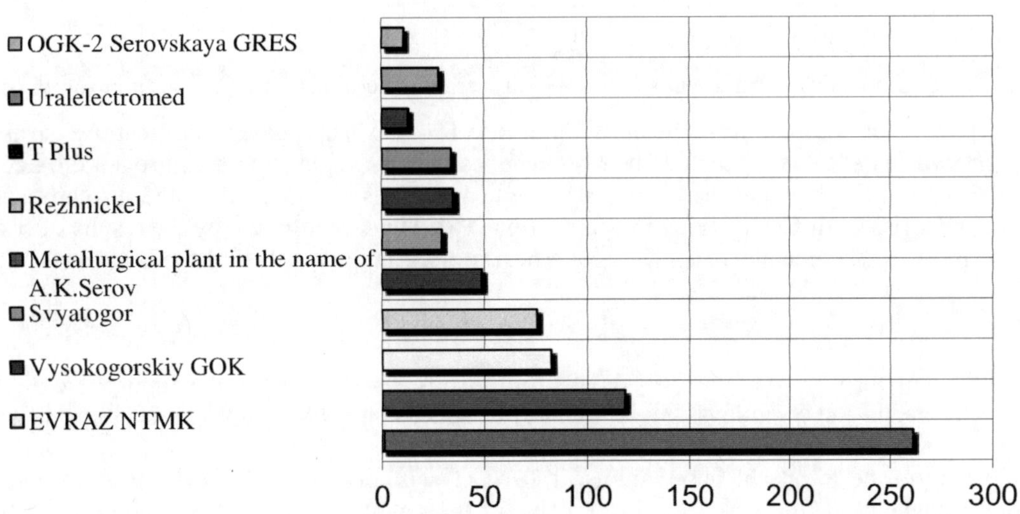

Figure 2. Emissions of the main polluting enterprises of atmospheric air, thousand tons/year

It should be noted that various environmental measures were carried out at the enterprises in order to minimize the negative impact of their activities on the environment in the Sverdlovsk region in 2016. The diagram (figure 3) shows methods of reducing emissions, which have been mostly used:

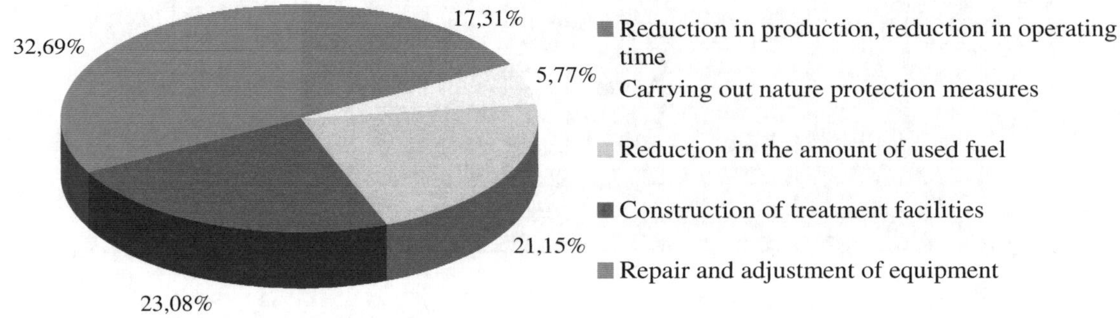

Figure 3. Environmental measures which are aimed at reducing emissions of pollutants

When we analyzed the information from the state report "On the state and protection of the environment of the Sverdlovsk region in 2015" it was established that 8,308.6 thousand tons of pollutants were formed from all sources of pollution in 2016. From this amount, 7,535.5 thousand tons were received from dust and gas treatment facilities. The average degree of pollutants catching in gas cleaning facilities was amounted to 89.1 %, including solids – 97.9 %, gaseous and liquid substances – 60.2 % [17]. Unfortunately, not all pollutants which were formed during the production process were supplied to dust and gas treatment facilities (figure 4).

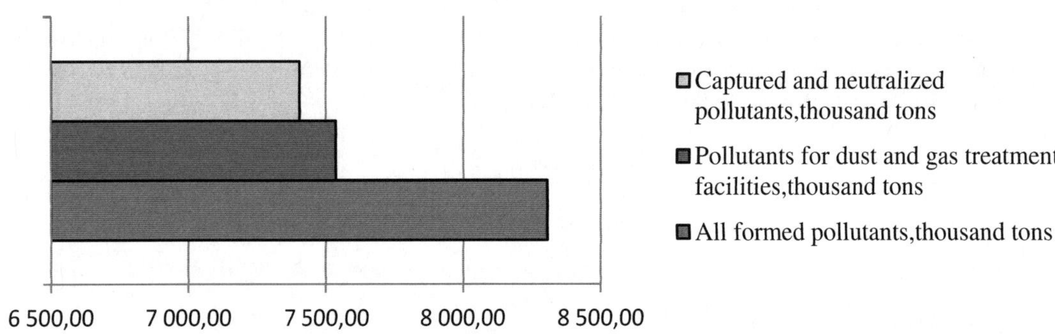

Figure 4. The amount of harmful substances which were cleaned in comparison to their total quantity

Therefore, even despite a sufficiently high level of exhaust gases purification, suspended substances with high toxicity, such as benzopyrene, as well as sulfur dioxide, nitrogen dioxides and formaldehyde, enter the atmosphere. Increased content of these substances in the air can cause an increase in the risk of diseases among the population [16]. This is confirmed by the results of research work on assessing the chemical risk for public health in 14 municipalities of the Sverdlovsk region, which has been conducted since 2003 [17].

3.2. Atmospheric pollution in Yekaterinburg

The next step of our work was the analysis of harmful substances content in the atmosphere of a large industrial center in the Sverdlovsk region namely Yekaterinburg and graphic processing of literary data (figures 5–7).

According to the results of observations in 2016, it was found that the level of atmospheric air pollution in the city, taking into account the value of the standard index (SI), should be classified as “high”. The concentrations of suspended particles, nitrogen dioxide, formaldehyde, benzopyrene and ammonia determined the value of the complex index of air pollution (IPA = 6). SI = 9.5

of ethylbenzene; the highest repeatability (HR) = 6 %. The level of air pollution also relates to a "high" category in assessing the level of formaldehyde TLV which was canceled in 2014, IPA = 8.

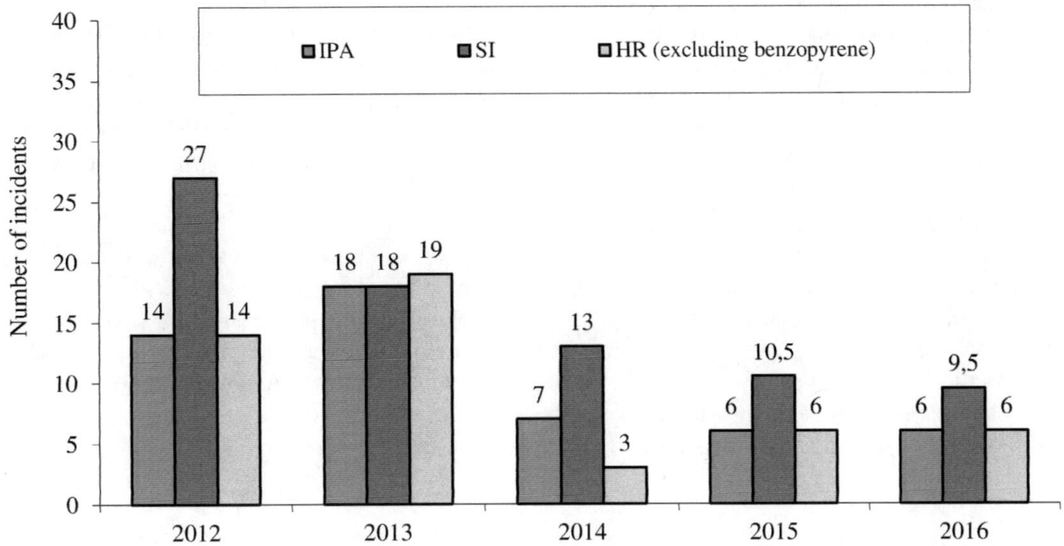

Figure 5. Dynamics of atmospheric pollution parameters in Yekaterinburg in the period 2012–2016

As we can see in figure 5, the level of atmospheric air pollution in the Yekaterinburg in 2012–2013 was "very high" and was determined by the values of complex IPA and SI parameter. In 2014–2016 the pollution level was decreased to "high" (according to the values of IPA and SI). For the period 2011–2013 in Yekaterinburg, the maximum repeatability was the repeatability of formaldehyde TLV and in 2014 was the repeatability of suspended substances. During the period 2012–2015 the SI parameter values were determined only by "very high" concentrations of ethylbenzene [17]. In 2016 the SI parameter value was decreased to "high".

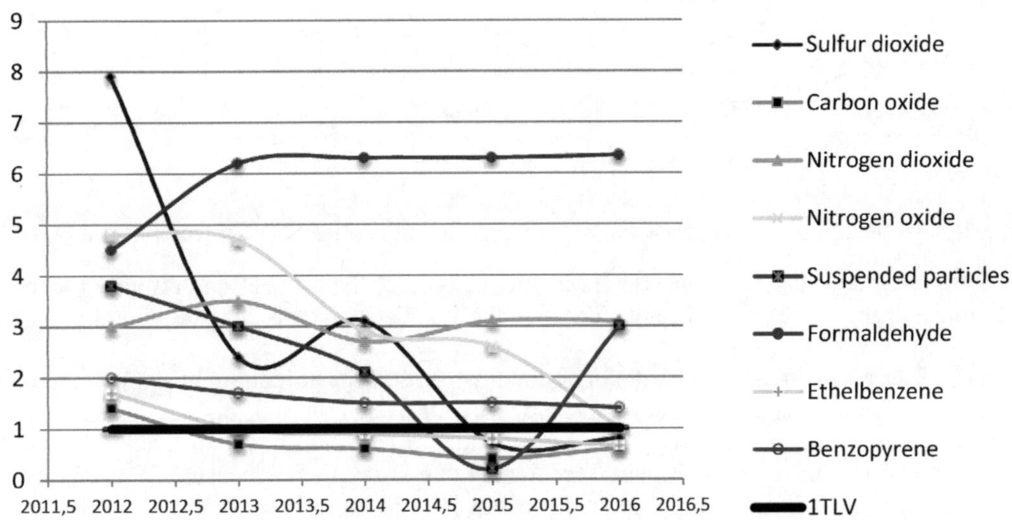

Figure 6. Maximum daily average concentrations of pollutants (in fractions of TLV)

Moreover, over the past five years the tendency to increase air pollution by suspended particles, nitrogen dioxide, phenol, benzene, and copper has been formed in the atmosphere of Yekaterinburg.

Basing on the data of automatic stations for the period 2012–2016, given in the official site [17], we constructed a diagram. As we can see, the level of air pollution in the city is "very high". This level exceeds several times the average daily threshold limit value (TLV) of the main pollutants (figure 6).

In 2016 the excess of nitrogen dioxides TLV was observed in 34.3 % of incidents, which corresponds to a high level of atmospheric air pollution by nitrogen dioxides. The maximum single concentration of nitrogen dioxides exceeded the established standards by 1.5 times. Moreover, the content of nitrogen oxide in the air increased by 3.1 times, and suspended particles – by 3 times.

3.3. The population morbidity

It should be noted that according to the WHO, 80 % of human diseases were caused by an environmentally unfavorable state of the environment and unsatisfactory living conditions in cities with large industrial enterprises. For the last 5 years the birth rate in Russia has fallen by 30 %, the death rate has increased by 15 %. 23 % of children remain healthy by the age of 7, and only 14 % – by the 17th anniversary. Since the 70s, the incidence of cardiovascular and oncological diseases has increased by 50 %. Children under the age of one die in Russia twice more often than in the United States. In some regions, more than a half of the children population suffers from the allergic diseases.

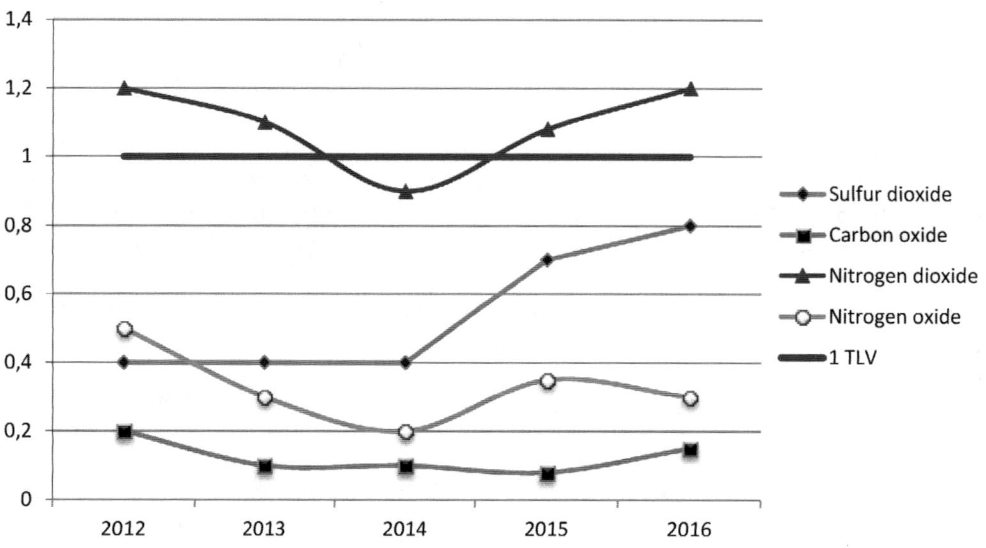

Figure 7. Average annual concentrations of pollutants in atmospheric air (in fractions of TLV) [17]

In this regard, the next step in our work was the analysis of the correlation between the growth in air pollution and the increase in the incidence of the population, based on the diagrams (figures 7 and 8).

We have found that if the level of contaminating substances is continuously increased in the atmosphere, in certain time, the population morbidity by individual types of diseases is increased.

We would like to note that in order to reduce the incidence of the population and to improve the ecological situation in the region, it is necessary to continue to strengthen measures not only to reduce, but also to prevent the formation of pollutant emissions into the atmosphere due to the development of low-waste and non-waste technologies. Moreover, people must pay special attention to their widespread implementation of these technologies in a large number of enterprises.

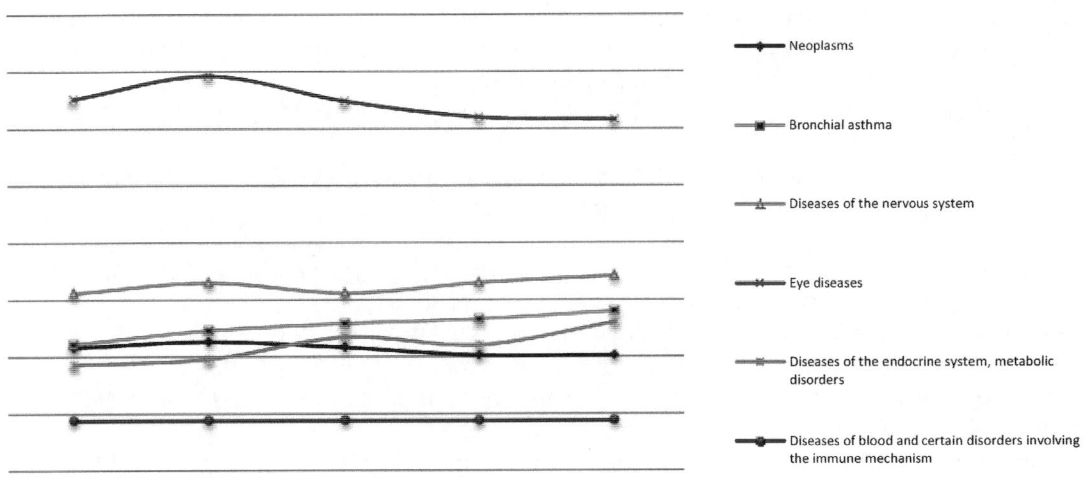

Figure 8. Dynamics of changes in the number of newly registered diseases per 1000 people [16, 18]

In addition, the creation of an environmental certification system for enterprises polluting the environment and equipping vehicles with exhaust gas neutralization systems can create a positive effect, since it is known that the contribution of road transport in air pollution in large cities reaches 90 %.

4. Conclusion

According to the results of observations during 2012–2016 the main stationary sources of pollution in the Sverdlovsk region are located in such cities as Yekaterinburg, Nizhny Tagil, Kamensk-Uralskiy, Pervouralsk etc. In this article we have investigated dynamics of pollutants' emission into the atmosphere by the industrial enterprises of Sverdlovsk region for the definition of population health effect. In 2016 (in comparison to 2015) emissions of pollutants into atmospheric air from stationary sources were decreased by 77.5 thousand tons (7.9 %), including solids – 33.9 thousand tons (20.4 %); emissions of liquid and gas-different substances – by 43.6 thousand tons (5.3 %), including reduction of emissions of sulfur dioxide – 28.9 thousand tons (10.9 %); nitrogen oxides – 10.4 thousand tons (7.1 %); volatile organic compounds – 1.2 thousand tons (10.8 %); hydrocarbons – 3 thousand tons (2.4 %), and carbon oxide – 0.6 thousand tons (0.2 %). However, over the past five years the tendency was formed to increase air pollution by suspended particles, nitrogen dioxides, phenol, benzene, and copper in the atmosphere of Yekaterinburg. Besides, we have found that if the level of contaminating substances is continuously increased in the atmosphere, in certain time, the population morbidity by individual types of diseases is increased.

References

[1] Gore Al 2006 *An Inconvenient Truth: The Planetary Emergency of Global Warming and What We Can Do About It* (Emmaus, Pennsylvania, USA: Rodale, Inc.)
[2] Liua L and Wub G 2017 The effects of carbon dioxide, methane and nitrous oxide emission taxes: An empirical study in China *Journal of Cleaner Production* (Beijing, China) **142 (2)** 1044
[3] 2007 Exposing Canada's chemical valley. An Investigation of Cumulative Air Pollution. Emissions in the Sarnia, Ontario Area *An Ecojustice Report* (Toronto, Ontario)
[4] 2001 Environmental Outlook for the Chemicals Industry (France: OECD)
[5] 2012 Environmental Outlook to 2050. The consequences of inaction (France: OECD)
[6] 2015 Air quality in Europe *EEA report* **5** (Demark: European Environmental Agency)
[7] Official website of EU Commission. Environment

IOP Conf. Series: Earth and Environmental Science **177** (2018) 012002 doi:10.1088/1755-1315/177/1/012002

[8] Zhu T, Melamed M, Parrish D, Gauss M, Klenner L, Lawrence M, Konare A and Liousse C 2012 WMO/IGAC. Impacts of megasities on air pollution ans climate *Report* (Geneva 2, Switzerland: World Meteorological Organization) **205**
[9] 2003 Climate change and human health: risks and responses *ed* A McMichael (Geneva: World Health Organization)
[10] 2003 Health Aspects of Air Pollution with Particulate Matter, Ozone and Nitrogen Dioxide *Report on a WHO Working Group Bonn* (Germany)
[11] Marchwinska-Wyrwal E, Dziubanek G, Hajo I, Rusin M, Oleksiuk K and Kubasiak M 2011 Impact of Air Pollution on Public Health *The Impact of Air Pollution on Health, Economy, Environment and Agricultural Sources* (Croatia: INTECH)
[12] Kilbane-Dawe I and Clement MA L 2014 The Impacts Of Air Pollution On Health. A Summary Of The State Of Current Knowledge *Parliament Hill Research* (London, UK)
[13] Kampa M and Castanas E 2008 Human health effects of air pollution *Environmental Pollution* (Elsevier) **151** 362
[14] Jane R and Palwa K 2015 Air Pollution and Health *Discussion Paper by The Energy and Resources Institute* (New Delhi: TERI)
[15] 2015 Outdoor Air Quality and Health. Care Process Model (Utah, USA: Intermountain Healthcare)
[16] Official website of Ministry of Healthcare of the Russian Federation
[17] Official website of Ministry of Natural Resources and Environment of the Sverdlovsk Region
[18] Official website of Office of the Federal State Statistics Service of the Sverdlovsk Region and the Kurgan region
[19] Byval'tseva S G and Belykh L A Differentiation of competence of environmental oversight and prosecution in the field of environmental protection *Business, Management and Law*
[20] The Criminal Code of the Russian Federation No. 63-FZ of June 13, 1996
[21] The Code of the Russian Federation on Administrative Offenses No. 195-FZ of December 30, 2001
[22] The Civil Code of the Russian Federation No. 51-FZ of November 30, 1994
[23] The Labor Code of the Russian Federation No. 197-FZ of December 30, 2001

Smart Institutions for Smart Cities

H Wiesmeth[1], O Fiala[2], E Stegareva[3], D Häckl[4] and I Weinhold[4]

[1] Chair of Economics, TU Dresden, Dresden, 01062, Germany, and Graduate School of Economics and Management, Ural Federal University, 19 Mira street, Ekaterinburg, 620002, Russia

[2] Chair of Economics, TU Dresden, Dresden, 01062, Germany

[3] Siberian State Medical University, Tomsk, Russia

[4] WIG2 GmbH, Wissenschaftliches Institut für Gesundheitsökonomie und Gesundheitssystemforschung, Nikolaistr. 6-10, Leipzig, 04109, Germany

e-mail: hans.wiesmeth@urfu.ru

Abstract. Smart cities employ creativity of the population for innovations supporting social and economic development. In this context, this paper explores the role of framework conditions on special supply effects of university hospitals, which can invite further research institutions for intense collaboration, thereby stimulating innovations. The case study, comparing a hospital in Russia with one in Germany, is based on the concept of the employment multiplier. The results show that exogenously given, but, more importantly, also modifiable framework conditions lead to large differences regarding the employment multiplier. Thus, it should be the concern of smart cities to make smart use of their institutions, such as university hospitals, by adjusting the conditions, under which they are operating.
Keywords: smart city, smart institution, university hospital, economic impact, supply effect.

1. Introduction

Smart cities are making use of knowledge and creativity to stimulate innovations, which help to better respond to the social and economic needs of the people [1–3]. In addition to information technologies, appropriate framework conditions are required to support the development of smart cities, for example, through optimal use of institutions, which are present in large cities anyway. In this context, this paper refers to university hospitals with their wealth of disciplines possibly attracting and inviting other research institutions and spin-offs for collaboration. Clearly, nature and extent of this collaboration depend on a variety of framework conditions, some of them given exogenously, some of them, however, determined by factors, which could, at least in principle, be modified and adjusted.

Thus, relevant research questions are: how to measure and how to increase the economic impact of these research hospitals by means of suitable framework conditions? For a first approach a university hospital in Germany (University Medicine Leipzig: UML) is compared to a university hospital in Russia (Siberian State Medical University Tomsk: SSMU). Since the early 1990s there have been various reforms to the Russian health care system with significant changes in the framework conditions [4, 5]. Similar to the German health care system [6, 7], the performance of the Russian system has been monitored over the last decades [8, 9], pointing to deficiencies and providing recommendations for further changes to increase efficiency [10, 11]. Thus, a thorough comparative

Content from this work may be used under the terms of the Creative Commons Attribution 3.0 licence. Any further distribution of this work must maintain attribution to the author(s) and the title of the work, journal citation and DOI.
Published under licence by IOP Publishing Ltd

economic appraisal of renowned university hospitals in Germany and Russia should allow some interesting conclusions.

The analysis of economic effects of universities and institutions of higher education has been addressed in the literature. Research on the regional economic impact of universities concentrates thereby mainly on the generation of income and employment through the universities' and their students' expenditure within a region [12–15]. Various studies estimated the economic impact of universities either by using the Keynesian multiplier model or by means of sector-specific input-output analysis [12, 13].

Less attention has so far been given to the role of an academic institution as a generator of scientific output, an academic institution teaches and trains students, and delivers executive education. In addition to the provision of human capital through well-educated graduates, research results, patents and spin-offs stimulate and accelerate innovation activities, which are of importance for further regional economic growth. There is, in particular, a need for appropriate measures, and for cross-national comparisons regarding these effects [15, 16].

Given this situation, this paper focuses on a more detailed analysis of special supply effects, associated with the two university hospitals located in different countries. The analysis allows some conclusions regarding the relevant framework conditions, also those, which can be modified. The ultimate purpose of this investigation is, thus, to provide some ideas on the concept of a smart institution, or a smart infrastructure, of relevance for a smart city [1–3].

2. Conceptual Framework

2.1. The economic impact of university hospitals

Apart from their role as providers of health care and higher education, university hospitals have been predominantly considered as expense factors, in particular from the point of view of the public administrations. The primary focus of these impact models [12–14] is the calculation of direct effects (income, consumption and employment effects, which are directly related to the university), indirect effects (generated by expenses of the institution with respect to construction and materials; third-party funded positions, effects of research institutes, effects caused by students and visitors) and induced effects ("induced" by the consumer spending of all employees). The ratio of direct effects to the sum of all effects can be interpreted as multiplier.

This focus is too narrow and neglects diverse economic and spillover effects, which are, admittedly, occasionally difficult to measure. Additional impacts to the benefit of a region include the education and training of a qualified labor force and, in particular, the provision of knowledge, supporting regional innovation activities and spin-offs. The main findings in the literature on the economic impact of universities always point to a significant supply effect of the research efforts on commercial innovative activities [15–18]. The geographical dimension of these knowledge spillovers seems to be limited, and the framework conditions have to be in favor of the transfer of scientific knowledge into the private sector of the economy. Moreover, the type of university (e.g., a focus on technology or medicine) as well as the characteristics of a region decisively affect the intensity of the interaction of universities within the regional economy [16, 19].

The attractiveness of a university hospital for such collaborations as given by the number of employees in external research institutions and spin-offs will be used as a measure for the effectiveness of the relevant framework conditions stimulating innovations. This is the main contribution of this paper to the existing literature, thereby extending and detailing the results obtained earlier [15–18].

2.2. Characteristics of UML

The impact model focuses on UML in 2009, which comprises the Faculty of Medicine (i.e. the Medical School) of the Leipzig University as well as the Teaching Hospital. The Faculty of Medicine of the University of Leipzig was founded in 1415, shortly after the university was established in 1409. The University of Leipzig is one of the oldest universities in Europe.

Today, the 5,177 staff members of the Faculty of Medicine and the university hospital, as well as the material and construction expenditures constitute the basis for the calculation of the different effects. The 3,129 students of human and dental medicine represent another mainstay of the model.

Some research centers settled in Leipzig because of the existence of the university hospital, and additional spin-offs were established. The 780 staff members of these institutions, comprising among others two Max Planck Institutes and one Fraunhofer Institute, are equally included in the analysis. Further indirect effects result from spin-offs and outsourced institutions that are directly related to the university hospital, such as the Cardiac Center Leipzig, which was established in 1994.

Moreover, visitors of medical congresses and fairs are considered in the calculation of the effects. Medical fairs are held in Leipzig primarily because of its reputation as a medical city with rich tradition shaped by the university hospital and the various non-university research centers.

Further institutions providing offers for students – e.g. libraries or student services – are, however, not included in the analysis even though they are being used by the 3,129 students of medicine. Given that the students of medicine constitute only 8.7 % of the total number of students, this effect has been neglected, as – in contrast to an evaluation of the whole university – the study focuses on the effect of the medical school and teaching hospital.

The region under study is the Free State of Saxony, one of Germany's sixteen federal states with around 4.2 million inhabitants and a size of some 18 thousand square kilometers. All employment and income effects investigated in this study are calculated for the Free State of Saxony.

2.3. Characteristics of SSMU

Each Russian region has a general hospital for adults (500–1000 beds) and a general hospital for children (300–600 beds) that accept referrals of complex cases from district hospitals and polyclinics, as they are intended to provide services for the entire population of the region. All specialties and subspecialties are represented, and the qualifications of staff and the care offered are more sophisticated than at the municipal level. The regional hospital often serves as the teaching unit of the local medical school.

Beyond that there are the medical institutions of higher education in the regions of the Russian Federation. One of them is SSMU in Tomsk, serving the region of Tomsk with a population of a little more than one million inhabitants and a size of almost 317 thousand square kilometers.

Today, SSMU is one of the leading medical universities in Russia. Its history dates back to 1878, when the Emperor Alexander II founded the first university in the Asian part of Russia. Imperial Tomsk University opened in 1888 with only one faculty – the Faculty of Medicine. Interestingly, the university was originally established to function as a research and educational center.

SSMU hosts various research centers focusing on molecular medicine, bioengineering, laser technology and others (cf. http://www.international.ssmu.ru/ru/research_innovation/research_centers/). The research centers promote the application of modern methods to conduct clinical and academic research and aim at creating an integrated system of research and innovation with the help of international scientists.

The basis for calculating the various economic effects is provided by the 2,369 staff members, and the 2,564 students of human and dental medicine.

3. Methodology and Data

The analysis is based on a case study approach with data from UKL and SSMU. The data were collected in close collaboration with the management of the hospitals.

We adopt in this paper – following the multiplier approach – a special impact model, an incidence analysis, which observes the flows of expenditures and their distributing impacts and helps to determine direct income, consumption and employment effects. The model will be likewise applied to study the demand side effects of UML in Germany and SSMU in Russia, with a special view on the supply side issues mentioned above. In the model, the respective institution is considered a consumer of various inputs. These inputs consist of the expenses for construction, material, personnel, etc. within one year. The goal is then the analysis of the regional economic impact of these expenses. These calculated demand effects can be divided into direct, indirect and induced effects. Although there are no consistent definitions in the literature, direct effects refer to primary income, consumption and employment effects originating from the analysed institution. Indirect effects arise from the university's material and construction expenditures. Moreover, the term comprises effects of third-party funds, outsourced establishments, research centers and spin-offs as well as students and visitors of medical fairs. Induced effects, for example in form of increasing employment, result from the staff's consumption expenditures and the related demand of goods and services. Income effects describe direct, indirect and induced incomes resulting from the existence of the analysed institution. Consumption effects describe the staff's consumption expenditures separated into different sectors.

This analysis focuses mainly on employment effects that are composed of direct employees (university hospital staff), indirect employees (university hospital staff paid by third-party funds, staff in supply firms as well as staff in research centers and spin-offs) and induced employees, the latter being employed because of the staff's consumption expenditures. Special attention will be given to the supply effects associated with external collaboration.

4. Results of the Impact Analysis

The analysis is based on the calculation of direct and indirect employment, income and consumption effects and the induced effects, which have to be broken down to the area of study (Free State of Saxony resp. the Region of Tomsk). Indirect employment effects arise from third-party funded projects and from other research institutions, which settled in the area because of the existence of the university medicine.

Again, in order to calculate direct income effects for the employees of the hospitals, internal staff statistics are used. A similar procedure is applied regarding the indirect income effects from employees of the research institutions and from the third-party funded projects. For the estimation of direct and indirect consumption effects of employees, students, and visitors of the fairs recourse is made to special surveys [20].

The induced effects are basically calculated by means of a multiplier analysis: a complex multiplier is used in order to ensure the consideration of separate sectors. This requires the determination of a regional consumption rate since the induced employment and income effects consider only the proportion of consumption expenditures that take effect in the region under study.

4.1. Results for UML

In 2009 there were 3,604 full time employees at both the medical school and teaching hospital of Leipzig. Another 339 full-time employees were paid by third-party funds. During the period of study 714 full-time employees were listed in the associated research centers and spin-offs, as well as 1,002 in the Cardiac Center. Additionally, the analyzed institution created and maintained jobs in Saxon supply firms due to its material and construction expenditures and expenditures for qualifying measures. 156 employees were identified by means of the method described above. Altogether the analyzed institution ensures directly and indirectly 5,817 employees.

The direct income effects, thus the employees' gross incomes (excluding third-party funded employees) amount to 158,097,524 € per year. Employees paid by third-party funds had indirect income effects of 15,493,400 €. Regarding the employees of the Cardiac Center the income effect was determined to 48,956,432 €, for associated research centers and spin-offs it amounted 22,868,051 € and 4,327,497 € for the suppliers' employees.

The consumption expenditures of the employees of the medical school and teaching hospital, the associated research centers, spin-offs and supply firms as well as the students and visitors of medical fairs amounted to 198,292,193 € per year. 25.5 % of this sum was spent on housing, housing maintenance and energy, 13.6 % on foodstuffs, drinks, tobacco products and 17.3 % on transportation.

Due to the re-spending of a part of the consumption expenditures in Saxony, another 1,131 jobs were created in enterprises in different economic sectors. These employees received a gross income of 32,958,895 €, of which 23,663,788 € were spent on consumption. The indirect process can be repeated seven times until no further consumption expenditures and thus employments are induced. However, 89 % of the effects are already realized after the first run.

6,947 direct, indirect and induced full-time employees were financed due to the analyzed institution in Saxony in 2009. This corresponds to an employment multiplier of 1.93 implying that each (full-time) employee of Leipzig's medical school and teaching hospital supports another 0.93 employees in the region. The actual number of employees (in the sense of created jobs) is higher than the calculated number of full-time jobs: the number of direct, indirect and induced employees is 9,385. This corresponds to an employment multiplier of 1.81 (Figure 1).

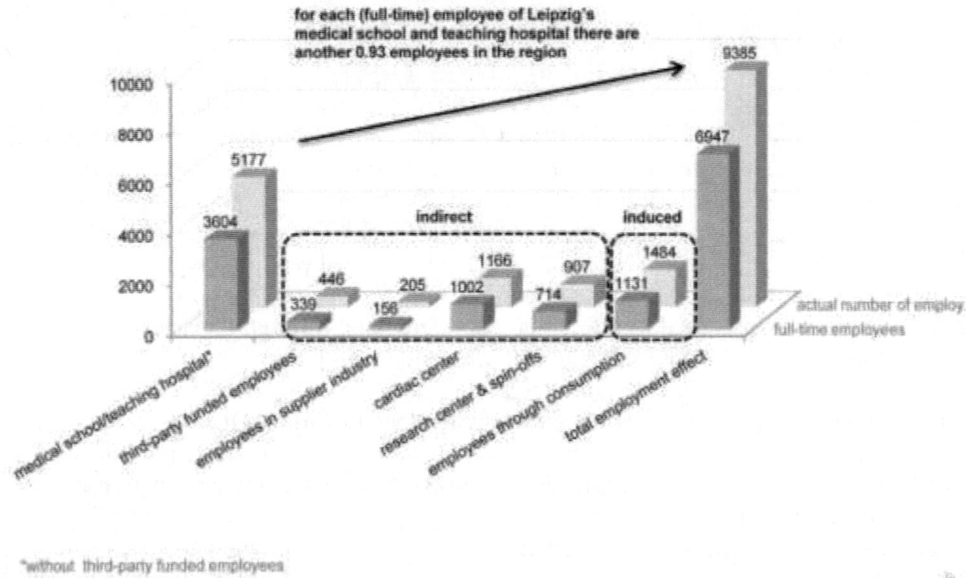

Figure 1. Employment multiplier and further results of the case study for UML

The gross income multiplier is calculated to 1.49. The difference between the employment and the income multiplier can be ascribed to the more detailed contemplation of earnings-related consumption expenditures and special sectors.

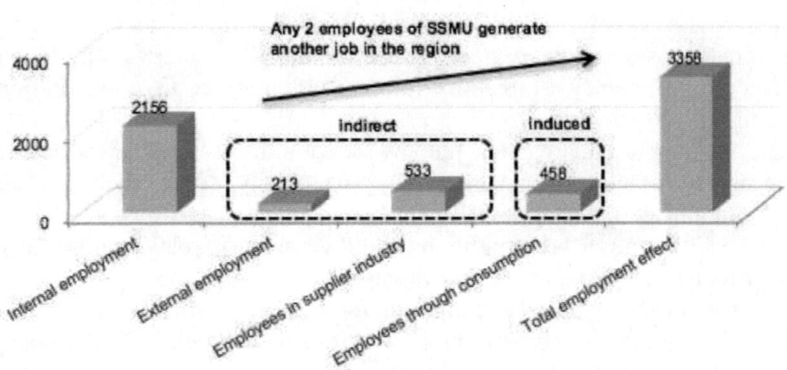

Figure 2. Employment multiplier and further results of the case study for SSMU

4.2. Results for SSMU

In 2014 there were 2,155.8 internal full-time employees and were 2,564 medical students at SSMU. Moreover, 212.5 were listed as external employees and expenses of the Medical University (483.5 million Ruble for material and another 47.3 million for investments) led to 304 employees in supplier industry. In total 516.5 indirect employees are found in the study. The number of induced employees amounts to another 458. Altogether the analyzed institution ensures directly and indirectly 3,358 employees. This amounts to an employment multiplier of 1.56, which is detailed in Figure 2.

The direct income effects, thus the employees' gross incomes (of the 2,155.8 internal employees) amount to 623.0 million Ruble per year. Due to additional gross income of external employees, employees in the supplier industry and student incomes, the analysis yields then an income multiplier of 1.74. Figure 2 shows the various components of the direct and indirect income effects for SSMU.

4.3. Comparison of the results

The analysis reveals employment multipliers of 1.93 for UML and 1.56 for SSMU, and income multipliers of 1.49 for UML and 1.74 for SSMU. The attractiveness of the university hospitals for external research institutions and spin-offs deserves, however, special attention: For UML, the cardiac center, research centers and spin-offs account for an additional 1,716 full-time positions. Adding the corresponding induced effects of 365 positions associated with these employees, one obtains 2,081 positions resulting from the scientific attractiveness of UML. This leads to a scientific employment multiplier, characterizing the effectiveness of the institution in view of the requirements of a smart city of 1.58 for UML. For SSMU, the corresponding numbers show 213 external employees, inducing additional 41 positions with a total of 254 employees generated through the scientific attractiveness of SSMU. Thus, the scientific employment multiplier for SSMU is 1.12.

The differences result obviously from differences in the relevant framework conditions. Some of the relevant framework conditions are, of course, exogenous in nature. One has to mention the density of the population, the size of the region under study, but also the differences regarding gross regional product (GRP) per capita. GRP per capita is a measure of regional income, which has an effect on demand for medical services, but also for innovations [21].

However, the higher scientific employment multiplier for UML likely results also from factors, which are related to organizational issues, maybe cultural issues. There is, in particular, the aspect of openness of the scientific environment. How easy is it, for external companies or spin-offs to cooperate with a large public-funded research institution? How easy is it, to setup a spin-off in the vicinity of such an institution? These are some of the questions, which result from this cross-nation study of university hospitals.

5. Conclusions

The results of this cross-national case study reveal substantial differences regarding the economic impact of the university hospitals under study. In particular, the scientific employment multipliers, pointing to the level of efficiency of the institutions with respect to collaboration with external companies, shows significant variability.

These differences result from the relevant framework conditions. Some of these conditions, such as demographics and density of population, are given exogenously, some, such as a general openness towards this collaboration, could be gradually further developed and modified. However, a closer investigation of the specific framework conditions and their specific relevance for university hospitals in Russia and in Germany is required for further research.

The results obtained in the paper are important for making optimal use of the knowledge and creativity of research institutions for stimulating innovations, which help to better respond to the needs of the people living in the cities: smart institutions for smart cities.

References

[1] Albino V, Berardi U and Dangelico R M 2015 Smart cities: definitions, dimensions, performance, and initiatives *Journal of Urban Technology* **22** 3–21

[2] Komninos N 2011 Intelligent cities: variable geometries of spatial intelligence *Intelligent Buildings International* **3** 172–88

[3] Hollands R G 2008 Will the real smart city please stand up? *City* **12** 303–20

[4] OECD 2012, OECD Reviews of health systems, Russian Federation, OECD Publishing

[5] Popovich L, Potapchik E, Shishkin S, Richardson E, Vacroux A and Mathivet B 2011 Russian Federation: Health system review *Health Systems in Transition* **13**

[6] Buchner F, Göpffarth D and Wasem J 2013 The new risk adjustment formula in Germany: implementation and first experiences *Health Policy* **109** 253–62

[7] Göpffarth D, Henke K-D 2013 The German Central Health Fund – recent developments in health care financing in Germany *Health Policy* **109** 55–64

[8] Chubarova T 2010 The health-care system in Russia: economic problems of theory and practice *Problems of Economic Transition* **52** 24–42

[9] Svetlichnaya S 2008 Monitoring of the health care quality (based on the data of a health insurance company) *Materials of the IV Research to Practice Conference "Medicine and quality"* (11–12 December 2008, Moscow)

[10] Roik V D 2013 Ways to improve medical insurance in Russia *Problems of Economic Transition* **55** 50–60

[11] Sheiman L M and Shishkin S V 2010 Russian health care: new challenges and new objectives *Problems of Economic Transition* **52** 4–49

[12] Armstrong H 1993 The local income and employment impact of Lancaster University *Urban Studies* **30** 1653–68

[13] Robson B, Deas I and Whyte D 2002 Measuring the economic impact of higher education in the North West Region (Centre for Urban Policy Studies, University of Manchester)

[14] Felsenstein D 1996 The university in the metropolitan arena: impacts and public policy implications *Urban Studies* **33** 1565–80

[15] Drucker J and Goldstein H 2007 Assessing the Regional Economic Development Impacts of Universities: A Review of Current Approaches *International Regional Science Review* **30** 20–46

[16] Robin S and Schubert T 2013 Cooperation with public research institutions and success in innovation: evidence from France and Germany *Research Policy* **42** 149–66

[17] Fritsch M and Slavtchev V 2007 Universities and innovation in space *Industry and Innovation* **2** 201–18

[18] Fritsch M and Aamoucke R 2013 Regional public research, higher education, and innovative start-ups: an empirical investigation *Small Business Economics* **41** 865–85

[19] Boucher G, Conway Ch and Meer E van der 2003 Tiers of engagement by universities in their region's development *Regional Studies* **37** 887–97

[20] Statistical Office of Saxony 2008 Einnahmen und Ausgaben privater Haushalte im Freistaat Sachsen *Statistischer Bericht O II 3 – 5j/08*

[21] Saviotti P P and Pyka A 2013 The co-evolution of innovation, demand and growth *Economics of Innovation and New Technology* **22** 461–82

IOP Conf. Series: Earth and Environmental Science **177** (2018) 012004 doi:10.1088/1755-1315/177/1/012004

Budget optimization modelling for sustainable development of the university research: the example of US

A Bogomolova[1, 2], I Balk[2] and N Ivachenko[1]

[1] Lomonosov Moscow State University, GSP-1, Leninskie Gory, Moscow, 119991, Russia
[2] Global Innovation Labs LLC, 258 Harvard str. #352 Brookline MA 02446 USA

e-mail:info@innovationlabs.net

Abstract. Sustainable development is an important condition for the successful and effective development of most organizations and projects. Universities as organizations are also interested in sustainable development of their activities. In this article, one form of university activity is considered, namely R&D activities and the possibility of its sustainable development over the long term. To achieve the objective optimization model is used, which gives the most favourable option for the use of own funds of universities to smooth and increase the growth rate of the financial flow. The model uses mathematical and machine learning methods. The model has some degree of simplification and abstraction of reality. But the application of the model on real US indicators provided interesting results and conclusions.

1. Introduction

From an economic point of view, the further planning horizon of organization, the more it has a chance of stable growth. The ability to maneuver under adverse economic conditions is a sophisticated task that requires a carefully mature strategy, too. And in this case it is necessary to initially have sufficient resources and have the confidence that the company will survive in the current period before building long-term plans. Therefore, we can conclude that the more unstable and unpredictable the nearest economic conditions, the more attention is paid to the nearest planning.

For universities, long-term strategy is especially important, because they are engaged in scientific R&D projects that last from 2 to 6 years. Also, the usual education programs (bachelor, master, doctorate) last for an average of 2 to 5 years. Since the university does not have the ability to crucially change its activities and react to external economic shocks, universities are highly vulnerable to this problem.

The society has always striven for stability and risk reduction. Stability and security are some of the basic needs not only of man, but also of the organization for development under normal conditions."Sustainability" was first used in a United Nations document in 1978 [1].However, this definition has several meanings. Charles Kidd finds at least 6 different sources (or roots, as he calls them) of the interpretation of this term. They are the "no growth"/"slow growth", the carrying capacity, the resources, the ecodevelopment, the critique of technology and the biosphere roots [1]. B J Brown also considered different ways of interpreting the term stability [2]. She also mentions sustainable development and the sustainable economy. Thus, this topic is very relevant among researchers, especially in recent times.

Content from this work may be used under the terms of the Creative Commons Attribution 3.0 licence. Any further distribution of this work must maintain attribution to the author(s) and the title of the work, journal citation and DOI.
Published under licence by IOP Publishing Ltd

Higher education also needs sustainable development. Typically, the authors highlight the complexity of adapting education to new global conditions, reducing quality and increasing demand, inadequate funding and the need to form long-term processes [3]. Sustainability in this article is seen as the absence of shocks and recessions in the flow of funding over the years. Accordingly, sustainable growth is a smoothed growing trend of investment in the RD, which is predictable and does not contradict expectations.

Building an optimization model is a way to more accurately plan for managing limited resources. One of the successful examples of building such a model can be found in the work of EM Modiano [4]. He builds an optimization dynamic model for the management of delectable resources and applies it on the energy industry using historical values.

One of the most important resources for research is financing. Interruptions in the flow of funds may adversely affect the development of the R&D project. Continuity and predictability of the volume of core resources should be provided for the sustainable R&D development. As we already discussed in the previous article "Impact of the macroeconomic factors on university budgeting: the US and Russia" [5], we found a correlation between macroeconomic downturns and the amount of investment in university R&D. So we continue to study this case and the goal is to build an optimization model of university budgeting. First of all, the model should provide a stable flow of financing, which will not strongly fluctuate under the influence of economic shocks.

2. Methodology and data

In this article we use methods of statistical analysis, logical generalization, mathematical programming and machine learning to find the optimal vector of solutions. The basis of the study is a constrained optimization problem [6].

The constrained optimization problem has some important components:

1) Optimization function F(x), which has the maximization or minimization criterion of the search.

2) A set of vector$x_{1...n}$ satisfying a certain set of constraints.

The solution of the problem consists in finding a vector x* such that the function F(x) would has minimum or maximum value, depending on the criterion of the search [6]. For a simple linear function, one could use the simplex method [7]. But if the function turns out to be non-linear, then you'll have to turn to machine learning. There are many automatic ways to find a solution by enumeration of allowable values.

The verification of the model was carried out on statistics collected by the National Science Foundation's National Center for Science and Engineering Statistics [8]. The data set contains University Science & Engineering R&D Funding by Source taken in billions of 2017 US dollars. The total R&D investments are divided into federal, state and local, universities own funds, industry and other. Constant dollar conversions based on OMB's GDP deflators from the FY 2018 budget [9].

Indicators of inflation in the US over the last 20 years have also been used. Rates of inflation are calculated using the current Consumer Price Index published monthly by the Bureau of Labor Statistics (BLS) [10]. We also took S&P 500 Historical Annual Returns [11]. They are calculated as a percentage change of the last trading day of each year to the last trading day of the previous year. These two indicators are needed to calculate the possible redistribution of funds through time periods and opportunities to multiply them. All data are taken from 1997 to 2016 and do not have missing values.

3. Optimization model

Let's take a look at the data that we have in the context of the future model. We have indicators of the activity of funding sources by years. They are federal, state and local, universities own funds, industry and other sources. Since we act on behalf of the university, the universities own funds are the only source of investment that we can influence, and all the rest are permanent. Together they form a panel data. Each year the university allocates some own funds for R&D. The university has two options how to use its money: to invest them in the R&D in the current period or to invest on the exchange to use in the future. As a simplification, we will assume that the decision on the allocation of funds is made at the beginning of each academic year.

The following variables are present in the model:
t – time, number of years in the survey data in increments of one year.
$R_{1...t}$ – total R&D Funding, historical values.
$C_{1...t}$ – constant, includes the historical values of funds that we cannot influence: federal, state and local, industry and others.
$U_{1...t}$ – university own funds, historical values of money that the university allocates for the year from its own funds.

Where $R_{1...t} = C_{1...t} + U_{1...t}$

The main problem consists in smoothing fluctuations of total R&D Funding (R). So first we need to build an ideally smoothed function (R*) to aim for. Thus, each next value of R_{i+1} should be proportional to the previous R_i.
Ri+1* = Ri * (1 + α), where α is the chosen growth rate.

Since we know the values of R_0, the resulting vector will be fully known and within the model will be a constant. To bring the actual values of $R_{1...t}$ closer to the normative $R_{1...t}$ *, we need to reduce the deviations($\delta_{1...t}$). Since the deviations can be either positive or negative, then we will use the square deviation to avoid mutual compensation ($\delta_{1...t}^2$). Thus, the objective function is

$$\min F(U_i) = \sum_{i=0}^{t} \delta^2 = \min \sum_{i=0}^{t} (R_i - R_i^*)^2$$

But the only variable we manage is our own R&D funds U_i. Thus, actual U_i consists of U_i^* that we spend in the current period and U'_i that we invest for use in future periods. So, we can rewrite objective function:

$$\min F(U_i) = \min \sum_{i=0}^{t} (C_i + U_i^* - R_i^*)^2$$

Now consider constraints,

1) $U_i^* \geq 0$, because we cannot take out foreign investments.

2) $U_i^* \leq I_i$ where I_i is investments that constitute the amount of deferred own funds and interest accrued on them. I_i is indicator of the state of the account at the beginning of a period *i* at the time of making a decision on investment in R&D.

$I_{i+1} = I_i*(1+\beta) + U'_i$ where β is return on investments rate.

The solution is a vector $U_{1...t}^*$ satisfying all the constraints and for which the value of the function $F(U_i)$ is minimal. Moreover, we get the estimated value of $R_{1...t}'' = C_{1...t} + U_{1...t}^*$, which was obtained after applying the optimization measures.

4. Model application on real data

Let's look at a specific case for the R&D Funding in the US between 1997 and 2016.

As a return on investments rate we are using the average S&P500 index returns as an indicator of the average market return:

β = 7.31% – S&P500 index returns, represents historical average annual profitability of S&P500 index, average for the period under study (1997–2016).

We take growth rare (α) for the second variable for the optimization model to see the best potential of the model. Use the minimax algorithm to find the optimum, since we must maximize the value of α. We will have to introduce one more restriction on the non-negativity of α. Following the solution algorithm based on the model, we use a method of machine learning to select the optimal vector $U_{1...t}^*$. Now we can compare historical $R_{1...t}$ with model $R_{1...t}^*$ and new estimated $R_{1...t}''$

The sum of the deviations turned out to be rather small (21.50), and on the graph we see that the line of estimated R'' is smoothed and has no negative shocks. Thus by managing own funds universities can make R&D funding flow smooth and more stable over the years.

The results can be seen in the graph below (fig. 1).

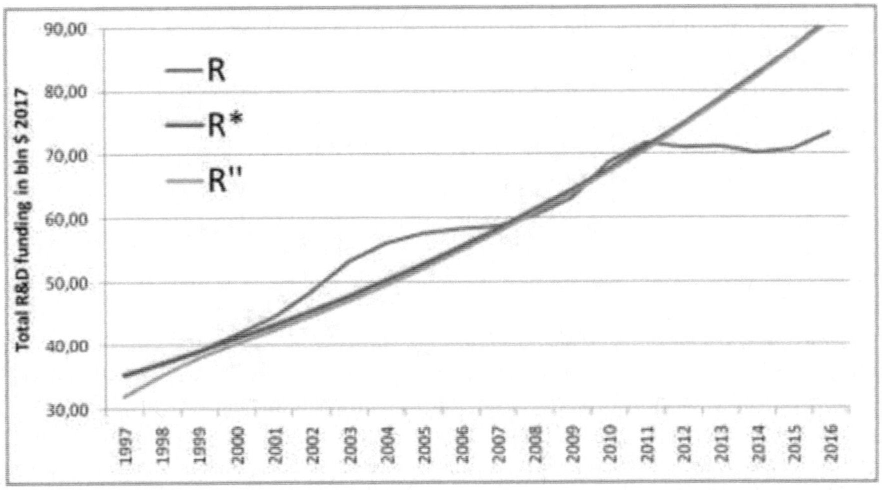

Figure 1. Comparison of the vector of historical total R&D funding (R) with the model (R*) and estimated (R") under the condition of an optimal growth rate

The optimal value of α in this case is 5.1%. Now we see that the result is much better. The lines of the modal (R*) and estimated (R") are almost completely coincides. Due to small savings in a period of abundance, it would be possible to completely compensate for the failures in recent years.

The average growth of historical total R&D findings is 4% on the background of the optimal α = 5.1%. Thus, by managing own funds universities can make R&D funding grow faster on average. For investments of this scale, 1% is a significant amount of money. If we count in absolute numbers, then the sum of all historical total investments in R&D for the last 20 years was 1151 bln dollars, but on the optimization model, they could be 1169 bln dollars and the difference is 18 bln dollars, which appeared due to returns on the investment of deferred funds.

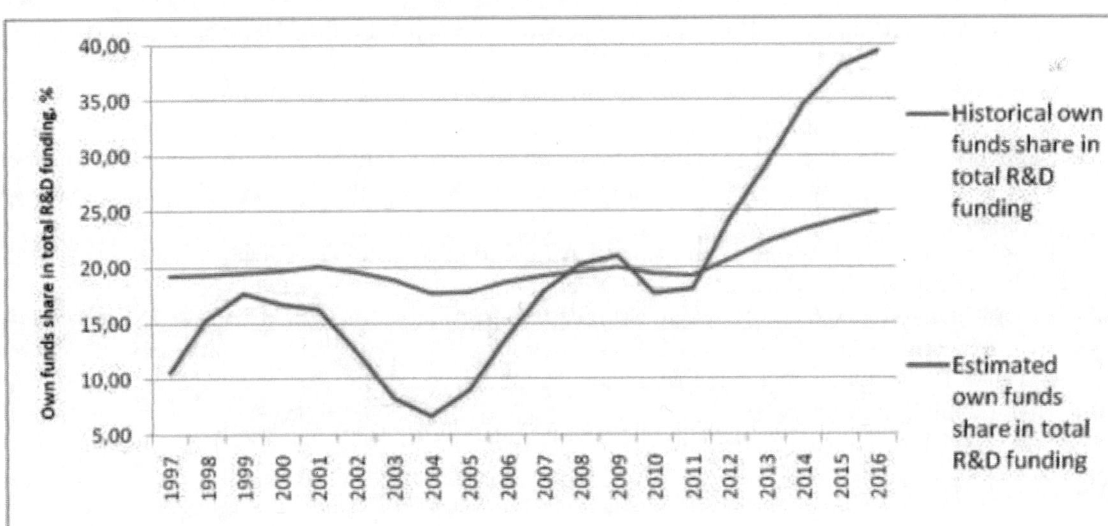

Figure 2. Own funds share in total R&D funding

Another effect that can be exaggerated by comparing the model and historical investment behavior of universities is significantly larger fluctuations in investing own funds in model situation, which, nevertheless, repeat the extremes of historical values. Let's look at own funds share in total R&D funding and compare historical and estimated values (fig. 2). If universities in terms of relative stability saved their money, now the share of own funds in general funds would be about 40% against the actual 25% in 2016.

One more argument in favor of this model. Since under the conditions of the model, universities compensate for insufficient funding from business and the government, they also depend on macroeconomic fluctuations. However, in periods of economic downturns the government and business invest less, but the universities, on the contrary, invest more to compensate this fluctuation. In this case, it is also correct in the opposite direction, the universities save and invest own funds on the stock exchange when the economy is growing, which means that it is likely to receive good interest returns.

5. Conclusion

Using the optimization model and machine learning, it is possible to smooth out the R&D investment flowing universities that make it more stable. Moreover, by investing surpluses in the years of abundance according to the average exchange rate coefficient and using them in future periods, it is possible to increase the average rate of investment growth. At the same time, the allocated money remained the same every year, but they were not invested instantly, but partly postponed for the future. Thus, the stability and efficiency of investments is improved. If we take into account the correlation between macroeconomic fluctuations and the investment behavior of the state and industry, we can predict future funding and optimize own funds of universities using the algorithm offered in this article. This is a challenge for future research.

References

[1] Kidd Ch V 1992 The Evolution of Sustainability *Journal of Agricultural and Environmental Ethics* **5(1)** 1–26
[2] Brown B J, Hanson M E, Liverman D M and Meredith R W 1987 Global Sustainability: Toward Definition *Environmental Management* **11(6)** 713–19
[3] Dimitrova E 2014 The 'sustainable development' concept in urban planning education: lessons learned on a Bulgarian path *J. Clean. Prod.* **62** 120–27
[4] Modiano E and Shapiro J 1980 A Dynamic Optimization Model of Depletable Resources *The Bell Journal of Economics* **11(1)** 212–36
[5] Bogomolova A, Balk I, Ivachenko N and Temkin A 2017 Impact of the macroeconomic factors on university budgeting the US and Russia *Journal of Physics: Conference Series* **913**
[6] Snyman J A 2005 Practical Mathematical Optimization. An Introduction to Basic Optimization Theory and Classical and New Gradient-Based Algorithms, *Springer Science + Business Media, Inc.*
[7] Dantzig G B 1947 Maximization of a linear function of variables subject to linear inequalities *Activity Analysis of Production and Allocation* 339–347
[8] Federal Funds for R&D National Science Foundation's National Center for Science and Engineering Statistics
[9] FY 2018 BUDGET
[10] Historical Inflation Rates: 1914–2018
[11] S&P 500 Historical Annual Returns

IOP Conf. Series: Earth and Environmental Science **177** (2018) 012005 doi:10.1088/1755-1315/177/1/012005

Monitoring of risk of multidimensional stochastic system as tools for a research of sustainable development of regions

A N Tyrsin[1, 2] and A A Surina[3]

[1] Ural Federal University named after the first President of Russia B. N. Yeltsin, Yekaterinburg, 620002, Russia
[2] Science and Engineering Center "Reliability and Resource of Large Systems and Machines", Ural Branch, Russian Academy of Sciences, Yekaterinburg, 620049, Russia
[3] South Ural State University (national research university), Chelyabinsk, 454080, Russia

E-mail: at2001@yandex.ru, dallila87@mail.ru

Abstract. New approach for a research of sustainable development of regions and cities is offered. It is based on the risk model of multidimensional stochastic system. In article the risk model of multidimensional stochastic system with interdependent factors is described. The hypothesis which consists that the risk can be managed by changing probabilistic properties of a component of multidimensional stochastic system is the cornerstone of the offered risk model. At the same time the multidimensional stochastic system is modeled in the form of a random vector which components in generally are mutually correlated. The questions of formation of multidimensional areas of dangerous states and calculation of risk are described. The representation of risk function is shown. For regions of the Ural Federal District on group of risk factors numerical characteristics of a multidimensional Gaussian random variable – a covariance matrix and a vector of mathematical expectations are found. Results of calculation of probability of a dangerous outcome and risk depending on the found numerical characteristics are given.

1. Introduction

The regions and the cities which are regions basis are the complex socio-economic systems consisting of huge number of the interacting elements [1]. This interaction in many respects is characterized by non-deterministic, stochastic character. Many authors point out the necessity of uses of systems approach for studying of such regional systems [2–4]. Systems approach involves the unification of all regional community in the system consisting of the interrelated elements of infrastructures [5] (environmental protection, ecology, economy, education, health, culture, policy etc.) the functioning of which is directed to achievement of sustainable development.

In this situation the multidimensional stochastic system is often modeled in the form of a random vector. Each component of this vector is the one-dimensional random value characterizing functioning of the corresponding element of the system. In [6] vector entroic model operation for the description of regional social and economic systems is used. The vector entroic model which components are the randomness and self-organization entropies allows to investigate system from the perspective of system regularities and to reflect the main trends of its development. But for development of concrete

Content from this work may be used under the terms of the Creative Commons Attribution 3.0 licence. Any further distribution of this work must maintain attribution to the author(s) and the title of the work, journal citation and DOI.
Published under licence by IOP Publishing Ltd

recommendations about improvement of a system condition it is necessary to add system-wide presentation with the quantitative assessment of a contribution to its deterioration in each of its infrastructures.

This problem can be solved with the help of risk analysis [7]. Some authors note that growth rates of damage considerably exceed growth rates of economy [8, 9]. It can be explained with constant increase of risk in the conditions of a scientific and technical revolution and the forced development of a technosphere [10]. Let's formulate a hypothesis: Sustainable development of territorial socio-economic system is inextricably linked with the risk of its functioning, if the risk level is lower, the development of region is steadier. Therefore diagnosis of the sustainable development of system can be carried out on the basis of a monitoring its risk.

Usually modeling a risk comes down to selection of dangerous outcomes, the quantitative assignment of consequences from their occurrence and estimation of the probabilities of these outcomes [11]. For relatively simple objects when it is possible to specify a priori all dangerous outcomes, in the presence of statistical information or expert estimates on chances of their emergence in general this approach yields the results acceptable in practice. However for many complex systems to allocate all these dangerous outcomes is not possible.

The risk model of multidimensional stochastic systems according to which the system is presented in the form of a random vector with mutually correlative components is offered in [12]. The aim of the article is the description of new approach to assessing the sustainability of regional systems on the basis of this model.

2. Model of risk analysis

Let's present a condition of the composite system in the form of some multidimensional stochastic system *S*. Let's mark out in this system risk factors of $X_1, X_2, ..., X_m$. The result is a representation of a system in the form of a random vector $\mathbf{X} = (X_1, X_2, ..., X_m)$ with some probability density of $p_{\mathbf{x}}(\mathbf{x})$.

Instead of the conventional selection of concrete dangerous situations we will set geometrical areas of failures. Its can look arbitrarily depending on a specific objective, and are determined on the basis of the available a priori information. For definiteness, we will describe the offered approach on the example of the common conception of dangerous conditions as large and improbable deviations of a random variable. Then we will consider dangerous situations larger and improbable deviations of selective values x_{ij} of any of the component X_j the best in sense of safety values θ_j, $j = 1, 2, \ldots, m$. If the prior information about the values θ_j is absent, then we consider that they are equal to expectations $\mu_j = M[X_j]$ of random values of X_j, i.e. $\theta_j = \mu_j$, $j = 1, 2, \ldots, m$. Then we will define probability of an unfavorable outcome for each of the component X_j as

$$P(D_j) = P(X_j \in D_j) = P(X_j \notin \overline{D}_j), \quad \overline{D}_j = \{x : d_j^- < x < d_j^+\},$$

where d_j^-, d_j^+ – the set left and right boundaries of admissible values ($d_j^- < d_j^+$), i.e. the area of the favorable outcomes is limited to range $(d_j^-; d_j^+)$.

Let's enter the lower b_j^- and upper b_j^+ threshold levels of permissible variations concerning values θ_j as $b_j^- = \theta_j - d_j^-$, $b_j^+ = d_j^+ - \theta_j$, at the same time the area of the favorable outcomes $\overline{D}_j$ for each component of X_j is described by range $(\theta_j - b_j^-; \theta_j + b_j^+)$.

If only the right boundary d_j^+ of admissible values is set, then we consider $d_j^- = -\infty$ and $\overline{D}_j = \{x : x < d_j^+\} = \{x : x < \theta_j + b_j^+\}$, with only a certain left boundary d_j^- we have $d_j^+ = +\infty$ and

$\overline{D}_j = \{x : x > d_j^-\} = \{x : x > \theta_j - b_j^-\}$. Expression $d_j^- = -\infty$ ($d_j^+ = +\infty$) means that, values of risk factor X_j less (more) θ_j are same safe as well as $X_j(\theta_j)$.

Now it is necessary to describe multidimensional area of dangerous situations of *D*, having considered the mutual influence of the component on the emergence of failures. It is equal $D = \mathbf{R}^m \setminus \overline{D}$ where $\overline{D}$ – area of admissible values of risk factors. Let's describe area $\overline{D}$. It can be done in various ways. The most justified from the geometrical point of view are represented to set it in the form of internal area of *m*-axis ellipsoid

$$\overline{D} = \left\{ \mathbf{x} = (x_1, x_2, \ldots, x_m) : \sum_{j=1}^{m} \frac{(x_j - \theta'_j)^2}{b_j^2} < 1 \right\}$$

with the center at point $\boldsymbol{\theta}' = (\theta'_1, \theta'_2, \ldots, \theta'_m)$, and $\forall j = 1, 2, \ldots, m$

$$\theta'_j = \begin{cases} \theta_j, & d_j^- = -\infty \vee d_j^+ = +\infty, \\ (d_j^- + d_j^+)/2, & d_j^- > -\infty \wedge d_j^+ < +\infty, \end{cases} \quad b_j = \begin{cases} (b_j^- + b_j^+)/2, & d_j^- > -\infty \wedge d_j^+ < +\infty, \\ b_j^-, & d_j^+ = +\infty, \\ b_j^+, & d_j^- = -\infty. \end{cases}$$

Then for a random vector *X* probability of unfavorable outcome will be equal

$$P(D) = P(\mathbf{X} \in D), \quad D = \left\{ \mathbf{x} = (x_1, x_2, \ldots, x_m) : \sum_{j=1}^{m} \frac{(x_j - \theta'_j)^2}{b_j^2} \geq 1 \right\}. \tag{1}$$

Let's notice what in (1) area *D* of unfavorable outcomes represents external area of *m*-axis ellipsoid which has semiaxes on each of coordinates are equal b_j respectively, i.e. on each *j*-th axis this area corresponds to a one-dimensional case of D_j. Obviously, when the outcome does not lie on one of axes, the event ($\mathbf{X} \in D$) can be implemented and in the absence of risk deviations on all component (situations $\mathbf{X} \in D$ and $\forall j\ X_j \notin D_j$ are possible).

Setting the function of consequences from dangerous situations (risk function) in the form of *g(x)*, we will receive model for the quantitative assessment of risk

$$r(\mathbf{X}) = \int\int_{\mathbf{R}^m}\ldots\int g(\mathbf{x}) p_{\mathbf{X}}(\mathbf{x}) d\mathbf{x}. \tag{2}$$

If in (2) to accept $g(\mathbf{x}) = 1\ \forall \mathbf{x} \in D$ и $g(\mathbf{x}) = 0\ \forall \mathbf{x} \notin D$, that $r(\mathbf{X}) = P(\mathbf{X} \in D)$, i.e. the risk is estimated as probability of unfavorable outcome. If at an early stage of system analysis is difficult to describe enough precisely the *g*(**x**) function, then the formula (2) becomes assessment of *P(D)* and is a convenient initial approximation of risk model.

To define a function *g*(**x**) requires quantitative assessment of consequences for the studied system depending on values of risk factors. It demands carrying out a separate research. Let's offer one of the variants of setting risk function. Let's make the following assumptions.

1. We consider that the *g*(**x**) function is nonnegative and the continuous everywhere on $\mathbf{R}^m$ function, and $g(\boldsymbol{\theta}') = 0$.

2. We consider $\forall \mathbf{z} \in \mathbf{R}^m$ and $\forall \alpha > 1$ $g(\boldsymbol{\theta}' + \alpha \mathbf{z}) \geq g(\boldsymbol{\theta}' + \mathbf{z})$, i.e. the *g*(**x**) function does not decrease in any direction from point $\boldsymbol{\theta}'$.

3. We consider that on each risk factor there is information at least on one of limit values: D_j^- is more to the left of θ_j and D_j^+ is more to the right of θ_j, at which achievement consequences become

IOP Conf. Series: Earth and Environmental Science **177** (2018) 012005 doi:10.1088/1755-1315/177/1/012005

almost uncontrollable or irreversible. If $d_j^- = -\infty$ ($d_j^+ = +\infty$), then we consider that $D_j^- = -\infty$ ($D_j^+ = +\infty$).

4. $\forall D_j^- > -\infty$ $g(\theta'_1, \theta'_2, ..., D_j^-, ..., \theta'_m) = 1$ and $\forall D_j^+ < +\infty$ $g(\theta'_1, \theta'_2, ..., D_j^+, ..., \theta'_m) = 1$.

Then the risk function can be set, for example, as

$$g(\mathbf{x}) = \sum_{j=1}^{m} \alpha_j (x_j - \theta'_j)^2 , \tag{3}$$

or

$$g(\mathbf{x}) = \begin{cases} \sum_{j=1}^{m} \alpha_j (x_j - \theta'_j)^2, \mathbf{x} \in D, \\ 0, \qquad \mathbf{x} \in \overline{D}, \end{cases} \tag{4}$$

where $\alpha_j = \begin{cases} \dfrac{1}{(D_j^- - \theta'_j)^2}, x_j < \theta'_j, \\ \dfrac{1}{(D_j^+ - \theta'_j)^2}, x_j \geq \theta'_j. \end{cases}$

It is apparent that if $D_j^- = -\infty$ and $x_j < \theta'_j$ or $D_j^+ = +\infty$ and $x_j \geq \theta'_j$, then $\alpha_j = 0$.

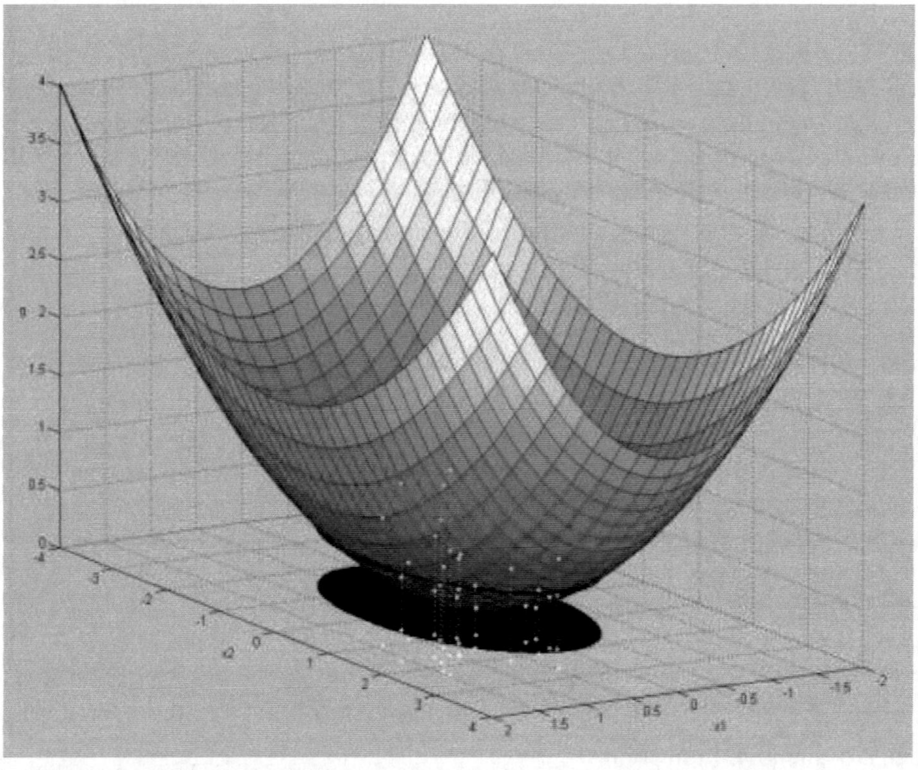

Figure 1. An example of a two-dimensional risk functions (3)

By way of illustration in figure 1 the example of the risk function set on a formula (3) for a case $m = 2$ is shown. The ellipse describing area $\overline{D}$ of admissible values of risk factors and lying on the

Ox_1x_2 $(r = 0)$ plane is shown by black color. The paraboloid above the plane represents possible values of risk $r(\boldsymbol{X})$. White points on the plane are values of risk factors, to them there corresponds points on paraboloid surface which set risk values; the image of border of an ellipse $\overline{D}$ is shown in the form of the black line. All corresponding couples of points (values of risk factors and risk values) are connected among themselves by vertical dashed lines.

In the problems of risk monitoring, Along with risk assessment $r(\mathbf{X})$ on all risk factors of $X_1, X_2, \dots, X_m$ of multidimensional system is expedient to estimate the contribution of each factor to total risk. We introduce a random vector $\mathbf{X}_k^- = (X_1, \dots, X_{k-1}, X_{k+1}, \dots, X_m)$. Then the absolute change of risk of multidimensional system due to addition of factor X_k is equal

$$\Delta r(X_k) = r(\mathbf{X}) - r(\mathbf{X}_k^-) \tag{5}$$

Dividing $\Delta r(X_k)$ of the risk $r(\mathbf{X}_k^-)$, we will receive the relative change of risk of multidimensional system by the addition of factor X_k

$$\delta r(X_k) = \Delta r(X_k) / r(\mathbf{X}_k^-) \tag{6}$$

Let's note that along with a contribution to the common risk of one factor, formula (4) and (5) allow us to estimate influence and groups of factors.

Monitoring of risk on the basis of model (1)–(6) consists in serial estimation in time of the actual values of $r(\mathbf{X})$, $\Delta r(X_k)$, $\delta r(X_k)$, $j = 1, 2, \dots, m$, and also dynamics of their change.

3. Practical application of the risk model of multidimensional stochastic system for monitoring of sustainable development of regions of the Ural Federal District in 1999–2016

Let's consider the most common case when $\mathbf{X}$ has joint normal distribution with a probability density

$$p_{\mathbf{X}}(\mathbf{x}) = \frac{1}{\sqrt{(2\pi)^m |\mathbf{\Sigma}|}} \exp\left\{ -\frac{1}{2} (\mathbf{x} - \mathbf{a})^T \mathbf{\Sigma}^{-1} (\mathbf{x} - \mathbf{a}) \right\},$$

where $\mathbf{a} = (a_1, a_2, .., a_m)^T$ – a vector of expectations, $\mathbf{\Sigma} = \{\sigma_{ij}\}_{m \times m}$ – a covariance matrix, $\sigma_{ii} = \sigma_i^2$ – dispersion.

Use of a Gaussian random vector is based on the central limit theorem [13]. It should be noted that this simplification isn't so critical and if there are any bases to consider that density of probabilities a component of a vector of $\mathbf{X}$ have more extended tails, then this can be corrected by setting the $g(\mathbf{x})$ function accordingly.

Let's carry out approbation of the offered risk model of multidimensional stochastic system for monitoring of sustainable development of four regions (Kurgan, Sverdlovsk, Tyumen and Chelyabinsk) the Ural Federal District after the default of 1998. Consideration of regions instead of the cities is explained by more representative statistical data [14]. The general assessment of probabilities $P(D)$ of macroeconomic risk factors of regions of the Ural Federal District in 2001–2015 was carried out in [15]. Below we will execute monitoring of sustainable development of these regions on dynamics of their macroeconomic risk factors, having taken as the analyzed time slice of 9 years. Let's use brought in [15] macroeconomic risk factors and threshold risk levels, having made some refinement:

1) make changes to the risk factors: remove from consideration the index of the industrial production in view of its close correlation with other factors on lesser intervals and instead of dependency ratio we use more informative indicator – unemployment rate;

2) reduce number of threshold levels of risk factors to three: K1 – an unstable state; K2 – a pre-crisis state; K3 – crisis state. These threshold levels correspond brought in [15] pre-crisis developing, crisis unstable and crisis extreme states, respectively.

Risk factors, its threshold levels and the limiting values are given in table 1

Table 1. Macroeconomic risk factors of the regions

Risk factor	Threshold levels			Extreme values
	K1	K2	K3	
X_1 – real income movement, in % to previous year	85,9	79,93	75	$D_j^- = 50$
X_2 – the ratio of the average size of pension to subsistence minimum of pensioners	0,86	0,66	0,5	$D_j^- = 0{,}25$
X_3 – morbidity on 1000 people of the population	920	960	1000	$D_j^+ = 1500$
X_4 – mortality from external causes, number of the dead on 100000 people of the population	287,6	322,1	350	$D_j^+ = 700$
X_5 – a wear of fixed assets on the end of the year, %	66,79	71,33	75	$D_j^+ = 150$
X_6 – the volume of budget revenues per capita, in the prices of 2016, thousand rubles	29,18	21,75	15	$D_j^- = 7$
X_7 – quantum index of gross regional product, % to previous year	92,62	88,4	85	$D_j^- = 40$
X_8 – unemployment rate, in %	15	18	21	$D_j^+ = 40$

In figures 2–4 results of calculation of probability of unfavorable outcome P(D) and risk of r(X) for threshold levels of risk factors K1, K2, K3, respectively are shown. The risk function was set on a formula (4). Calculation of integral (2) was carried out by means of a method of statistical tests of Monte Carlo.

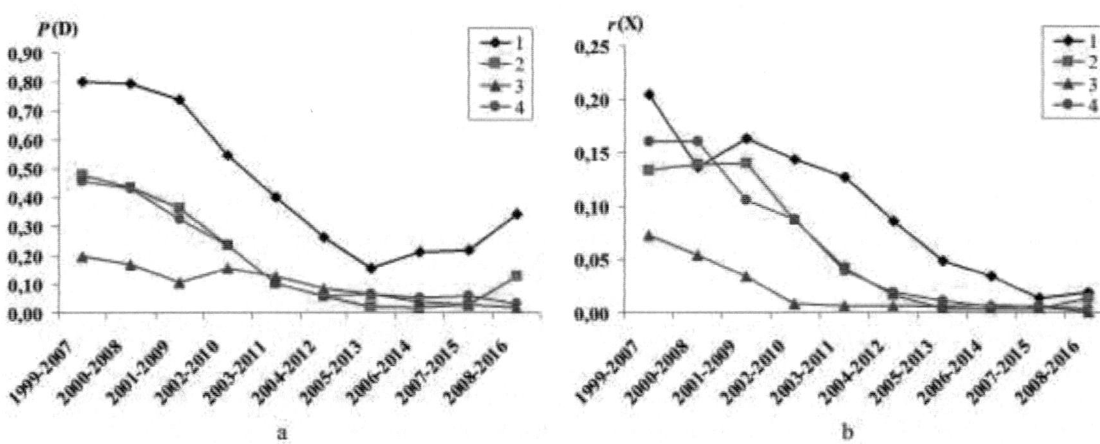

Figure 2. Estimates of P(D) and r(X) for threshold levels of risk factors K1:
1 – the Kurgan region, 2 – the Sverdlovsk region, 3 – the Tyumen region, 4 – the Chelyabinsk region

IOP Conf. Series: Earth and Environmental Science **177** (2018) 012005 doi:10.1088/1755-1315/177/1/012005

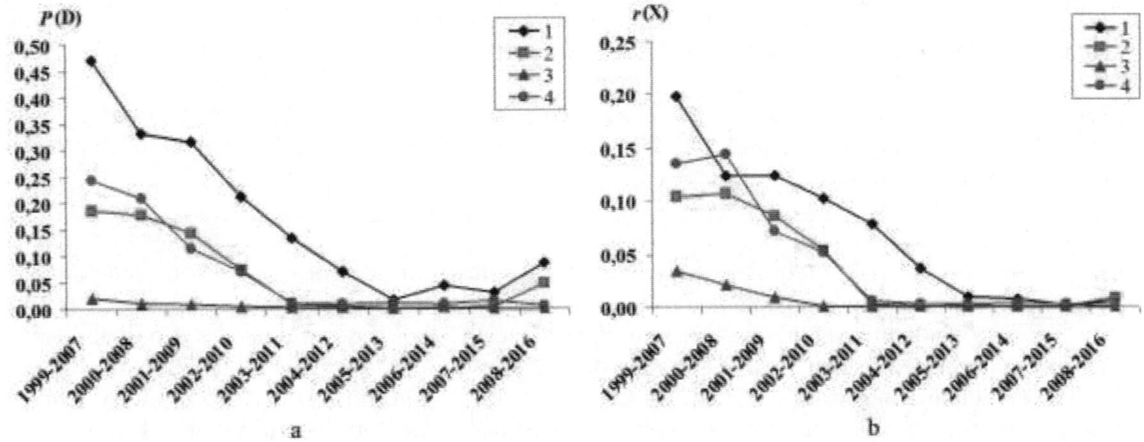

Figure 3. Estimates of P(D) and r(X) for threshold levels of risk factors K2:
1 – the Kurgan region, 2 – the Sverdlovsk region, 3 – the Tyumen region, 4 – the Chelyabinsk region

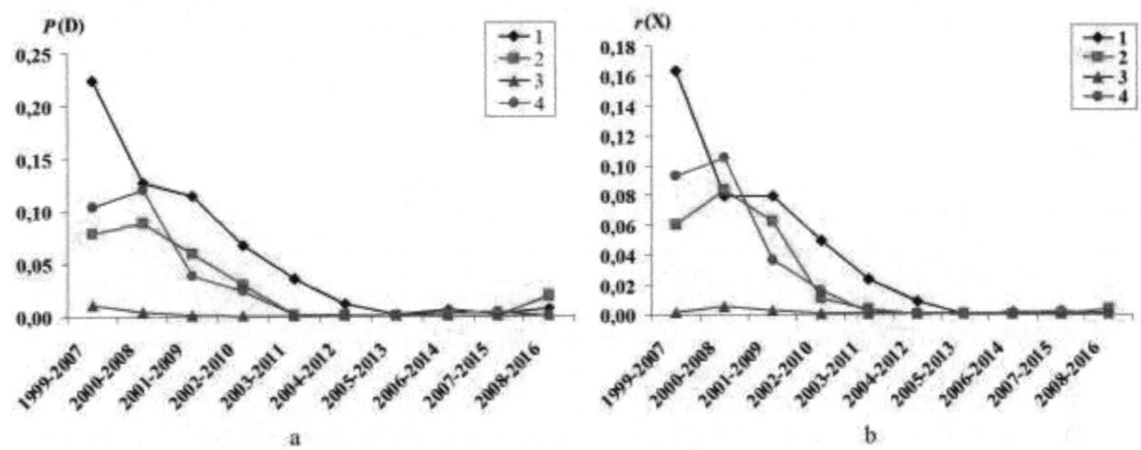

Figure 4. Estimates of P(D) and r(X) for threshold levels of risk factors K3:
1 – the Kurgan region, 2 – the Sverdlovsk region, 3 – the Tyumen region, 4 – the Chelyabinsk region

Analysis of the results of monitoring the sustainable development of regions showed the following:

1. During the initial period (1999–2007) on all four regions the greatest values of both probabilities of a failure of *P(D)*, and risk of *r*(**X**) on all three levels of danger K1, K2, K3 were observed. It corresponds to the minimum stability level of each of the regions. The Kurgan region had the worst level of stability, and Tyumen region – the best among four regions.
2. In dynamics stability gradually increased (decrease of *P(D)* and *r*(**X**)). However in the last three years for the Kurgan and Sverdlovsk region the tendency to decrease in stability of development was outlined again.
3. Analysis on each of regions:
   3.1. In the Kurgan region the greatest contribution to instability of functioning untill 2013 was made by X_6 factor, and then the situation significantly changed – practically all factors began to exert approximately similar impact.
   3.2. In Sverdlovsk region untill 2010 the main contribution to instability of functioning was made by a factor X_6, then X_2 became such factor, and since 2015 factor X_7 began to make the main contribution to instability.

3.3. In the Tyumen region untill 2011 the greatest contribution to instability of functioning was made by factor X_2, then such factors became several (X_1, X_2, X_4, X_6), and during the last two periods X_1 and X_2 belong to such factors.
3.4. In Chelyabinsk region untill 2011 the main contribution to instability of functioning was made by a factor X_6, then such factors were X_2, X_6, X_7, and since 2014 the main contribution to instability began to make factor X_7.

4. Conclusion

1. A new approach to risk analysis of the complex systems is offered. It is based on modeling the system as a multidimensional random variable, which components are risk factors.
2. Two options of the risk analysis are considered. In the first case evaluated the probability of dangerous states of the system, and the second – directly the risk based on the risk function.
3. The carried out approbation of the offered risk model of multidimensional stochastic system on actual data showed its adequacy for monitoring of sustainable development of regions. The received results are well interpreted and correspond to the actual situation in general.

Acknowledgements
Present study was carried out under financial support of the Russian Fund of Fundamental Research grant No. 17-01-00315 “Development of models and methods of monitoring, forecasting and control optimization of multidimensional regional socio-economic systems based on entropy and minimax approaches”.

References

[1] O'Sullivan A 2012 *Urban Economics* 8th edition (McGraw-Hill/Irwin)
[2] Forrester J W 1970 *Urban dynamics* (Pegasus Communication)
[3] Satzinger J W, Jackson R B and Burd S D 2008 *Systems Analysis and Design in a Changing World* 5th edition (Course Technology)
[4] Popkov Yu S 2014 *Mathematical Demoeconomics: Integrating Demographic and Economic Approach* (Berlin: De Gruyter)
[5] Rinaldi S M, Peerenboom J P and Kelly T K 2001 Identifying, Understanding, and Analyzing Critical Infrastructure Interdependencies. *IEEE Control Systems* **12** 11–25
[6] Tyrsin A N, Gevorgyan G G 2017 Entropy modeling of sustainable development of megacities *IOP Conference Series: Earth and Environmental Science* **72** 012010
[7] Mun J 2010 *Modeling Risk* 2nd edition (Wiley)
[8] Vorobyov Yu L, Malinetsky G G and Makhutov N A 2000 Management of risk and sustainable development: Human measurement *News of higher education institutions. Applied nonlinear dynamics* **8** (6) 12–26
[9] Gore A 1992 *Earth in the Balance. Forging a New Common Purpose* (Earthscan Publications Ltd., Copyright – Senator Al Gore)
[10] Porfiryev B N 1991 *Management in emergency situations* Results of science and technology. Security concerns: emergency situations **1** (Moscow: VINITI)
[11] Vishnyakov Ya D and Radayev N N 2008 *Common theory of risks* 2nd edition (Moscow: Academy)
[12] Tyrsin AN and Surina A A 2017 Modeling of risk in multidimensional stochastic systems. *Bulletin of the Tomsk state university. Management, computer facilities and informatics* **2(39)** 65–72
[13] Gnedenko BV 2005 *Probability theory course* 8th edition (Moscow: URSS editorial)
[14] Regions of Russia *The main social-economic indicators of cities* Federal State Statistics Service
[15] Kuklin A A, Pecherkina M S , Tyrsin A N and Surina A A 2017 Methodical tools of diagnostics of risks for welfare of the person and the territory of accommodation *Region economy* **13** 4

Published under licence in *Journal Title* by IOP Publishing Ltd.

IOP Conf. Series: Earth and Environmental Science **177** (2018) 012006 doi:10.1088/1755-1315/177/1/012006

The principles of the green economy in the context of agglomeration: evidence from Big Ekaterinburg

N A Vukovic[1], V A Larionova[2] and V V Biryulina[2]

[1] Ural State Forest Engineering University, Ekaterinburg, Russian Federation
[2] Ural Federal University named after the first President of Russia B. N. Yeltsin, Ekaterinburg, Russian Federation

e-mail: shpak17121978@gmail.com, viola-larionova@yandex.ru, vivvvv99@gmail.com

Abstract. The authors consider an urban agglomeration as a more integral informal economic subject uniting economic interests of its municipalities and expressing interests of the territory which can become a driver of the "green economy" strategy. High level of social and economic development of an agglomeration kernel, and high capabilities related to minimization of environmental risks and multiplying natural capital of the small cities allows synchronization of economic growth, optimal use of natural resources and decreasing decoupling effect. Using Big Ekaterinburg as an example of urban agglomerations the authors have shown that strategic planning documents on the municipal level require cardinal revision in the context of complex development of the territory, "green" economy principles implementation and harmonization of relations among the economy, society, and environment.

1. Introduction

Creating modern habitat environment and extensive program of Russian Federation spatial development, including development of the cities and other settlements relates to complex resolution of multiple issues. City environment renovation is tightly connected with compliance to the high standards of ecological prosperity, and is a priority task with regard to stability and sustainable development of the country, which was reflected in the Address to the Federal Assembly by the President of Russian Federation, dated March 1, 2018 [1] and determines actuality of this study.

It is important that urban development becomes motive force in developing the country as a whole. Active and dynamic life of Russia with its vast territories cannot concentrate within the several metropolitan cities. Large cities shall distribute their energy, and become cornerstones for the balanced and harmonic spatial development of the whole country. Well developed communications of the centers allow small cities enjoying all opportunities of the modern services while integrating into the common economic and social space, and at the same time preserving distinctiveness, and abilities to unveil their unique potential.

Nowadays it is possible to achieve sustainable socioeconomic development and safe habitat only through increasing efficiency of using public resources, and transition to the innovative development path, use of scientific achievements and transition to new technologies.

Content from this work may be used under the terms of the Creative Commons Attribution 3.0 licence. Any further distribution of this work must maintain attribution to the author(s) and the title of the work, journal citation and DOI.
Published under licence by IOP Publishing Ltd

2. Methodology

"Green" economy is a strategically important direction providing stability and sustainable growth of national economy [2]. In terms of modern definition of the "green" economy the authors consider necessary to base upon the synthesis of economic, industrial, technological, and civilization approaches presented by V. Bochko [3]. That would allow defining "green" economy as a sustainable growth economy with domination of environmentally-friendly industries that use alternative sources of energy, and resource-conserving technologies. In this type of economy public environmental and economic policies actively stimulate economic growth and improving environmental culture of the population in order to preserve human civilization on Earth [4]. Schematic representation of the "green: economy is presented on Fig. 1 [2].

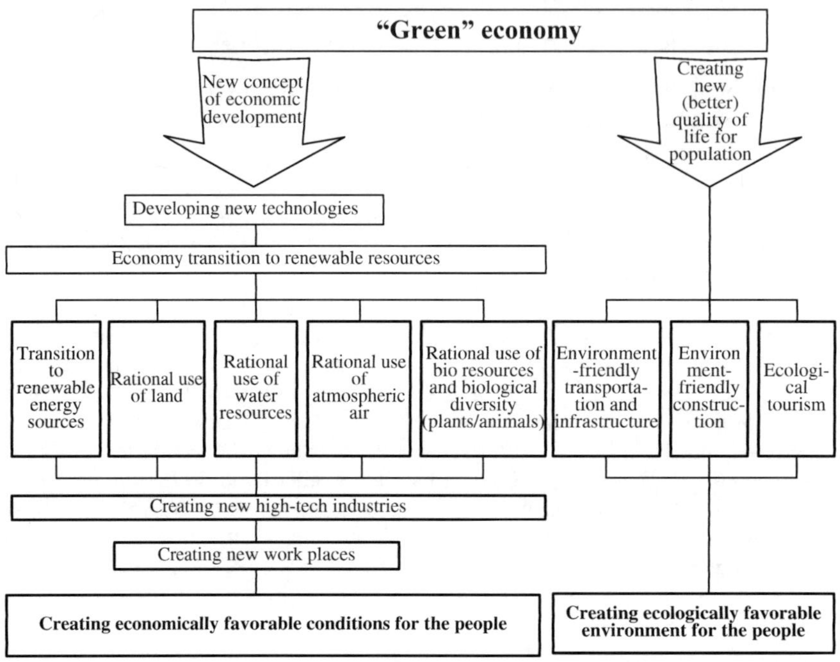

Figure 1. Environmental and economical model of the "green" economy [2]

Technological approach suggests "...transition of all production to the technologies that provide creation of environmentally friendly industrial products and foods" [3]. This approach does not cover tourism and new "green" industries, for example, "green" construction. At the same time priority positions in transition to the "green" economy in this approach are given to business, which contradicts the main business goal of profit generation, and the state role is neglected and not considered to be critical. However the state is a key stakeholder of a "green" economy, because in today's definition the "green" economy is aimed primarily at achieving public goals.

In authors' opinion the weakness of an industrial and technological approaches is in neglecting "green" economy development interdependencies with development of the "green" eco-cities, which actively generate demand for "green" technologies, and act as key drivers for "green" economy development. Important role of eco-cities in developing "green" economy, along with importance of developing efficient "green" budgeting and financial policy, is stressed by modern authors, such as I. Monasterolo and M. Raberto [4] and P. Baranova and F. Paterson [5].

P. Newton and P. Newman [6] state that "green" agenda for the cities and economy as a whole is a key direction for international organizations' efforts, and national and municipal priority of increasing value. Using Australia as an example, authors illustrate interdependencies and mutual effects of "green" city infrastructures, eco-cites and "green" sector of economy. The survey results demonstrate that 85% of the participating companies consider "green" growth to be among the priorities. Also the companies included in the survey recognize opportunities for increased involvement of the industries into a transition to a low-hydrocarbon ("green") economy that would appear in case of increased stimulation by the government [7].

Criteria for assessing "green" economy in international practice were reviewed by A. Nahman, B. K. Mahumani and W. J. De Lange [8]. These criteria formulated with regard to "green" economy declared principles are listed in Table 1.

Table 1. Declared principles of the "green" economy with corresponding assessment criteria

Item	"Green" economy principle	Assessment criteria
1	A tool for achieving sustainable development	Inter-sector principles
2	Measuring principles outside of GDP use bounds (corresponding indicators)	GDP indicators
3	Respect for planetary boundaries, ecological limits and deficits (absolute compartmentalization)	
4	Requirement for good governance (inclusion; democracy; participation; responsibility; transparency, and stability) and use of integrated ecological, social and economical basis for decision-making	Efficient government Complex decision making
5	Internationalization of external effects, stimulation of investments and innovations, sustainable development, behavior and way of living	Ceasing harmful subsidies Internationalization of external effects "Green" investments, "Green" innovations
6	Investments into environment-friendly economy and infrastructure	Sustainable production (agriculture, wood industry, fishing, mining, and manufacturing) Green infrastructure (power, buildings, cities, and) Sustainable consumption
7	Actions aimed to improve energy use efficiency, and resource conservation	Water use efficiency Materials use efficiency
8	Low carbon content, low emissions, zero waste, and generally low-carbon character	Zero waste Air quality Water quality
9	Protection of biological diversity and ecological systems, investments into natural capital	Biological diversity of ecosystems
10	Lowering poverty level and increasing prosperity	Fighting poverty Satisfying key demands Human capital investments Subjective well-being
11	Creating decent jobs and "green" vacations	High quality of employment
12	Fairness among the countries and generations	Fairness within the generations Fairness among the generations Reform of international institutions

Nowadays new economic reality dynamically all over the world, where ecological priorities turn from formal requirements into real conditions required for transition to sustainable development, and new economy model, and for development of new projects. These issues need to be reflected in the national ecological policy. Economic and legal mechanisms of Russian economy shall consider ecological factors correctly, in order to mitigate risks of transferring to sustainable development and leaving the export-oriented raw materials export development model.

International practice presents a certain amount of "green" economy development experience. It is necessary to thoroughly analyze the multitude of methodologies recommended by international organizations and used in international agreements related to regulation and monitoring of anthropogenic environmental impact occurring in course of international economy globalization.

The understanding of sustainable development, "green economy", and ways of forming said economy differ from country to country. Though any countries implement crisis-response programs based upon the principles of "green" economy, methodological approaches for reflecting "green" economy parameters and indicators in forecasts substantially differ for European countries, North America, Asia-Pacific, and CIS states.

Success of dispersing sustainable development ideas based upon "green" economy principles and active participation in formulation of best practices requires adoption of the methods that considers national specifics. At the same time general goal remains the same – providing and improving complex approach based upon three cornerstones of sustainable development – i.e. economic growth, social protection, and environmental protection.

Sustainable development methodology outlines four groups of the indicators showing the role of "green" economy as an engine of economic growth that is reflected, among other parameters, by the increase of GDP value [9].

First group of indicators describe potential opportunities for "green" economy transition, including productivity increase for natural resources (forest resources, fisheries, arable land); efficient capital management (decreasing economic damages related to the loss of capital by improving management of environmental risks); increasing human potential quality, specifically by improving environmental conditions (morbidity decrease and increase of longevity).

Second group includes indicators for increasing efficiency of key industrial sectors, including power, construction, utilities, etc., such as energy efficiency improvement, lowering greenhouse gas emissions, and introduction of alternate energy sources.

Third group describes investments into "green" ecology development, including water supply and sewer systems, alternative-fuel oriented public transportation, etc.

Fourth group of indicators stimulates innovations in the field of "green" economy development (including company-level innovations) aimed at creating favorable competitive environment, and obtaining innovation impact by implementing related standards and regulations.

At the same time evaluation of these indicators cannot take place at municipal, and regional levels due to lack of systematic approach, and contradiction of sustainable development goals for these public management subjects.

3. Results and Discussion

As it was shown in [10] a contradiction exists in the very concept of "region", which denotes a certain territory with historically formed and legally fixed administrative borders, but which does not have signs of integrity from the viewpoint of economic activity subject. As a rule, the economic potential and competitiveness of the region is determined by one or several large cities located on its territory and producing a large part of the gross regional product (GDP), as well as the degree of agglomeration processes occurring in this territory. Small cities around the agglomeration kernel are experiencing strong problems connected with outflow of population, lack of qualified personnel, decreasing the economic growth rate, deteriorating the investment climate and quality of living environment. And only a small part of these cities can follow a sustainable development strategy by choosing one of the alternative favorable scenarios, which include, in particular, the development strategy of a small city as part of an agglomeration with positioning and clear definition of its role and functions.

In this sense, an agglomeration is a more integral economic entity uniting economic interests of its municipalities and expressing interests of the territory as a whole. In the opinion of the authors the best segment for evaluation of "green" economy maturity level is the level of urban agglomeration.

Modern understanding of complex systems requires new approach to the functioning of their subsystems – when these subsystems are viewed not as elements subordinate to the whole system, but as synergistically interacting parts that create the system. Synergy law is expressed by the fact that any complex dynamic system tries to maximize its effect by means of integrity and cooperation. Thus synergy effect is displayed not only in the favorable combination of resources, but also in form of consistent behavior, connections, and relations, i.e. by the set of parameters that describe complex and developing system [11].

With regard to agglomeration processes synergism reveals itself in form of a skill of assessing and producing joint effects related to higher-scale development of a territory. Agglomerative thinking allows considering tight economic, social and cultural interactions that exist among the territorial units forming a complex object of management. Complete overall agglomeration management structure helps aligning interests of its territories.

At the same time the position of considering interests of separate territories or personalities instead of the ones of the agglomeration, causes chaotic development of agglomeration processes, and substantially damages social and economic life of the municipality, region, and country.

In order to set organizational, economic, and institutional conditions for transitioning to the "green" economy which in turn would become a local level sustainable development tool, it is necessary to reveal existing ecological, and economic interests of the whole agglomeration and its municipal entities, and interactions of various elements of the economy, society and environment that are required to fulfill these interests, and provide ground for the development of the mechanism for synchronizing interests [12].

In a practical sense "green" economy is tightly connected with resource preservation strategies, and in a methodological sense it is related to expanding the traditional parameter set that describes economic dynamics. It is possible to assess production modernization results considering environmental imperatives using the so-called decoupling effect. The effect occurs in situations when the processes that were previously demonstrating a certain degree of conjugation, start changing development trajectories. Therefore, analysis of the data describing these processes that did previously demonstrate correlation ort other dependencies, starts showing antagonistic trends. Speaking of environmental and economic problems of state development, decoupling factor characterizes the phenomenon of separating trends for economic parameters, and use of natural resources. Whatever are the decoupling factors in economy-environment sphere, its occurrence is favorable for environment and population [13].

High level of social and economic development of an agglomeration kernel, and high capabilities related to minimization of environmental risks and multiplying natural capital of the small cities assist in appearance and decrease of decoupling effect, which, in turn, would assist to economic growth of an agglomeration as a whole (not of its single subjects), and allows synchronization of economic growth and use of natural resources.

In case of Ekaterinburg agglomeration, that is the fourth in size in Russia after Moscow, Saint-Petersburg, and Samara-Togliatti, it is possible to track the following trends.

Agglomeration kernel is Ekaterinburg – one of the largest economic centers on Earth, and a member of top-600 largest cities that produce 60% of global GDP [14]. Satellites of the first belt include Aramil', Berezovskiy, Verkhnyaya Pyshma, and Sredneuralsk. Satellite-cities of the second belt are Degtyarsk, Zarechny, Novouralsk, Pervouralsk, Polevskoy, Revda and Sysert'. Third belt peripheral cities are Rezh, Asbest, and Nizhniye Ser'gi.

Revealing synergic effect of agglomeration thinking assist in equalizing levels of subjects' social and economic development and organize mutually beneficial collaboration in the area of distributing economical and ecological benefits of the kernel and satellites.

Emergence principle for the kernel and small cities in case of developing "green" economy can be implemented by developing recreational territories in satellite cities and implementing development projects with administrative support provided by the kernel (touristic projects in Revda, Sysert' and Pervouralsk, natural reserve in Rezh, farmers market and gastronomy festival in Sysert', providing waste processing facilities by small cities – Berezovskiy and Polevskoy) if the ecological and economic interests of the subjects are fulfilled. As it is illustrated in Table 2 a synergy effect can be achieved in the project of tourism development in the satellite cities of the second belt around Yekaterinburg if.

IOP Conf. Series: Earth and Environmental Science **177** (2018) 012006 doi:10.1088/1755-1315/177/1/012006

Table 2. Synergy effect in case of agglomeration subjects interaction and parameters of fulfilling their economic and ecological interests

Project	Responsibilities of the parties	Parameters describing ecological and economic interests of the subjects
Tourism development in the second-belt satellites (Revda, Sysert', Pervouralsk)	Agglomeration kernel – Ekaterinburg: providing administrative support to municipalities, ensuring the small cities' population ability to meet the needs of environmental well-being Small cities – Revda, Sysert, Pervouralsk: provision of the environmental resources (optimization of an access to natural assets), income generation, the ability to preserve the identity of the cities and open an unique potential, while integrating into a single economic space of agglomeration	Tourism development in the second-belt satellites (Revda, Sysert', Pervouralsk) Ensuring the best quality of life of the agglomeration population by maximizing the income received by municipalities

Strategic goals of providing ecological security by means of transition to "green" economy that define modern trends in national environmental policy were listed in the Concept of the Russian Federation long-term social and economical development for the period until 2020, in the National Security Strategy of Russian Federation, in the Basics for state policy in the area of Russian environmental development until 2030 and other strategic planning documents that exist on federal, regional and municipal levels [8].

The future long-term challenges mentioned in the Concept of the long-terms social and economic development include strengthening the environmental impact upon all aspects of life. For example, global economical environment features acceleration of technological challenges, and transition to a new technological base featuring nanotechnologies, biotechnologies, energy and resource saving technologies. "Green" economy sector is expanding rapidly, and environmental requirements for technological processes and production organization become stricter. All these factors reflect in transformation of priorities, goals and tasks related to social, economic, and environmental policies. Harmonization of ecological, economic, and social goals during the innovative modernization of Russia becomes a key goal of state policy.

Federal level strategic planning documents describe providing environmental safety and "green" economic growth as key priorities of territories development. There is a required list of strategic planning documents that has to be developed at each planning level. For example, developing strategies for social and economic development of each macro region and subject of Russian Federation is mandatory.

The value of economic factors for sustainable development of a territory makes actual the analysis of Sverdlovsk oblast and Ekaterinburg strategic planning documents describing kernel and satellites aimed at defining the importance of introducing "green" economy in the strategies of social and economic development, and degree of coordinating the listed events.

Sverdlovsk oblast strategy of social and economic development for 2016–2030 is based upon the following political priorities:

- Creating competitive conditions for accumulation and preservation of human potential;
- Creating conditions for increasing competitive abilities of the Sverdlovsk oblast economy industrial, innovative, and entrepreneurial;
- Balanced development of Sverdlovsk oblast territories.

Paragraph 27 of the said strategy specifies a special direction ion social and economic policy of Sverdlovsk oblast for 2016–2030 named "Sustainable environmental development of the Sverdlovsk oblast territory" aimed at lowering the scale of negative environmental effects caused by industrial and other activities.

Strategy states that it is possible to harmonize economic and environmental goals, lists the tasks related to making economy more ecological, performing transition to the model of "green" economic growth, for example, by supporting implementation of power and resource saving technologies in manufacturing, agriculture, housing and utilities, introducing economic stimulation measures in the area of environmental protection aimed at limiting emissions and dumping of pollutants and microorganisms into the environment, increasing the share of processes and neutralized production and consumption waste.

Ekaterinburg strategic development plan that was actualized up to 2030 with target values set until 2035 defines strategic goal of city development to provide stable, high and improving life quality for current and future generations of citizens. Strategic development directions and programs list features Item 6 – "Forming comfortable, environmentally favorable city environment". Key tasks include stabilization and improvement of the city ecology, crating favorable conditions of habitation, increasing levels of ecological literacy and culture of the citizens, providing territory cleanness, and improving public amenities. However this direction is not listed among the priorities and it is evident that there is no synergetic interaction among the mentioned strategic documents.

Municipal level strategic documents are still being developed, however published intermediary results allow making a conclusion that strategic planning level of satellite cities can be characterized by the lack of goals related to harmonization of relations among the economy, society, and environment. For example, the strategy of social and economic development for Berezovskiy city municipality lists a strategic priority development direction named "Ecology. Developed Environment. Recreational zones", however, environmental aspects are not considered when defining key strategic goals for development of economy and social sphere. There are no activities aimed at implementation of "green" economy. Therefore, this strategy does not reflect strategic goals of "green" economy development that were defined on national level, and the level of agglomeration at the whole.

Positive dynamics in the system of setting strategic goals on federal level aimed at harmonization of ecological, economic and social goals and providing "green" economic growth is not developed further on municipal level. Besides that there is a contradiction related to achievability of parameters related to "green" economic growth. The absence of possibility of jointly using natural assets by municipalities and organizing mutually beneficial cooperation leads to creating and increasing decoupling effect.

There is a right time to resolve these problems now keeping in mind the current stage of developing strategic documents on municipal level. That would require further improving of strategic planning system, and increasing degree of its alignment in terms of goals and priorities.

4. Conclusion

While forming institutional conditions of regional transfer to the "green" economy that would become a tool for local sustainable development, and considering unequal implementation of green economy principles over the country, agglomeration becomes the most efficient level of assessing "green" economy development status.

Recommendations regarding the system for assessing the "green" economy development level would help forming complex approach for providing "green" economic growth at all levels.

Key directions of maximizing profits in terms of ecological and economic interests of agglomeration subjects include:

- Improving consumer qualities of resources and assets, increasing their long-term use efficiency;
- Efficient collaboration among the municipalities with regard to optimizing the mode of accessing to natural resources and benefits;
- Considering interests of all agglomeration subjects in course of "green" economy implementation, and preserving agglomeration thinking in course of developing strategic documents.

Acknowledgments

The authors would like to express their deepest gratitude to the Russian Foundation for Basic Research (RFBR) for the support of the research within the project No. 17-22-07001 "The complex algorithm

IOP Conf. Series: Earth and Environmental Science **177** (2018) 012006 doi:10.1088/1755-1315/177/1/012006

of culture-based regeneration of minor industrial cities in the context of agglomeration processes in Russia and Europe".

References

[1] *Address of the President of Russian Federation to the Federal Assembly* on March 1, 2018
[2] Vukovic N 2018 Green Economy: Definition and Contemporary Ecological-Economic Model *Bulletin of Ural Federal University. Series Economics and Management* **17(1)** 128–45
[3] Bochko V S 2014 Green economy: the second eternal problem of mankind *Bulletin of Ural Federal University. Series Economics and Managemen* **3** 113–9
[4] Monasterolo I and Raberto M 2018 The EIRIN flow-of-funds behavioural model of green fiscal policies and green sovereign bonds *Ecological Economics* **144** 228–43
[5] Baranova P and Paterson F 2017 Environmental capabilities of small and medium sized enterprises: Towards transition to a low carbon economy in the East Midlands *Local Economy* **32(8)** 835–53
[6] Newton P and Newman P 2015 Critical Connections: The role of the Built Environment Sector in Delivering Green Cities and a Green Economy *Sustainability* **7(7)** 9417–43
[7] Newton P 2014 City transitions, infrastructure innovation, green economy and the eco-city *Resilient Sustainable Cities: A Future* ed L Pearson, P Newton, P Roberts (New York: Routledge) 91–104
[8] Nahman A, Mahumani B K and De Lange W J 2016 Beyond GDP: towards a green economy index *Development Southern Africa* **33(2)** 215–33
[9] Matova N I 2016 Contents of the "green" economy term in conditions of todays Russia *Proc. Russ. Sci. Pract. Conf. "Russian regions in new economic conditions" (Sochi, Russia)* 66–75
[10] Vukovic D B 2016 Regional competitiveness: misconceptions *Proc. Int. Conf. Problems of development and enhancing the competitiveness of the economy in a multipolar world (Cholpon-Ata, Issyk-Kul, Kyrgyztan) Karaganda economic university bulletin* **3(37)** 20–3
[11] Biryulina V V and Yurasova I I 2016 Features of agglomeration thinking: national and foregn experience *Proc. XI Int. Conf. Russian regions in focus of changes (Ekaterinburg, Russia)* **2** 217–26
[12] Rodionova I A and Lipina S A 2015 Green ecomomy in Russia: model and development forcast *Fundamental research* **2** (part 24) 5462–6
[13] Samarina V P 2015 Green economy of Russia: some issues of theory and methodology *National Interests: priorities and security* **2(287)** 2–9
[14] Mishulina S I 2017 Environmental safety in the Russian system of strategic goal-setting *Environmental law* **1** 39–43

Published under licence in *IOP Conference Series Earth And Environmental Science* by IOP Publishing Ltd

The dynamic index of urban environment quality as a tool for sustainable urban development

D M Suvorov[1], L A Suvorova[2], I V Pestova[2] and T V Baibakova[2]

[1] Department of Heat Engineering and Hydraulics, Vyatka State University, 36 Moskovskaya str., Kirov, 610000, Russia
[2] Department of Economics, Vyatka State University, 36 Moskovskaya str., Kirov, 610000, Russia

E-mail: dmilar@mail.ru

Abstract. Based on the analysis of domestic and foreign experience, the authors of the article have developed and tested the methodology for constructing a dynamic index of urban environment quality (DIUEQ) of a single municipal entity in Russia. The main methodological principle of building DIUEQ is a hierarchical approach, the main method is the method of summary indicators, and the main features of the calculation methodology of the index are its simplicity, clarity and transparency. Each of the four aggregated groups of indicators includes three sub-groups of the same name–indicators of provision, indicators of sociological measurements and economic indicators. The dynamics of DIUEQ and the dynamics of its structural components reflect the nature of changes in the city indicators both in retrospect and in forecasting. The authors give an example of calculating the index according to the proposed method for the city of Kirov for 2016. Moreover, the authors show how promising it is to assess by DIUEQ the dynamics of the quality of the urban environment in general, of certain groups and subgroups of indicators for the temporal dynamics of the corresponding indices included in the DIUEQ, as well as forecasting and setting the tasks of sustainable urban development by both municipalities and public groups.

1. Introduction. Domestic and foreign experience in measuring the quality of the urban environment

The quality of urban (city) environment can be defined as "...the ability of the urban environment to meet the objective needs and demands of the city's residents in accordance with generally accepted norms and standards of living" [1]. Integral quality assessment by virtually all domestic researchers is proposed in the form of a dimensionless index that takes into account all the factors presented in each specific methodology (usually environmental, economic, socio-demographic, infrastructural) [2–5]. At the same time, groups of factors and indicators taken into account in each group, methods of calculation, methods of translating the initial values of the indicators into the index form, including methods for rating the indicators, differ significantly in different methods.

The objectives of calculating the quality index of the urban environment include, as a rule, the definition of the current state of the urban environment, the development of activities for its improvement, and monitoring processes in the sphere of urban development [6]. But one of the purposes in many methodologies is to compare the quality of the urban environment or the quality of

Content from this work may be used under the terms of the Creative Commons Attribution 3.0 licence. Any further distribution of this work must maintain attribution to the author(s) and the title of the work, journal citation and DOI.
Published under licence by IOP Publishing Ltd

life in different municipalities or regions, which implies the construction of an appropriate rating by the magnitude of the index [5, 7 and 8].

Ratings of cities or regions on the quality of the urban environment, quality of life and other aspects developed by different state structures or expert communities plays an important role in Russia and abroad. In Russia, such indices are calculated and published in Internet resources, for example, by the rating agency Expert RA [8], consulting company Strelka KB [9], etc. The declared goals when constructing the quality index of the urban environment are not always realized. This is most typical for the indices used in the construction of ratings according to the ministerial methods of the Ministry of Regional Development of the Russian Federation [10] and the Ministry of Construction of the Russian Federation [6]. The current official methodology of the Ministry of Construction [6] is based on the consulting company Strelka KB [9]. It has developed the city's quality index, which makes it possible to evaluate each city in comparison with other cities. It is clear from the text of the document [6] that calculations of most indicators require obtaining and processing of large amounts of inaccessible information from cartographic services, geoinformation systems, special Internet services, and it is practically impossible to reproduce these calculations for verification or forecasting at the level of the leadership of a municipality or research group. The city index is determined by methods [6, 9] only in comparison with other cities for each year, that is, every year it is determined on a new basis for comparison.

The influence of the urban environment on the quality of life of people and on the sustainability of urban development is also revealed by various international comparisons. The UN Research Group has developed the City Prosperity Index (CPI) [11], one of the closest to the quality index of the urban environment. The most valuable for this work is the methodical approach of constructing the CPI index [12] based on four principles. The first principle is having three levels of nesting indices: dimension, sub-dimension, and indicator. The second is when an integral index, and all indices of groups, subgroups and indicators have the same limits of change – from zero (0) to hundred (100) when 0 is the worst value of the index and 100 is the best. The third one is when all the indices of the groups for the integral index, the indices of the subgroups for the group indices, and the indices of the dimensions for the index of the subgroup are in equilibrium, that is, all higher order indices are determined as the arithmetic mean of the indices of the preceding nesting level. The fourth is when all indices of indicators are subjected to the procedure of linear normalization.

The second important method is the Urban Sustainability Index (USI), used for the cities of the PRC. It was developed by the McKinsey Global Institute (MGI), and the Urban China Initiative (UCI) research center. The latest report on this index was published using a methodology modified in 2016 [13]. This report presented a rating for 185 cities of China of different sizes and at different stages of their development in terms of their sustainability in the period from 2006 to 2014 according to 23 indicators broken down into four groups covering the areas of the economy, social sphere, resources, and environment. A special feature of the report [14] is a deep analysis of the dynamics of the index change calculated by this method over the period under consideration for different groups of cities in the PRC, including consideration of index ratios between groups, between subgroups and within groups of indicators. Calculation of USI and CPI indices is carried out on the basis of a single standardization for several years in retrospect, which makes it possible to track the dynamics of changes in the state of the city for each indicator and for each group of indicators by the magnitude of the change in the corresponding index.

The third method, widely used in the world, was developed by Mercer Human Resource Consulting. Its experts use the world's quality of life index, which takes into account 39 factors grouped into 10 categories [15]. In 2017, this index was calculated for more than 450 cities around the world and based on its results a rating of 231 cities was compiled [16]. The peculiarity of this methodology is the presence of only subjective assessments based on relative differences for each factor between any two cities that respondents estimate in pairs.

The fourth method, which has a fairly wide popularity, is the Sustainable Cities Index (SCI) from ARCADIS [14]. The SCI index in 2016 includes 100 global cities estimated by experts. The city receives an estimate for each of the three sustainability indicators, and the overall index score is equal to their average value.

In all considered and the overwhelming majority of other foreign methods (for example, [17, 18]), all indices are calculated on the basis of a methodical approach based on the use of the second and fourth principles from the CPI index technique (see above).

2. Methodological foundations for constructing a dynamic quality index of the urban environment

Based on the analysis of domestic and foreign experience, the purpose of this studies to develop and test the methodology for constructing the quality index of the urban environment of a single municipal entity in Russia, the temporal dynamics of which (and the dynamics of its structural components) would accurately reflect the nature of the changes in the indicators of this city both in retrospect and in forecasting, which would allow municipalities and public groups to set and solve the tasks of the sustainable city development. To distinguish it from alternative construction options, the newly developed index of urban environment quality is called dynamic (DIUEQ).

Methodological principles of this work are based on the use of the hierarchical approach, described in the monograph [19] and used in [3], and the method of summary indicators described and applied to form the index of the integral estimation of the life quality of the population in two Russian megacities in [4]. The composite indicator method is used as the basis of the methodology for constructing international indices CPI, USI, SCI; the methodology based on it is presented in detail in the document [12]. Also, the base for the development included the principles formulated in [5, 20]: the relative simplicity of the methodical apparatus; use of statistical [21] and other data from public sources; reliability, balance and comparability of the results of estimates; the ability to reproduce and verify the results of other researchers.

The work of Perm researchers [3], which describes the approach to the formation of the index of urban environment quality (IUEQ), served as the basis for the formation of DIUEQ in this study. Despite the necessary complexity of the assessment according to the methodology [3], expressed in the fact that it identifies nine structural elements of the urban space (groups of indicators), it has, in our opinion, excessive subjectivity in arbitrarily set scales of indicators and groups of indicators. In addition, the methodology [3] poorly presents economic aspects of the functioning of the urban environment, including indicators of budgetary provision.

When constructing DIUEQ, it is proposed to use a significant number of indicators used in [3]. These indicators, having dimensional or relative (specific) values, determined directly or by simple calculations based on objective data obtained from public sources (statistical reporting, reports of municipalities, the Ministry of Internal Affairs, the State Traffic Safety Inspectorate, other organizations) are called further indicators of provision. While forming DIUEQ, they form a subgroup of indicators of provision in each group of indicators. In addition, each group of indicators has two more subgroups, namely, objective economic indicators and subjective indicators of sociological measurements. The selection of the last subgroup is not mandatory and is carried out with the availability of data from measurements, obtained and processed according to a certain open methodology.

The selection in the methodology of the DIUEQ formation from three similar subgroups of indicators in each group allows expanding analytical and prognostic possibilities of the methodology by comparing the indices of subgroups within each group and the indices of the same subgroups of different groups of indicators. It is especially valuable when considering the time dynamics of all indicators.

3. Methods of forming a dynamic index of urban environment quality and an example of its implementation

With reference to the purpose of this work and in accordance with the stated methodological principles and positive international experience, the number of groups of indicators is minimized in such a way as to ensure their a priori equilibrium on the one hand and cover in aggregate all aspects of the urban environment relevant to the analysis and sustainable development, on the other hand. Therefore, the groups of indicators for which DIUEQ is formed in this study are enlarged, in comparison with the methods closest in content [3, 6].

This methodology uses four groups of urban environment quality indicators: "Safety and Ecology", "Architectural and Landscape Environment", "Social and Leisure Environment", "Engineering and Transport Infrastructure". The methodology was first applied to the formation of the DIUEQ of the municipal entity "The city of Kirov", which is the regional center of the Kirov region, the subject of the Russian Federation.

The temporal discreteness of the determination of all indicators is established in one calendar year. The calculation example gives the values of the indicators determined for 2016, but the methodology assumes that the dynamics of the indicators change both to a depth of 3-5 years in retrospect, and to continue calculations based on the results of 2017 and subsequent years.

The dynamic index of urban environment quality has four levels: a) integral DIUEQ, denoted as I_k, reflects all the low-level indicators without exception and is calculated on their basis; b) group indices of urban environment quality, calculated on the basis of the indices of the subgroups entering into each group; c) subgroup quality indices of the urban environment, calculated on the basis of the indices of all the indicators included in each subgroup; d) indices of all indicators included in the subgroup, calculated on the basis of the values of the corresponding indicators (private indices). Each indicator can be used to calculate only one particular index and be counted only in one subgroup.

The values of all the partial indices, subgroup indices, group indices and DIUEQ have the same possible limits of quantitative change, namely: the lower limit is zero (0.0), the upper limit is 1.0. The lower limit of the index corresponds to the worst value of the indicator, the upper limit to the best. These limits for private indices are established by their normalization (setting the limits of the indicator for which the numerical value of the index cannot be below zero and its limits for which the numerical value of the index cannot be higher than unity, and also the mathematical formulas for calculating the value of the index of the indicator in range of its values corresponding to the value of the index between zero and one). Such a valuation is carried out by linear mathematical transformations of the first type [4, 12] for parameters for which the value of the index equal to 1 corresponds to the highest value of the index under normalization, or by transformations of the second type [4, 12] for parameters for which the index value, equal to 1, corresponds to the lowest value of the indicator at normalization.

DIUEQ of municipal formation I_k is calculated on the basis of indices of all n groups of indicators I_i, taking into account their weight coefficients B_i, the sum of which is equal to unity, according to the formula (1):

$$I_k = \sum_{i=1}^{n}(I_i \cdot B_i) \quad (1)$$

In determining subgroup indices and group indices, the formula (2) is used, where z is the number of indicators included in a given subgroup (from 4 to 12) or in a given group (z = 3, if there are all three subgroups, or z = 2, if there is no subgroup of sociological indicators):

$$I_k = \sum_{i.j,k=1}^{z} I_{ijk} / z \quad (2)$$

When applying formula (1), the weights of the individual groups of indicators should be justified. When calculating DIUEQ, formula (1) is used with equal values of the weights of all groups. In this case, the index I_k is defined as the arithmetic average of the group indices by formula (3), analogous to formula (2):

$$I_k = \sum_{i=1}^{n} I_i / z \quad (3)$$

The method of rationing of each particular index is of particular importance. Normalization of each indicator of provision and the economic indicator is carried out once. Rationing, if possible, is performed on the basis of the natural limits of the minimum and maximum values of the indicator.

If this is not possible, either legislative norms or local government regulations are used for it. In the absence of such standards, rationing is carried out with a single use of the values of the corresponding indicators for a representative group of large and largest cities of the Russian Federation (with a population of more than 250 thousand people). Rationing of each indicator on the upper and lower limit of the index is carried out strictly individually and once; for this purpose, the absolute values of the corresponding indicators are used, and not the places of the cities of the representative group in the ranking by the indicator.

The standardization of quantitatively determined indicators of sociological measurements is also carried out by linear mathematical transformations of the first or second type [4, 12]. Respondents chose the answers mostly on a five-point or a ten-point scale. To convert the first type to the minimum value on the response scale (1 point), the index value equal to 0.0, the maximum equal to 5 or 10, the index value equal to 1.0 was put in correspondence. Intermediate scores for the index calculation were normalized by linear transformation in the range from minimum to maximum. The resulting value of the index of indicators of sociological measurements was formed as the arithmetic mean of the respondents' indices; the total number of questionnaires of a representative sample of residents of the city of Kirov was more than a thousand.

Further, Tables 1 to 4 are given as an example of methodology, the numerical values for the security indicators for Kirov in 2016 (in total 36 indicators in four groups).

Table 1. Indicators of the quality of the urban environment of Kirov in 2016 for the group "Safety and Ecology"

Number, formulation of the indicator and units of measure	Indicator value	Indicator index
1.1.1. The quantity of pollutants from all stationary sources per capita, kg/person	292.017	0.5133
1.1.2. The density of air emissions of pollutants from stationary sources, t/sq. km	49.637	0.7518
1.1.3. The density of emissions of pollutants into the atmosphere from road transport, t/sq. km	27.351	0.5442
1.1.4. The quantity of solid household waste exported per capita, m^3/person.	2.796	0.6990
1.1.5. Crime rate per 10 000 people	186.8	0.3160
1.1.6. Number of injured in an accident for 100 000 people	163.9	0.4537
1.1.7. Number of people killed in an accident for 100 000 people	6.064	0.3936
1.1.8. Number of accidents per 100 000 people	135.7	0.4572
1.1.9. Number of participants in the voluntary formations of the population for the protection of public order for 10 000 people	16.18	0.5393

Indicators of subgroups of sociological measurements are similarly defined (32 indicators, with the majority of indicators being aggregated, that is, they were determined by the results of respondents' answers to several questions), and for subgroups of economic indicators (24 indicators in total, formed mainly in size, in rubles per person, or share, as a percentage of the expenditure part of the budget of the municipality, budgetary security).

The components of DIUEQ for Kirov in groups and subgroups of indicators are given in Table 5 and in Figures 1 and 2.

DIUEQ for the city of Kirov in 2016, calculated according to formula (3) with equal weights of subgroups and groups of indicators, is $I_k = 0.4985$, as shown in Table 5; DIUEQ, calculated without regard to sociological indicators, is 0.5164. According to the indicators of provision and sociological indicators, the group "Social and leisure environment" has the highest value of subgroup indices, where these two subgroups of indicators are sufficiently balanced among themselves. The indicators of these two subgroups are even more balanced for the second and fourth groups, Table 5. The indicators of provision reflect mainly the quantitative characteristics of the urban environment, and the sociological indicators are predominantly qualitative, and therefore they complement each other well.

Table 2. Indicators of the quality of the urban environment in Kirov in 2016 for the group "Architectural and landscape environment"

Number, formulation of the indicator and units of measure	Indicator value	Indicator index
2.1.1. Provision of the population with housing stock, sq. m/person	22.9	0.2900
2.1.2. The share of old and dilapidated housing stock in the total housing stock, %	0.89	0.7527
2.1.3. Number of construction permits issued per 1000 inhabitants	0.12	0.0800
2.1.4. The share of green space in the total area of urban land, %	22.93	0.5733
2.1.5. Commissioning of the total area of housing, sq. m. per 1000 people	942.8	0.5893
2.1.6. Level of the city illumination, %	81.98	0.8198
2.1.7. Diversity of residential development, %	8.96	0.2987

Table 3. Indicators of the quality of urban environment in Kirov in 2016 for the group "Social and leisure environment"

Number, formulation of the indicator and units of measure	Indicator value	Indicator index
3.1.1. Coefficient of migration growth by 10 000 people	65.8	0.2700
3.1.2. Provision of the population with doctors for 10 000 people	29.3	0.7146
3.1.3. Proportion of children aged 1–6 years who receive pre-school educational services and (or) service for their maintenance in municipal organizations, %	76.5	0.8703
3.1.4. Proportion of children aged 5–18 years who receive supplementary education services, %	78.2	0.8026
3.1.5. Proportion of students in municipal general education institutions who study in the second (third) shift, %	33.1	0.6827
3.1.6. Level of actual provision of clubs from the regulatory requirements, %.	94.5	0.9450
3.1.7. Level of actual provision in cultural institutions of the club type in terms of the number of spectator seats from the normative requirements for 2016, %	16.91	0.1691
3.1.8. Retail area of retail trade facilities for 10 000 people, sq. m.	34250	0.8346
3.1.9. Number of places in public catering facilities for 10 000 people	666.6	0.4943

This trend of mutual complementarity was manifested to an even greater extent for the indicators of the group "Security and ecology". For this group, the overall group index is the lowest, which indicates that there are more significant problems in this area than in others. As for the indices of subgroups of economic indicators, on the whole they correspond to the indices of the other two subgroups.

Table 4. Indicators of the quality of urban environment in Kirov in 2016 for the group "Engineering and Transport Infrastructure"

Number, formulation of the indicator and units of measure	Indicator value	Indicator index
4.1.1. The share of vehicles with equipment for low-mobility citizens, %	11	0.3055
4.1.2. The share of the length of public roads of local importance that do not meet regulatory requirements, %	77.6	0.1378
4.1.3. Passenger turnover of public transport per capita, passenger-km / person	1037	0.6481
4.1.4. Average distance between public transport stops, m	305.6	0.9813
4.1.5. The share of heat networks need to be replaced, %	57.69	0.4231
4.1.6. The share of replaced heat networks, %	1.9662	0.2458
4.1.7. The share of a street water supply network need to be replaced, %	60.28	0.3972
4.1.8. The share of replaced street water supply network, %	0.8685	0.1086
4.1.9. The share of a street sewer network needs to be replaced, %	54.65	0.4535
4.1.10. The share of replaced sewerage network, %	1.2021	0.3005
4.1.11. State of the street gas network	0.0000	1.0000

Table 5. Components of DIUEQ of the city of Kirov in 2016

Number and name of the indicator group	Designation and value of the index of the group	Number and index of the subgroup of indicators of provision	Number and index of the subgroup of sociological indicators	Number and index of the subgroup of economic indicators
1. Safety and ecology	$I_1 = 0.4186$	$I_{11} = 0.5186$	$I_{12} = 0.4089$	$I_{13} = 0.3283$
2. Architectural and landscape environment	$I_2 = 0.5131$	$I_{21} = 0.4863$	$I_{22} = 0.4658$	$I_{23} = 0.5873$
3. Social and leisure environment	$I_3 = 0.5565$	$I_{31} = 0.6426$	$I_{32} = 0.5338$	$I_{33} = 0.4932$
4. Engineering and transport infrastructure	$I_4 = 0.5058$	$I_{41}= 0.4547$	$I_{42} = 0.4432$	$I_{43} = 0.6195$
Total	$I_k=0.4985$	$I_{sc} = 0.5256$	$I_{sl} = 0.4629$	$I_{ec} = 0.5071$

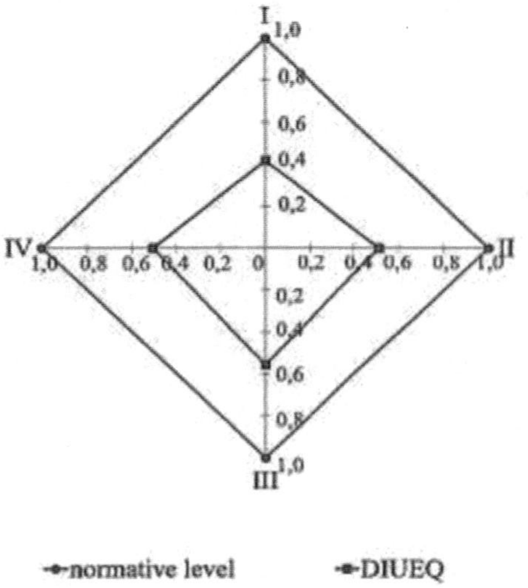

Figure 1. The square of the quality of the urban environment of the city of Kirov for 2016 by DIUEQ and by group indices

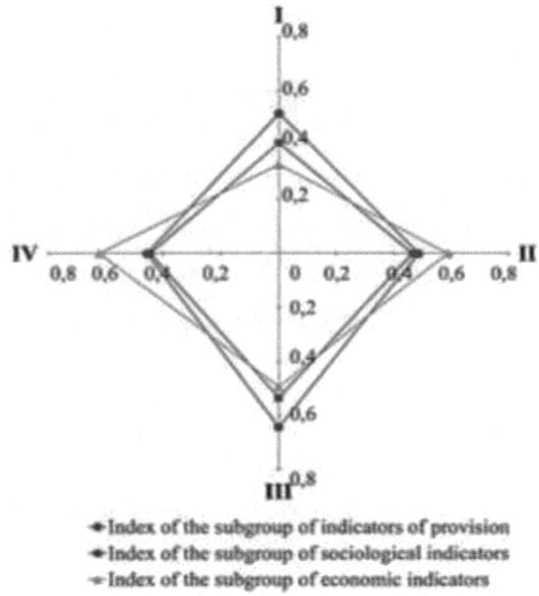

Figure 2. The square of the quality of the urban environment of the city of Kirov for 2016 by indices of subgroups of indicators

In Figures 1, 2: I – Safety and ecology; II – Architectural and landscape environment; III – Social and leisure environment; IV – Engineering and transport infrastructure

4. Conclusion

The authors have solved the following tasks according to the goal of the research.

1. The most simple, clear and transparent method for calculating DIUEQ, suitable for any city or municipal entity, is proposed. At the same time, there was used positive domestic and international experience of methods of constructing indices of urban environment quality, quality of life and sustainable development. One of the main elements of this methodology is the identification of four groups of indicators; each group includes three sub-groups of the same name – indicators of provision, indicators of sociological measurements and economic indicators.
2. The calculation of the index excludes indicators that are determined by inaccessible sources or by complex methods.
3. Comparability of all private, group indices and DIUEQ as a whole: the possibility of correction or replacement of indicators for a particular city is ensured.
4. Methodically, it is possible by means of DIUEQ to assess the dynamics of the quality of the urban environment in general and of certain groups of indicators on the temporal dynamics of the corresponding indices, including the assessment and prediction of sustainable development of cities. The effectiveness of the application of this technique for this purpose is confirmed by the use in China of the USI index.
5. When calculating and recalculating (with time variation) all the partial indices, there is no need to directly compare them with the indices of other cities due to a single normalization of all the partial indices in the initial construction of each of them.

The management of municipalities can use the proposed methodology for assessing the effectiveness of management actions for changing certain parameters of urban environment quality, both for the past time period and for planning. According to the authors, the sustainability of urban development can be achieved by ensuring the growth on a temporary basis of at least 5 years of both

DIUEQ itself and its group indices, if there is a significant reduction in the value of the index for no subgroup of indicators in each group. At the same time, economic indicators, expressed in absolute units per capita, are subject to recalculation taking into account the average annual inflation (GDP deflator).

The experience of applying the proposed methodology for calculating the DIUEQ for Kirov in 2016 showed its effectiveness and expediency of performing calculations using this method for Kirov both for the previous period (in particular, for 2012–2015) and for the accounting data in 2017. It is also planned to test this technique in 2018 in relation to other cities of the Kirov region.

References

[1] Sarchenko V I and Hirevich S A 2016 Determination of the economic essence of the formation of a quality urban environment *Problems of modern economy* **2(58)** 182–86
[2] Ilyina I N 2015 Quality of the urban environment as a factor of sustainable development of municipalities *Property relations in the Russian Federation* **5(164)** 69–82
[3] Kataeva U V and Lapin A V 2014 Formation of a methodical approach to the integrated assessment of the quality of the urban environment *Bulletin of Perm University* **2(21)** 31–9
[4] Boblakova L M and Dmitriev V V 2014 Integral assessment of the quality of life of the population of St. Petersburg and Moscow *International Journal of Experimental Education* **3-1** 91–5
[5] Ovsyannikova T U and Nikolaenko M N 2015 Assessment of the quality of the urban environment in urban areas: interregional comparisons *Bulletin of Novosibirsk State University Series: Socio-economic sciences* **15(2)** 120–31
[6] 2017 *Order of the Ministry of Construction and Housing and Communal Services of the Russian Federation* No. 1494
[7] Gurban I A 2015 Ranking of territories as a tool for measuring regional well-being *Economic Analysis: Theory and Practice* **42(441)** 36–51
[8] 2017 *Rating of regions of the Russian Federation for quality of life*
[9] *Urban environment Quality Index* (Strelka KB)
[10] 2013 Order of the Ministry of Regional Development of Russia No. 371
[11] CPI_2015 *Global City Report*
[12] 2016 *Measurement Of City Prosperity/ Methodology and Metadata*
[13] UCI Launch 2016 *Urban Sustainability Index (USI) Report*
[14] 2016 *Sustainable Cities Index*
[15] 2014 *Quality Of Living Worldwide City Rankings – Mercer Survey*
[16] 2017 Vienna Tops Mercer's 19th Quality Of Living Ranking *Mercer Global*
[17] Suter L and Orttung R 2016 Building an Arctic Urban Sustainability Index *Conf. Arctic PIRE Annual Meeting At* (Washington DC)
[18] *European Commission Indicators for sustainable cities*
[19] Perskiy U K 2008 *Interaction of micro- and macroeconomics: a hierarchical approach* (Ekaterinburg: Institute of Economics, UrB RAS)
[20] Skoufina T P and Baranov S V 2011 Interregional and intercity comparisons as a special type of scientific research: the theoretical aspect *Theory and practice of social development* **7** 322–5
[21] *Official site of the Territorial Body of the Federal State Statistics Service for the Kirov Region*

Published under licence in *Journal Title* by IOP Publishing Ltd

Delimitation and analysis of the spatial structure of the Kazan urban agglomeration

V A Fedorova, G R Safina and B Essuman-Quainoo

Kazan Federal University, 18 Kremlevskaya str., Kazan, 420008, Russia

E-mail: fva_14@mail.ru, safina27@mail.ru.

Abstract. Intensive urbanization is a feature of the modern stage of civilization development. The urban population of the planet is concentrated in new urban formations – urban agglomerations. The main goal of this work is to analyze the approaches to the delimitation of the Kazan urban agglomeration and its spatial structure. The isochronous method of transport accessibility of the agglomeration core was used. For the Kazan agglomeration in the delimitation of optimum orientation on 1-hour travel time band of accessibility. Comparison with the formal method of agglomeration delimitation (administrative-territorial) showed that the isochronous method more objectively reflects the agglomeration processes of the studied area. Realization of the "real" borders of the Kazan agglomeration will promote the adoption of the right management decisions and the implementation of measures in the field of regulation of social processes.

1. Introduction

The current stage of urbanization in various regions of the world is characterized by active education and the development of urban agglomerations. The urban agglomeration is a close territorial grouping of urban and rural settlements, which forms a complex local system characterized by various intensive links – industrial, communal, recreational, environmental, and general use of various resources of the range, while pendulum labor and cultural-household trips [1–3]. Along with the term "urban agglomeration" in modern scientific literature such terms as "local settlement systems", "areas of large cities", "group settlement systems", "urban clusters", "constellations of cities", "urban complex", "city-fields "," conurbations" etc.

Currently, issues related to the delimitation of urban agglomerations are relevant and actively discussed by both domestic and foreign urbanists. The definition of boundaries is essential to explain the main directions and prospects of development of urban areas, solving socio-economic and environmental problems arising in the process of their operation.

The purpose of this work is to distinguish the boundaries of the Kazan urban agglomeration by different methods and to analyze its spatial structure.

2. Methods

One of the approaches that make it possible to identify the territory subject to agglomeration processes and to indicate its boundaries is the method for determining the accessibility of the centre of agglomeration [3]. This method is implemented through isochronous accessibility, built based on the equal time required to achieve the agglomeration core. As the boundaries of agglomeration, different

Content from this work may be used under the terms of the Creative Commons Attribution 3.0 licence. Any further distribution of this work must maintain attribution to the author(s) and the title of the work, journal citation and DOI.
Published under licence by IOP Publishing Ltd

researchers propose to focus on 1-, 1.5 and 2-hour isochrones of the transport accessibility of the centre [4, 5].

We, with the help of this method, tried to implement the process of delimitation of the Kazan urban agglomeration. The centre of agglomeration, undoubtedly, is a large industrial, commercial, financial, tourist centre – Kazan. To build the isochrones of the centre's accessibility, it is necessary to focus on the main automobile highways and highways leading from the boundaries of the agglomeration core. With the help of the SAS Planet program and the Yandex map service, the main transport arteries connecting Kazan with the settlements located in the zone of influence of the centre were identified: P-239, Z-241, M-7 and A-245. Then, in the ArcGIS 10.2.2 program, we have digitized these transport routes and through the MapInfo Professional program, we built zones of 0.5-, 1.0-, 1.5-hour transport accessibility.

To build isochronous transport accessibility, such an indicator as the speed of movement of vehicles should guide it. As you know, under the rules of the road, the maximum permitted speed of vehicles within settlements should not exceed 60 km/h; outside the city, it can reach 90 km/h. In this study, we adopted the average speed is 75 km/h [5]. Thus, municipalities, located at a distance of 37.5 km from the borders of Kazan, on the existing roads form the Kazan agglomeration. In the zone of half-hour transport accessibility; at a distance of 75 km from Kazan-in the zone of 1-hour transport accessibility; 112.5 km – in the zone of 1.5-hour transport accessibility and at a distance of 150 km in the zone of two-hour transport accessibility.

3. Analysis

It should be noted the existing attempts to identify the boundaries of the Kazan agglomeration. Thus, in accordance with the Law of the Republic of Tatarstan No. 40-ZRT "On Approving the Strategy of Social and Economic Development of the Republic of Tatarstan until 2030", the borders of the Kazan agglomeration are conducted along the boundaries of 5 municipal districts: Verkhneuslonskiy, Vysokogorskiy, Zelenodolskiy, Laishevskiy, Pestrechinskiy, Kazan [6, 7]. Thus, the composition of the Kazan agglomeration includes the municipal districts surrounding its core or centre to it (figure 1). In addition, it is indicated that Volzhsk (the Republic of Mari El) is involved in the agglomeration processes.

Characteristics of the structure and population of Kazan urban agglomeration within the boundaries proposed in the "Strategy of socio-economic development of the Republic of Tatarstan until 2030" are presented in tables 1 and 2, figure 1.

Table 1. The characteristics of the settlements and residents of the Kazan agglomeration, as of January 1, 2016 (was compiled on the basis [8])

Types of settlements in the agglomeration	Number of settlements	Population size	
		In thousands	%
Cities	4	1324.19	87.2
Urban-type settlements	2	25.56	1.7
Rural settlements	109	168.23	11.1
Total	115	1517.98	100.0

Kazan agglomeration is located in the North-West of the Republic of Tatarstan, mainly on the left Bank of the Volga River. The total population as of January 1, 2016, is 1.5 million people; the average population density is 179.4 people/km2. Kazan agglomeration – monocentric post-industrial agglomeration, including the cities of Kazan and Zelenodolsk, it is in these two cities that the vast majority of the population of the agglomeration lives – 80.17 and 6.5 %, respectively (table 3).

Table 2. The structure of the Kazan urban agglomeration (was compiled on the basis [8])

Municipal districts and administrative centres	Area of the territory (km²)	Population size (in thousands)	Number of Settlements
Kazan	614.16	1216.97	1 city
Verkhneuslonskiy (Verkhniy Uslon)	1302.82	16.58	1 town (Innopolis), 73 villages as part of 19 rural settlements
Vysokogorskiy (Vysokaya Gora)	1701.20	47.34	124 villages as part of 25 rural settlements
Zelenodolskiy (Zelenodolsk)	1396.00	164.78	106 villages as part of 3 urban and 21 rural settlements
Laishevskiy (Laishevo)	2094.00	40.04	68 villages as part of 1 urban and 23 rural settlements.
Pestrechinskiy (Pestretsy)	1352.40	32.27	73 villages as part of 21 rural settlement:
The Kazan agglomeration	8460.58	1517.98	446

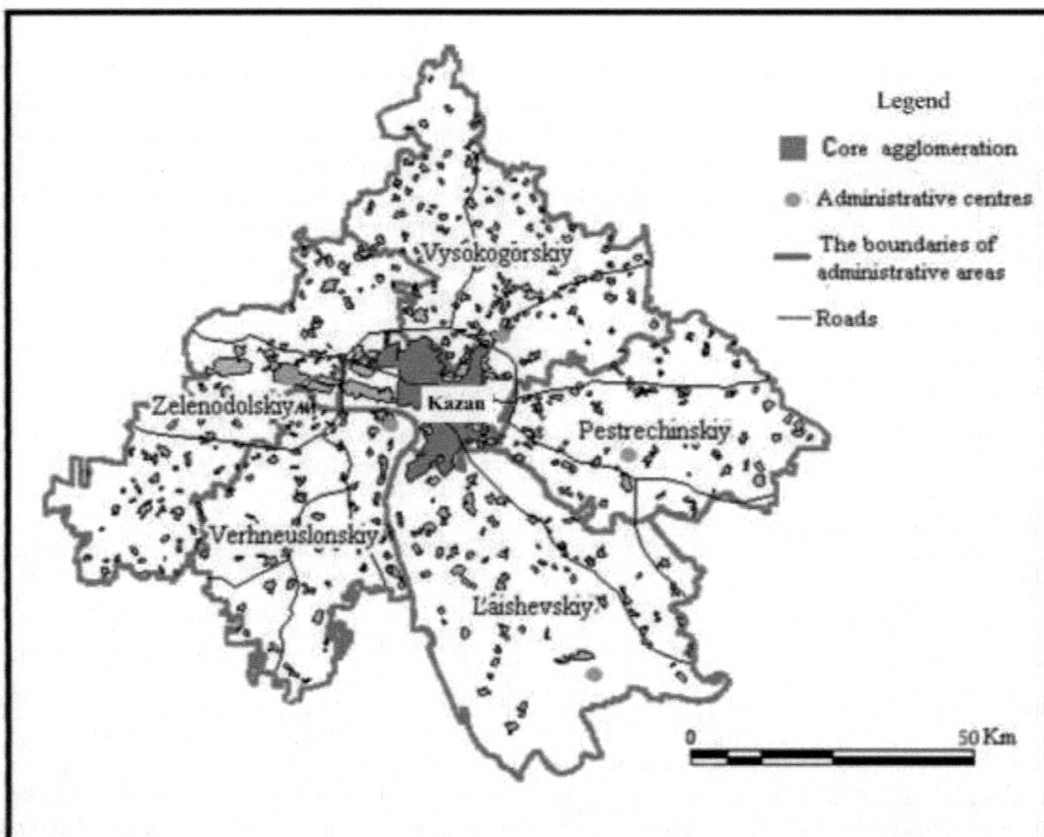

Figure 1. Map-scheme of the Kazan urban agglomeration in accordance with the "Strategy of socio-economic development of the Republic of Tatarstan until 2030"

Table 3. The Urban Population of the Kazan Agglomeration (was compiled on the basis [8])

Populated locality	Population size (in thousands)
Kazan	1216.97
Zelenodolsk	98.76
Laishevo	8.36
Innopolis	0.096
Vasilyevo	17.53
Nizhniye Vyazovye	8.02

The exhaustion of many resources within the core of the Kazan agglomeration and the impossibility of endless expansion of its urban boundaries causes the need to develop satellite cities that reduce the load on the agglomeration centre and make fuller use of the potential of the territories [9–11].

So, within the Kazan agglomeration (on the territory of Verkhneuslonskiy district, on the right bank of the Volga river, at a distance of 40 km from Kazan) at the present stage "from scratch" created such a satellite – an innovative science city, Innopolis city, designed for 155 thousand inhabitants. It is worth mentioning the Kazan SMART city project – this satellite city is planned to be located in the agglomeration zone closest to the core, at a distance of 15 km from Kazan, near the Kazan international airport. The estimated population of Kazan SMART city is 58.8 thousand people.

The construction of the isochron for the transport accessibility of the centre of agglomeration allows us to objectively and accurately identify the area of agglomeration processes and to delimit. The 0.5, 1.0, and 1.5-hour isochrons we constructed are presented in figure 2.

Generalized data on the population living in areas of 0.5-, 1- and 1.5-hour transport accessibility of the core of the Kazan agglomeration, as well as the degree of development of transport communications allow to identify the most objectively the territory and settlements, gravitating to the core of agglomeration and associated with it diverse connections (industrial, labor, municipal, recreational, etc.). The constructed isochrones of transport accessibility of the city centre reflect the intensity of daily pendulum migrations of the population that allows defining "real" borders of an agglomeration. Thus, the modern territory of the Kazan agglomeration is limited to the area of 1-hour transport accessibility, in which the satellite cities are located, which are closely connected with the centre of agglomeration by numerous factors (figure 2).

Within the first zone, namely, the area bounded the 0.5-hour isochronous transport, placed 3 cities (Kazan, Zelenodolsk, Innopolis) and 2 urban-type settlements (Vasilyevo and Nizhniye Vyazovye). In the second zone – the area between the 0.5- and 1-hour zones, there are 2 towns (Laishevo and Arsk) and 1 urban-type settlement (Rybnaya Sloboda).

In the third zone, encompassing the interval between 1- and 1.5-hour isochronous transport accessibility, there is 1 town (Chistopol) and 5 urban-type settlements (Kamskoye ustye, Apastovo, Baltasi, Bogatye Saby, Alekseevskoye) (figure 2). However, it should be noted that the population living in this zone, in addition to Kazan, there are other centres of attraction, so at this stage include the specified territory in the Kazan agglomeration is inexpedient. The 1.5-hour access zone of the core of the Kazan agglomeration is the territory of its future development. The vast majority of the agglomeration population (93.2 %) lives in a 0.5-hour transport accessibility zone and only 6.8 % in a 1-hour zone.

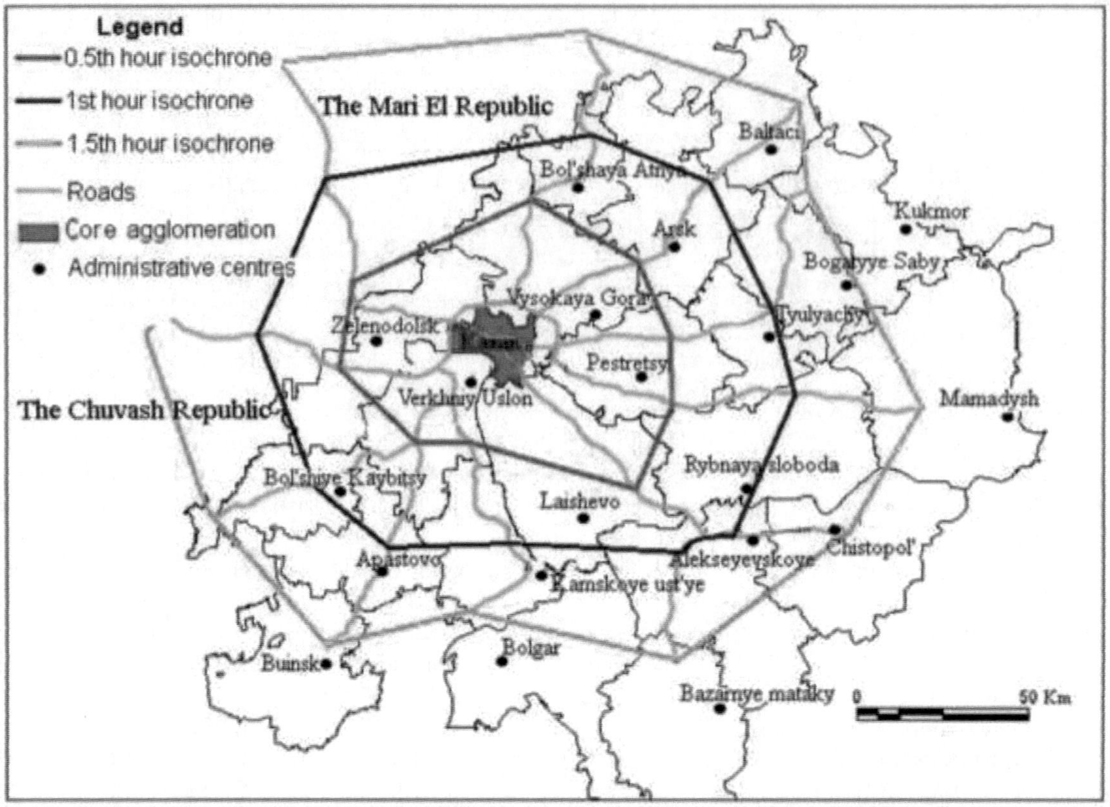

Figure 2. Isochrones of transport accessibility of the core of the Kazan agglomeration

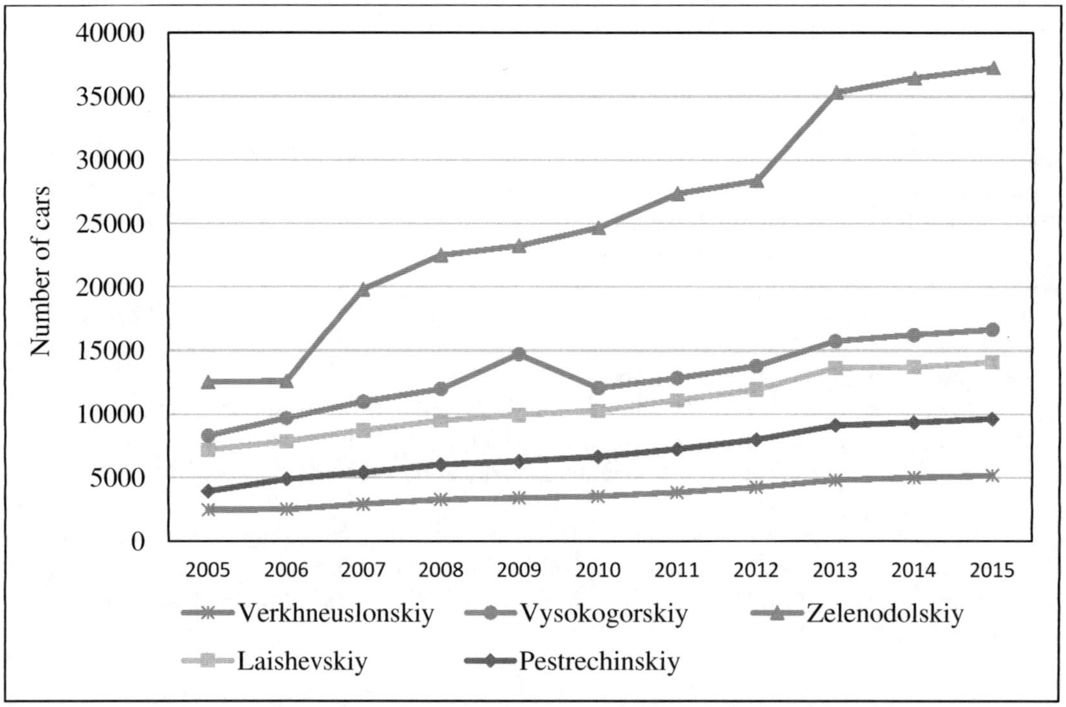

Figure 3. Dynamics of the number of vehicles within the Kazan agglomeration

"Motorization" of the population and the availability of vehicles at the present stage are one of the factors of strengthening the relationship between the settlements that make up the Kazan agglomeration. The number of vehicles reflects the intensity of pendulum migration, which is an inter-urban labour relation, expressed in the location of places of work and residence at a significant distance from each other, in different administrative-territorial formations. In recent years, the tendency of increasing the number of vehicles in individuals in the municipal areas that are part of the Kazan agglomeration (figure 3).

Thus, in Kazan, this figure has increased by 2 times (compared to 2005 and 2015), in municipal areas – by 2–3 times. This fact indicates an increase in the possibilities of pendulum migration, which affects the development of agglomeration and the extension of its borders.

A comparison of the results of the delimitation carried out by two different methods is presented in table 4.

Table 4. Characteristics of the Kazan agglomeration within different boundaries

The name of the method of delimitation		The area of the land/territory (thousand km^2)	Population size (in thousands)
A formal approach to the allocation of borders – on the boundaries of municipal districts surrounding the core (Strategy of socio-economic development of the Republic of Tatarstan until 2030 [7])		8.46	1517.98
A method based on the determination of the possibility of a pendulum migration	0.5-hour isochrone accessibility centre	4.01	1448.7
	1-hour isochrone accessibility centre	11.86	1554.1
	1.5-hours isochrone accessibility centre	25.55	1699.3

Thus, the area of the Kazan agglomeration, determined based on 1-hour isochronous transport accessibility of the centre, is 40 % higher than the figure established in accordance with the formal approach. In addition to the 5 municipal districts included in the Kazan agglomeration identified in the 2030 Strategy, the agglomeration processes cover Atninskiy, Arskiy, Tyulyachinskiy, Rybnoslobodskiy, Kamsko-Ustinskiy and Kaybitsky districts of the Republic of Tatarstan, as well as part of the territory of the Mari El Republic. The number of residents involved in the process of agglomeration within the Republic of Tatarstan, with an increase of 36.12 thousand people (2.4 %) compared to the Kazan agglomeration in the interpretation proposed in the Strategy 2030.

4. Conclusion

Thus, comparison with the formal method of delimitation of agglomeration (administrative-territorial) showed that the isochron method more objectively reflects the agglomeration processes of the investigated territory. For the Kazan agglomeration in the delimitation of optimum orientation on 1-hour travel time band of accessibility.

The obtained results indicate the need for a more detailed study of the peripheral zone of the Kazan agglomeration for the intensity of agglomeration processes, which will allow more objectively and accurately carry out its delimitation. The establishment of "real" boundaries of the Kazan agglomeration will facilitate the adoption of good management decisions and activities in the field of regulation of social processes.

Acknowledgements

The publication was prepared with the financial support of the Russian Foundation for Basic Research and the Government of the Republic of Tatarstan, Project No. 17-13-16003-ОГН\18.

References

[1] Pivovarov Y L 1999 *Fundamentals of georbanistics: Urbanization and urban systems* (Moskow: Vlados)

[2] Percik E N 2006 *District planning (territorial planning)* (Moskow: Gardariki)

[3] Lappo G M 1997 *Geography of cities* (Moscow: Vlados)

[4] Lappo G et al 2007 Agglomerations of Russia in the 21st Century *Bulletin of the Regional Development Fund of the Irkutsk Region* **1** 46–8

[5] Izhguzina N R 2014 Approaches to the delimitation of urban agglomerations *Discussion* **9** 44–52

[6] *The law of the Republic of Tatarstan of 17.06.2015 N 40-ZRT "On approval of the Strategy of social and economic development of the Republic of Tatarstan till 2030"*

[7] *The strategy of social and economic development of the Republic of Tatarstan for the period up to 2030*

[8] 2016 *Population of municipalities of the Republic of Tatarstan: Statistical bulletin* (Kazan: Tatarstanstat Publishing Center)

[9] Safina G R et al 2016 Territorial reserves of major cities: challenges, experience, solutions *International Journal of Pharmacy and Technology* **8** (3) 14864–71

[10] Medvedeva R A et al 2017 Urban densification: Features, environmental problems, and prospects *International Journal of Pharmacy and Technology* **11** (4) 868–71

[11] Medvedeva R A et al 2017 Ecological and social aspects of urban environment quality (on the example of the cities of the Republic of Tatarstan) *Proceedings of the National Academy of managing staff of culture and arts* **4** 140–4

Published under licence in* Journal Title *by IOP Publishing Ltd.

IOP Conf. Series: Earth and Environmental Science **177** (2018) 012009 doi:10.1088/1755-1315/177/1/012009

Technologies for optimization of ecosystem services and functions of soils under anthropogenic impact in urban areas

N Nevedrov and E Protsenko

Scientifically Research Laboratory of Environmental Monitoring, Kursk State University, 33 Radishcheva str., Kursk, 305000, Russia,

e-mail: 9202635354@mail.ru, protselena@yandex.ru

Abstract. The article presents the data of ecological monitoring and assessment of soils of the ecosystems represented in Kursk. On the basis of the description of the structure of 17 soil profile cuts laid out in different functional zones and geomorphological elements of urban landscapes, 7 main types of soils mainly represented in the ecosystems of Kursk were identified and depicted on the map-scheme. According to the monitoring of heavy metals, a tendency to the increase of the concentration of their gross forms in soils has been noted. The increase in their content reaches up to 48.9 % per year. Changes in the concentrations of mobile forms are determined by the dynamics of soil processes and, consequently, they do not always have increasing tendency. The soil remediation technologies, adapted for urban environment, are proposed. Phytoremediation technology and sorption method of detoxification helps to increase the environmental comfort in cities significantly.

1. Introduction

The effectiveness of the state policy in the sphere of ecology can be achieved only through the analysis of the full-fledged data on environmental objects and their integrated assessment. The necessity to evaluate all the components of natural and natural-anthropogenic objects of the environment is obvious. However, soil ecological monitoring in order to assess the quality of the soil cover is essential. Due to their natural specifics, the soils serve as a kind of depot that accumulates, stabilizes and neutralizes numerous products of anthropogenic emissions, fulfilling the role of a barrier on the way of migration of toxicants to adjacent environmental components. These characteristics of soils, as one of the most important components of the environment, determine the stability of ecosystems to anthropogenic impact and ensure their stability and comfort for humans [1].

Preservation of ecosystem functions of soils is a key priority for modern ecologists [2]. This is an integral component when forming an environmentally friendly zones in the cities. The urban soils fulfill a number of important ecological functions and are the basis of the ecological framework [1, 3]. Soils of modern cities are under powerful anthropogenic pressure, leading to their degradation [3]. Technogenesis supports numerous soil damages that results in the formation of various types of urban soils. Basic characteristics and development strategies of urban soils differ greatly from their background analoques [2–4].

Heavy metals (HM) pollution of the soils is one of the main environmental problems that limits a number of ecosystem services of soils. An important practical task of urboecology is the development and introduction of effective methods for the detoxification of soils with heavy metals pollution.

Content from this work may be used under the terms of the Creative Commons Attribution 3.0 licence. Any further distribution of this work must maintain attribution to the author(s) and the title of the work, journal citation and DOI.
Published under licence by IOP Publishing Ltd

Various selective treatment methods are currently proposed. However, their adaptation to the spatial heterogeneity and excessive anthropogenic changes of the environment requires a large number of studies and their approbation within the conditions of urban environment. For example, an environmentally safe and adequate phytoremediation method of treatment can be successfully applied to reduce toxic concentrations of HM in soils. It is sufficient to select the plants-hyperaccumulators of HM the use of which is not limited by the bioclimatic conditions, the level f soil pollution and metal specificity of a particular plant [5–7].

Sorption treatment methods, based on the ability of materials and fertilizers applied on the polluted soils to immobilize heavy metals in surface horizons, preventing their transfer into adjacent environments and trophic chains, are highly effective. Sorbents can save a significant amount of natural and financial resources required for ecological function recovery of urban soils.

The aim of the work was an environmental assessment of the soils polluted by the heavy metals in Kursk and testing of selective methods for their remediation.

2. Objects and methods of research

When choosing the key areas of the research and sampling sites, such factors as land-use category and topographic effect were taken into consideration. The study of soils and their diagnostics were carried out by laying soil profile cuts and bore holes in various functional zones of the city, as well as in various elementary geomorphological units of urbolandscapes. 17 soil profile cuts that characterize the up-to-date soil cover of Kursk were analyzed. Diagnosis and classification were carried out in accordance with the modern approaches to the classification of urban soils [8, 9]. The selection of soil samples in the profile cuts was carried out from each genetic horizon.

The data of ecological monitoring of heavy metals pollution in soils are obtained from three observation stations, located in the industrial and residential areas of the southern part of Kursk. Samples of wood and leaves of Tilia platyphyllos, Aesculus hippocastanum and Acer platanoides were also taken in the industrial and residential areas of the city in the autumn period. This approach allowed us to determine the maximum accumulation of heavy metals during the growing season. Based on the experimental data, the composition of the studied tree species of the remediation park was modeled, and practical recommendations were also made to introduce this method of remediation into the urban environment.

The detoxification capacity of the sorbent on basis of natural lime and sapropel was studied in laboratory experiments with model pollution of gray forest middle loamy soil with Plumbum ions.

The laboratory experience was laid in plastic vegetation vessels, where 2 kg of gray forest soil were placed. In each vessel, a pollutant element was introduced – Plumbum in the form of its oxide in a dose of 10 MPC (320 mg/kg). After soil system stabilization (after 20 days), four variants of the laboratory experiment were put into the vessels in 6-fold replication according to the scheme:

1) Test sample (background – 10 MPC for Plumbum, 640 mg/vessel);
2) Background + sapropel + lime – 5 g of mixture/vessel (ratio of components 1:5 by weight);
3) Background + sapropel + lime – 10 g of mixture/vessel (ratio of components 1:5 by weight);
4) Background + sapropel + lime – 15 g mixture/vessel (ratio of components 1:5 by weight).

The experiment lasted for 40 days. Every 5 days the actual soil acidity (pH) was measured by the ionometric method. Artificial watering, corresponding to the average annual rainfall for the Kursk region, was applied. Quantitative analysis of heavy metals in soils was carried out using atomic absorption spectrometry. Movable forms of heavy metals were extracted with ammonium acetate buffer (pH = 4.8).

3. Geo-information modeling of the soil cover of Kursk

Carrying out an ecological and geochemical survey of the soil cover of Kursk resulted in constructing a map-scheme of the city soils (Fig. 1).

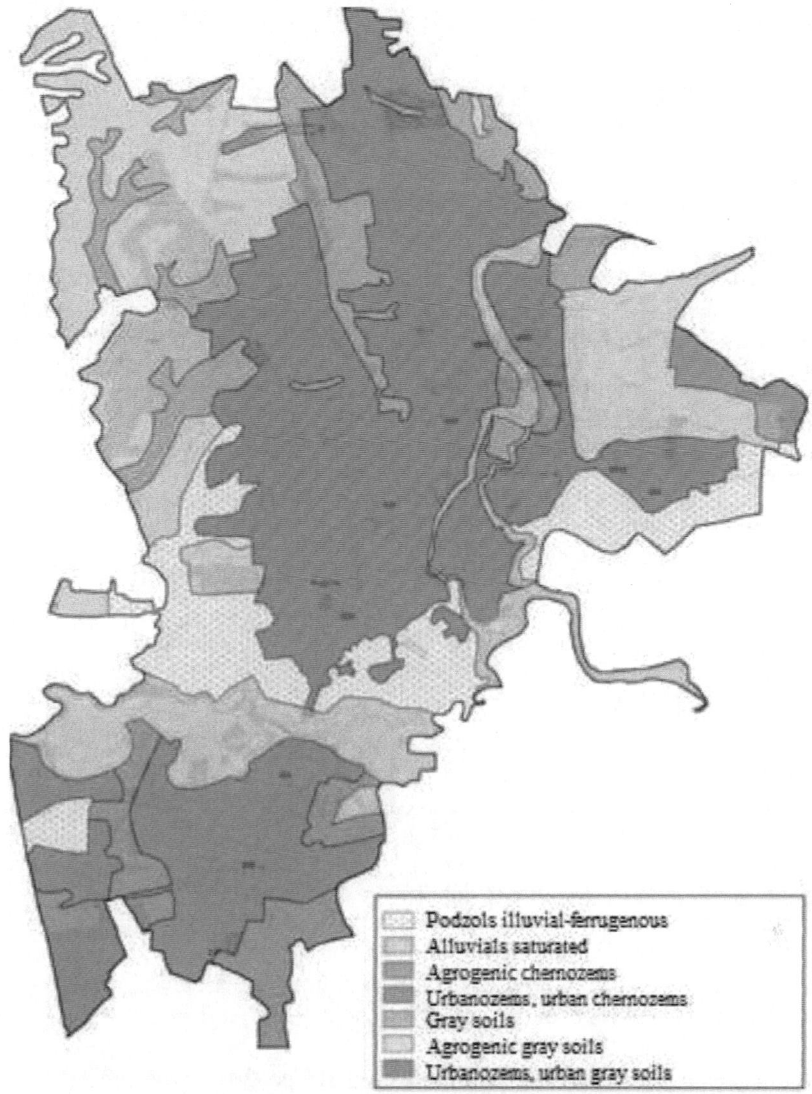

Figure 1. The map-scheme of the city soils of Kursk

The territory of the city is represented by four types of initial native soils historically formed during the Holocene – leached chernozems and gray forest soils on loess loams, alluvial soils of overflow lands on ancient alluvial sediments and sandy podzols formed on ancient alluvial and fluvioglacial sandy sediments in the terrace above the floodplain of the Seym River. During steps of the morphological description of soil profile cuts and the analysis of the physicochemical characteristics of the soils of the chosen sites, with reference to the cadastral schemes and the functional zoning of the territories, seven main soil groups have been identified. These 7 groups include soils of different nature and degree of anthropogenic transformation (Figure 1). The soils of industrial zones are highly fragmented. Disturbed soil profiles formed as a result of the functioning and construction of industrial sites of various purposes can be found there. Urbanozem with depth of urban horizons of more than 50 cm is the most presented type of soil. In such soils, the amount of anthropogenic inclusions in the Urbik horizons reaches up to 30–35 %. Heavy metals pollution that reaches the depth of 1 m is recorded.

Residential zones of the city have a great variety of anthropogenic changes in the soil cover. Along with urbanozems in urban wastelands urbogray forest soils and urbodark-gray soils are presented, characterized by the gradual intake and processing of urbotehnogenic sediments. Depth of profile transformation is less than 50 cm.

The soils of sanitary protection and recreational zones of the key sites under research have undisturbed profiles with a classical order of genetic horizons. The soils of sanitary protection zones adjacent to the industrial-active zones are polluted with HM. The soils have frank elements of physical degradation: soil consolidation and erosion, in the areas that are relatively distant from pollutant emission sources

4. Ecological monitoring and assessment of soils in different functional areas of Kursk

When analyzing the results of environmental monitoring obtained from observation stations in the southern part of Kursk, the increase in the gross form of Plumbum for 5 years was noted and totalled 47 mg/kg (Table 1).

Table 1. The average content of gross and mobile forms of heavy metal ions in the soils of Kursk in the period from 2012–2017 year

Element	Content 2012 (mg/kg)	Content 2015 (mg/kg)	Content 2017 (mg/kg)	Reproportion, (%) 2012–2015 year	Reproportion, (%) 2015–2017 year	MPC (mg/kg)
Zn_{gross}	86.3±5.1	98.2±9.1	79.7±8.4	13.7	-18.8	100
$Zn_{mob.}$	25.2±3.9	28.6±3.4	6.3±1.2	9.5	-77.8	23
Pb_{gross}	80.4±8.2	98.2±13.1	127.4±5.6	21.2	29.7	32
$Pb_{mob.}$	14.3±2.6	27.6±4.3	23.0±2.1	93.0	-	6
$Ni_{gross.}$	24.1±5.2	27.5±3.4	54.4±4.2	14.1	97.8	40
$Ni_{mob.}$	1.6±0.4	1.6±0.4	2.2±0.3	-	39.3	4
Cd_{gross}	2.1±1.3	2.1±1.1	2.7±0.9	-	32.8	1
Cd_{mob}	0.6±0.07	0.9±0.2	0.5±0.1	50.0	-44.4	0.1
Cu_{gross}	16.1±3.8	21.7±1.7	65.2±1.9	34.7	200.0	55
Cu_{mob}	0.9±0.1	1.4±0.1	1.1±0.2	55.5	-	6

The stable increase in concentration of Plumbum in soils is fed by to aerosol deposition of the polluted atmosphere and a constant input of accumulating urbosediment. An increase in the content of gross cadmium has been observed only for the last two years and is 32.8 % (0.34 mg/kg per year) (Table 1).

The maximum accumulation in the period from 2015 to 2017 was recorded in gross forms of cuprum and nickel, it figures up to 97.8 and 200 % respectively. Such an increase in the content of the pollutants could have occured as a result of peak industrial emissions. The content of the gross form of zinc has decreased noteworthily, apparently due to translocation into plants and deposition in wood, due to subsurface leaching. This is proved by a decrease in the concentration of the mobile form of the metal.

Concentrations of mobile plumbum and cuprum have been stable for the past two years, and the concentration of mobile cadmium has decreased by 0.4 mg/kg. This process goes on despite of the high rate of growth of its gross forms, that shows a high degree of their affinity for sorption on geochemical barriers (organic matter, clay minerals, magnetic phase of the soil) and high capacity of the soil absorbing complex with respect to these elements. However, the amount of mobile nickel is increasing (39.3 %). It seems that nickel in comparison with other metals under analysis has the lowest affinity for sorption on geochemical barriers and a considerable part of it stays in the soil solution.

Figure 2 shows the data of the total soil pollution of the main catenas in industrial, residential and sanitary protection zones in the southern part of Kursk, obtained from the analysis of six lots located in the eluvial and transsuperaqual geomorphological elements of the landscape. Urbanozems of lot No. 1 and No. 2 have a very high very dangerous level of pollution, the alluvial soil of lot No. 3 has high, dangerous level. The soils of lots No. 4–6 (alluvial soil, soddy-podzol sandy illuvial-ferruginous, urban soil) refer to low level of pollution. In general, it is possible to notice a decrease in the total pollution moving away from the active source of pollution to the Seim River (Figure 3). And a slight increase in the total soiling index (Zc) in urbanozems in the territory of the former machinery plant.

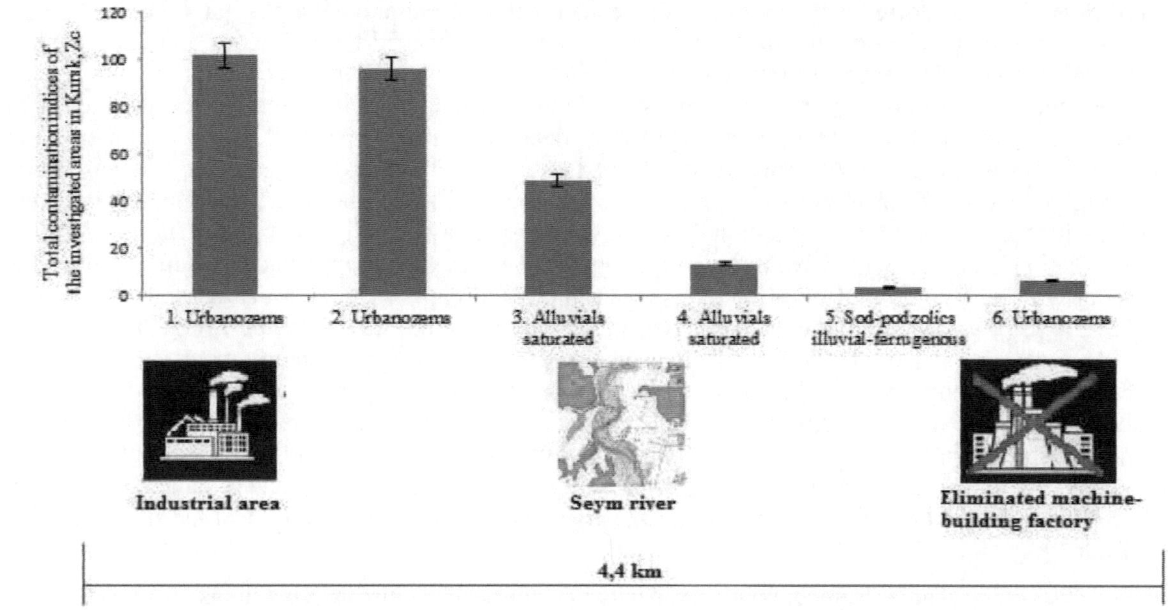

Figure 2. Distribution of total pollution in the soils of the catena under analysis

This can be explained by the prevailing wind and lateral migrations of HM in the soil catena from the alluvial elements (lots No. 1 and No. 2) to transsuperaqual (plot No. 3). Natural barriers for the further distribution of HM are the river and the forest range situated on its right bank. As a result, the transported pollutants produce a high level of pollution on the left bank of the river, and on the right a low level is observed. The territory of the former machinery plant were subject to reorganization (plot No. 6), the polluted soil was removed. This explains the low level of urban pollution at the moment.

5. Selective technologies for urban soils remediation with heavy metals pollution

Environmental comfort level increase in the cities can be achieved by reducing the toxicity of soils with heavy metals pollution. Figure 3 shows the proprietary model of urban microlandscape remediation.

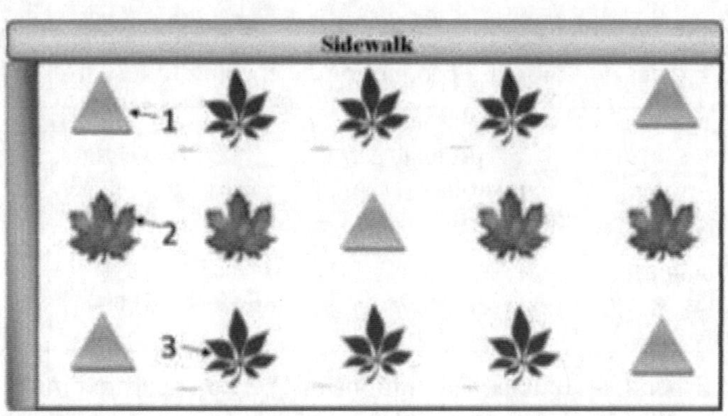

Figure 3. The proprietary model of tree planting
that forms a single unit of remediating urban microlandscape [10]
1 – Tilia platyphyllos, 2 – Acer platanoides, 3 – Aesculus hippocastanum

An important aspect of the patent is the composition of wood species, selected not only by the HM accumulation characteristics, but also by the period of their leaf abscission. A set of remediation species allows to prevent the transfer of HM to the soil during the leaf abscission of Aesculus hippocastanum and Acer platanoides, and also to ensure the supply of organic substance and the progress of the soil-forming processes due to the later leaf abscission of Tilia platyphyllos.

The proposed technology of trees-remediators planting enhanced the aesthetics and facilitated to increase the pollution export of heavy metals not only from the surface horizons, but also from deep layers due to the rooting depth. The volumes of produced wood of all trees-remediators will allow to reduce the concentrations of HM in the soil by depositing them in wood. The location of Tilia platyphyllos at the corners and in the center of the area allows the distribution of leaf abscission throughout the soil surface. The distance between the trees facilitates the collection of leaves and lawn care with the use of compact agricultural vehicles. The grass lawns will give a picturesque look to the remediation park and will prevent the migration of heavy metals deeper into the soil profile. This model will be a single unit of sustainable urban microlandscapes. The area of this unit is 1.76 Are (52 × 34 m). The number of the units will depend on the area of the site planned for remediation.

The effectiveness of the described phyto-landscape model in urban environment with increased spatially-nonhomogeneous technogenic burden is justified by the obtained empirical data on the accumulation of heavy metals by the vegetative organs of the species-remediators. The greatest accumulating capacity in relation to plumbum and cadmium has Aesculus hippocastanum, which deposits up to 856 mg/kg and 0.9 mg/kg of cadmium. Such remediator as Acer platanoides is highly effective in relation to zinc. The leaves of Acer platanoides can accumulate up to 1355 mg/kg of zinc (Table 2).

Table 2. Accumulating capacity of tree species in urban ecosystems (according to the research carried out in Kursk region), mg/kg

Element	Aesculus hippocastanum		Acer platanoides		Tilia platyphyllos	
	Leaves	Wood	Leaves	Wood	Leaves	Wood
Pb	732±2.4	856±6.7	122.2±1.3	124.1±1.8	22.7±0.1	23.4±0.2
Cd	0,6±0.1	0,9±0.1	0,29±0.01	0,32±0.02	0.12±0.01	0.18±0.01
Zn	878±7.6	956±8.4	1355.4±7.4	1687.5±8.7	137.2±1.4	143.4±1.7

Leaves of Tilia platyphyllos accumulate small amounts of the metals. This will limit the secondary soil pollution during the natural course of soil formation processes and the mineralization of leaf abscission.

Table 3 shows that the annual wood growth will allow the removal of a significant part of heavy metals from the soil cover. Thus Aesculus hippocastanum will accumulate up to 22.2 g of plumbum, 0.02 g of cadmium and 24.8 g of zinc per year.

Table 3. The dependence of the annual amount of accumulated heavy metals on the tree species and biomass production (according to research in Kursk region)

Tree species	Wood production by a unit of sustainable urban microlandscapes, m^3	Wood production dry weight, kg	The amount of HM, deposited in wood, g		
			Pb	Cd	Zn
Aesculus hippocastanum	0.04	26.0	22.24	0.023	24.86
Tilia platyphyllos	0.03	11.4	0.27	0.02	1.63
Acer platanoides	0.024	14.4	1.78	0.004	24.29

Thus, it was established that toxic concentrations of HM concentrated in the soil are transfered with leaf abscission of Aesculus hippocastanum and Acer platanoides and are "weakened" by migration and accumulation in wood.

Phytoremediation method can hardly be used in the epicenters of HM emission with "top-cut grade" of heavy metals. In such cases it is rational to use the means of physical and chemical recreation. The introduction of a sorbent based on natural lime and sapropel materials have a beneficial effect on the immobilization of plumbum by increasing the actual soil acidity and increasing the capacity of the soil absorbing complex. This method allows you fix plumbum in the humus-accumulative horizon of soils and prevents its subsoil migration.

Analyzing the results of the temporal dynamics of the acidity of the soil-water extract (monitoring the temporal changes in the pH (H_2O) of the soil solution in the soil-water extract) at 40 days' exposure, unstable changes of the index were observed during the first 25 days, but then the acidity became stable in the variants of the experiment with the introduced sorbent. In the test sample there was a high decrease in pH (H_2O) and on the 25th day it also stabilized. The highest pH value – 8 was noted in the experiment with a sorbent dose of 15 g/vessel, the lowest reaction of the medium (pH = 7.6) – in the test sample (Figure 4). The effect of the sorbent increases the alkalinity of gray forest soil, that definitely reduces the level of mobile forms of plumbum in it. It is known that in the alkaline environment the electric charges of ampholytoides are redistributed and the sorption of Pb^{2+} increases on mineral and organo-mineral soil colloids.

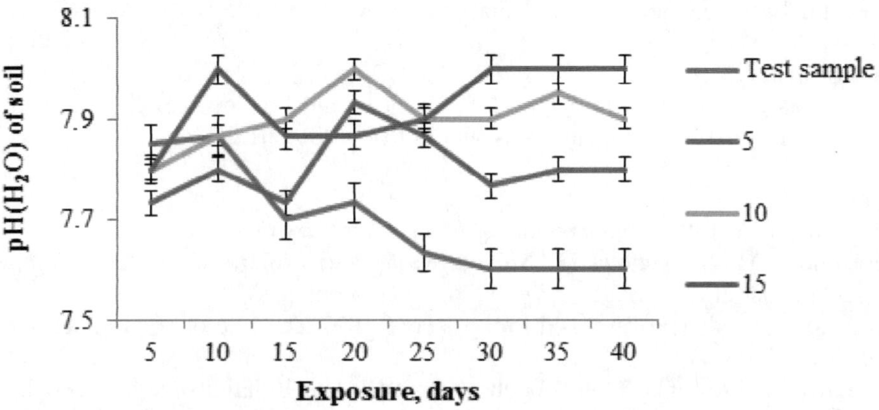

Figure 4. Dependence of (temporal) dynamics of soil acidity pH (H2O) of gray forest soil on the dose of the sorbent introduced (g/vessel)

During the analysis of the obtained data on the mass concentrations of plumbum, it was found that the mobility of plumbum reduces significantly when the sorbent is introduced, by 82.5–89.8 % to test sample. This is explained by the composition of the introduced sapropel, which increases the capacity of the soil absorbing complex (organic matter, clay minerals) thus provides additional sorption of plumbum on mineral and organo-mineral colloids. The minimum proportion of mobile forms of plumbum was noted at the dose of the recultivant – 5 g/vessel (Table 4).

Table 4. The effect of the dose of the sorbent on the mobility of plumbum in gray forest soil

Dose of sorbent (g/vessel)	Test sample (without introduction)	5	10	15
The content of mobile form of Pb	226.2±26.6	23.1±5.7	39.7±9.2	34.7±8.6
Decrease of mobility, %	-	89.8	82.5	85.7

The results obtained during the experiments convincingly prove that the use of a sorbent based on lime and sapropel effectively reduces the concentrations of mobile forms of plumbum in polluted soils.

6. Conclusion

The mixed character of the soil cover of Kursk is determined by the intensive development and multi-functionality of the industrial, town-planning and transport complexes. The functioning of existing and construction of new industrial economic entities and residential estates of the city is accompanied by a powerful impact on the soil. Increased spatial heterogeneity of the environment and the anthropogenic impact lead to the limitation of ecosystem services of the vast majority of urban soils. This has irreversible effect on the ecological balance and courses a decrease in the environmental comfort level. The annual increase in plumbum is about 7–14.8 %, cadmium 16.4 %, nickel to 48.9 %. This tendency results in the extension of urban soils of with a very high and very dangerous level of pollution that can be ranked to environmental distress. The data obtained from soil monitoring indicate the worsening of the ecological situation in Kursk, which indicates the need to introduce the practice of remediation and restoration of the ecological functions of the urban soils. The use of selective methods of phytoremediation made it possible to transfer up to 22.2 g of plumbum, 0.02 g of cadmium and 24.8 g of zinc from polluted soils per year. The introduction of a sorbent based on lime and sapropel into polluted soils reduces the ratio of the mobile form of plumbum by 82.5–89.8 %. Ecologically safe methods of soil remediation allow forming a sustainable comfortable environment and have obvious resource potential and economic and financial advantages.

Acknowledgements

The research was carried out with support of Grant of the President of the Russian Federation for Young Russian Scientists – Candidates of Sciences MK-4086.2018.5.

References

[1] Dobrovolskiy V 1998 *Fundamentals of biogeochemistry* (Moscow)
[2] Gerasimova M, Stroganova M, Mozharova N and Prokofieva T 2003 *Anthropogenic soils: genesis, geography, reclamation* (Oikumena, Smolensk)
[3] Stroganova M, Myagkova A and Prokofieva T 1997 The role of soils in urban ecosystems *Soil Science* **1** 96–101
[4] Nevedrov N, Protsenko E and Glebova I 2018 The Relationship between Bulk and Mobile Forms of Heavy Metals in Soils of Kursk *Eurasian Soil Science* **1** 112–19
[5] Galiulin R and Galiulina R 2008 Phytoremediation of soils polluted with heavy metals *Bulletin of the Russian Academy of Sciences* **3** 247–49
[6] Baker A and Brooks R 1989 Terrestrial Higher Plants which Hyperaccumulate Metallic Elements *A Review of Their Distribution, ecology and phytochemistry. Biorecovery* **1** 81–126
[7] Protsenko E, Nevedrov N and Zubkova T 2017 Selective Phytoamelioration of Soils in Kursk Oblast *Moscow University Soil Science Bulletin* **2** 78–83
[8] Rossiter D 2007 Classification of urban and industrial soils in the world reference base for soil resources *J. Soils Sediments* **7** 96–100
[9] Prokofieva T, Martinenko I and Ivannikov F 2011 Systematics of soils and soil-forming species of Moscow and the possibility of their inclusion in the general classification *Soil Science* **5** 611–23
[10] Nevedrov N, Protsenko E and Terekhov V 2018 Method for reducing heavy metal concentrations in soils in urban areas Patent 2642868 Russian Federation

Published under licence in Journal Title by IOP Publishing Ltd.

IOP Conf. Series: Earth and Environmental Science **177** (2018) 012010 doi:10.1088/1755-1315/177/1/012010

Sustainable development issues of Almaty as the largest metropolis in Central Asia

G Alibekova[1], A Panzabekova[1], Z Satpayeva[1] and N Abilkayir[2]

[1] Institute of Economics of Science Committee, Almaty
[2] Kazakhstan's medical university "KSPH", Almaty

e-mail: g_alibekova@mail.ru, aksanat@mail.ru, almatyszt_kz@mail.ru, almatyabilkaiyr.nazerke@gmail.com

Abstract. The research is aimed to examine the sustainability of Almaty city as a largest city in the Central Asia. The literature review revealed lack of methodology for city sustainability assessment in developing countries. Proposed methodology of Denevizyuk was explored to calculate the Sustainable City Index for Almaty and demonstrate it's dynamics in last 9 years. As expected, environmental issues are becoming worse while economic and social parameters have positive trends and smooth out the final index growth.

1. Introduction

More than 50% of the world population lives in cities. It is expected that in 2030 it will be 6 out of 10 people. Cities consume 75 % of the world's energy production and produce 80 % of CO_2 emissions. These figures are indicative of the colossal burden that cities face in the modern world. Therefore, the question of "sustainable" development of countries and cities sharpen in the world. The UN defines "sustainability" as "development without burdening future generations". One of the goals of sustainable development according to the UN is to ensure sustainable development of cities. Cities are one of the main urban, transport, architectural challenges of the 21st century.

Almaty is the largest metropolis not only of Kazakhstan, but of the whole of Central Asia.

From 2010 to 2017 the population of Almaty increased by 23.9 %. As of January 1, 2018, there were 1,801,713 people living in Almaty, the largest agglomeration of the first level is developing around Almaty. In 2014 part of the Almaty oblast with a total area of 23,200 hectares and a population of over 92,000 people were passed to Almaty city. Today the territory of the city is 683.5 square kilometers.

Sustainable development of the city is of great importance, because uncontrolled migration of the population can lead to an excessive burden on the capacity of housing infrastructure, social and economic situation in the periphery of the city, disproportional development of the city, negative environmental consequences and serve as an obstacle to economic diversification.

According to the indices of economy, the development of Almaty corresponds to the level of medium megacities. In accordance with "Financial Post" in 2015, the city entered the top 10 of 96 cities with the fastest growing economy.

By the social context, there is a moderate lag behind the level of medium-sized cities. So, according to "Mercer" – the quality of life index – Almaty is on 176 place among 230 cities (2016),

Content from this work may be used under the terms of the Creative Commons Attribution 3.0 licence. Any further distribution of this work must maintain attribution to the author(s) and the title of the work, journal citation and DOI.
Published under licence by IOP Publishing Ltd

and by "Economist Intelligence Unit" Index of life quality – on 100 place among 140 megacities (2015).

In terms of infrastructure and environmental issues, Almaty lags far behind the level of developed cities in the world. In the "Mercer" index of infrastructure Almaty is on 178 place among 231 cities, and in the index of "Numbeo" on level of pollution – 214 place from 297 cities.

According to the City Prosperity Index in 2012–2013, Almaty ranked 28th and was included in the second category of cities – "cities with solid prosperity ratios" with a coefficient of 0.83 along with cities such as Ankara, Mexico City, Guadalajara, Bucharest , Shanghai, San Paulo, Moscow, Seoul, Prague, Athens, Budapest and Lisbon. In 2015, Almaty dropped to 33rd place with a coefficient of 0.67.

In order to bridge the gaps in sustainable development, the Almaty 2020 Development Program has been developed and implemented. It covers all major areas of the city's development, including the economy, social sphere, public security and law, infrastructure, ecology and land resources, public services. The Program covers 7 priorities: a comfortable city, a safe city, a socially-oriented city, an economically sustainable city, a city for business and private capital, an integrated city, a city of active citizens.

However, Almaty does not participate in sustainable development indices. For example, ARCADIS, the world's leading consulting company, jointly with the Center for Economic and Business Research (CEBR) calculates the Sustainable Cities Index for 100 cities on the planet based on 32 indicators for People, Planet and Profit. The "green" cities index of Siemens is adapted mostly for the countries of Europe and North America. The United Nations Sustainable Development Index or the OECD Environmental Performance Review focuses on the national level statistics.

In this regard, many countries are developing adjusted sustainable development indices for their cities to monitor and decision-making. For example, the Special Committee for Human Settlements (SCHS) of Chinese Society for Sustainable Development proposed the following indicators for assessing urban sustainability (Urban Sustainability Index): resources, social context, environment, economics and innovation [1].

The analysis shows that most of the effort to determine the sustainability of cities focuses on establishing the necessary theoretical basis to measuring sustainability, resulting in a variety of models duplicating each other. At the same time, many methodologies are often focused on the national level or on developed countries. As a result, there is a lack of applicable methodology for measuring the sustainability of cities in developing countries, where the problem is often the most acute.

2. Methodologies for the assessment of sustainable urban development

The City Prosperity Index (CPI) of the UN-Habitat program as a tool for assessing the sustainability of cities is based on six key indicators of city development: 1) productivity; 2) infrastructure development; 3) quality of life; 4) equity and social integration; 5) environmental sustainability; 6) management and legislation. The index calculation takes the following stages: a) the standardization of variables; b) development of specific weights; c) compiling an aggregated index. Standardization converts a variable from its original unit of measurement to a dimensionless value that ranges between 0 and 100. There are different methodologies to develop weights. When developing the CPI, the recommendations Alkire and Foster on weighting scheme for multidimensional poverty index were used [2]. When summarizing their recommendations, the following weight scheme is used: a) the indicators have the same weight in the index b) the sub-indicators are the same weight in indicators c) the variables have the same weight in the sub-indices. This scheme of weights clearly shows the assumption that all selected measurements are equally effective in determining the prosperity of the city [3]. There are cases of specific weights calculation based on the fuzzy logic method [4].

Unlike the City Prosperity Index, the urban sustainability assessment model developed by a group of Iranian authors showed significant differences in the significance of subsystems (social, economic and environmental) sustainability. Researchers took into account seven indicators – existence, effectiveness, freedom of action, security, adaptability, coexistence and psychological need. The final susstability calculation showed that the "efficiency" indicator gives the highest value in the developed model. As methods of verification, Delphi and fuzzy logic methods and the Kruskal-Wallis criterion were used [5].

When developing the Urban Sustainability Index by the Chinese Society for Sustainable Development, researchers selected the most accessible indicators in developing countries and the most relevant ones. The Urban Sustainability Index is designed to measure the relative effectiveness of Chinese cities in the overall set of sustainability categories. The authors developed a comprehensive definition of sustainable development based on 18 separate indicators to assess not only the environmental sustainability of the cities analyzed, but also the level of services needed to serve the growing urban population and the effectiveness of the resources of each city. The index measures the effectiveness of the city with respect to five aspects of sustainable development: basic needs, resource efficiency, environmental well-being, engineering communications, commitment to sustainable development.

This Index is based on the demonstrated efficiency of each city, and not on the theoretical absolute potential. After processing the data for consistency, the cumulative score for the average indicators was derived. Weights for individual indicators were not assigned to the overall assessment in order to avoid a priori estimates of the relative importance of indicators that would be difficult to justify. However, the authors experimented with weighting measures based on data quality, but found minor effects, as the data set was rounded. To analyze any correlation between GDP growth and stability, once the index was calculated and adjusted, a series of f-tests were conducted. Although there are more sophisticated methods for weighing factors and validating the test, the study was limited to ANOVA testing, regression, and factor analysis.

Since 2012 the "SGM" Agency conducts evaluation and rating of sustainable development of the Russian Federation cities in accordance with the principles of sustainable development of the country, international organizations practice and scientific community. The research covers 185 cities of the Russian Federation with a population of more than 100 thousand people. The assessment of sustainable urban development is based on an analysis of 42 statistical indicators that characterize the city in three main categories: the state of the economy, the social sphere, and the environmental situation.

To construct the Sustainable Development Index, the indices inside the three blocks are subject to normalization operations. Some of the indices with an obviously asymmetric distribution before normalization were logarithmized in order to "stretch" along the scale small values of the indicator and, conversely, to bring each of its highest values closer to each other. The aggregated index is obtained from the sub-indices by summing up with adjustment coefficient – weights chosen by the expert evaluation method in accordance with international recommendations and the most common practices. The assessment of the "weight" of the group of indicators in the final rating was carried out on the basis of the principle of the main components equity: the social sector, environmental situation, the economy and the level of urban infrastructure development. The final integral index of sustainable urban development is obtained by calculating the mean value among particular indices and can theoretically take values from 0 to 1.

Russian scientists propose an integral index of urban sustainability. It is based on the concept and methodology of calculating the index of adjusted net savings. In order to assess the sustainability of urban development, taking into account economic, social and environmental factors, the authors propose to allocate three relevant sub-indices: gross fixed capital formation, human capital development expenditures and damage from environmental pollution in cities. The proposed author's system of indicators for Russian cities reflects the most hot topics of sustainable development of Russian cities and the quality of life of citizens, and is also adequate to the capabilities of Russian statistics. Key indicators were proposed reflecting important economic, social and environmental city priorities: economic indicators; energy efficiency; transport; social and institutional indicators; air and climate; water resources; waste; specially protected natural territories; noise effect [6].

To evaluate the sustainable development of Almaty, the authors of the given paper used the methodology proposed by Denevizyuk [7]. Integrated indicators of subsystems – economic, social, ecological – are built for its calculation.

The index of sustainable development of the city is calculated as follows:

$$I_{CSD} = \sum_{i=1}^{n} w_i I_i$$

$$I_i = \frac{1}{m+k}\left[\sum_{j=1}^{m} I_{ij}^{+} + \sum_{j=1}^{k} I_{ij}^{-}\right]$$

$$I_j^{+} = \frac{I_j - I_{min}}{I_{max} - I_{min}}$$

$$I_j^{-} = 1 - \frac{I_j - I_{min}}{I_{max} - I_{min}}$$

where

I_{CSD} - city sustainability index;

I_i – integral index of the corresponding subsystem (economic, social, ecological)

w_i – the weight coefficient of the corresponding subsystem satisfying the following conditions: $w_i \geq 0, \sum w_i = 1$;

m — number of positive indicators;

k — number of negative indicators;

I_j^{+} – standardized value of j-th positive indicator, i.e. at increase of which the subsystem improves;

I_j^{-} – standardized value of the j-th negative indicator, i.e. at increase of which the subsystem deteriorates;

I_{min} - minimum value of the j-th indicator;

I_{max} - maximum value of the j-th indicator.

This index was calculated on the basis of 33 statistical indicators characterizing sustainable development of the city in three main blocks: economic, ecological and social. At the same time, the system of indicators for each of these subsystems was constructed by the authors taking into account the following principles:

• availability of statistical materials. All used data are placed on official websites and databases of the Committee on Statistics of the Ministry of National Economy of the Republic of Kazakhstan, Statistics Department of Almaty.

• comparability of data. Only those indicators that can be compared with each other are used, which allows correct comparisons.

• Maximum relevance of indicators. Only those indicators that are relevant to the characteristic of sustainable development are used.

• The same proportion of all indicators forming an index of sustainable development. This is a common practice in the formation of multidimensional indices.

The proposed methodology for an integrated assessment of the city sustainable development is reliably applicable both for conditions with minimal initial data, and for conditions with a wide range of indicators [8].

The indicators for the period from 2008 to 2016 were used for research. The statistical base of the research can be increased. The index of sustainable development of the city was calculated in comparison by years, therefore the change of the period under consideration will lead to a change in the value of the index itself.

Thus, the city sustainable development index takes the following form:

$$I_{CSD} = w_{EcD} I_{EcD} + w_{SocD} I_{SocD} + w_{EnvD} I_{EnvD}$$

I_{EcD} – index of economic development of the city;
I_{SocD} – index of social development of the city;
I_{EnvD} – the index of environment development of the city.

The index of sustainable development lies in the range from 0 to 1. The interpretation of the results of the sustainable development assessment is based on the established limits of the allowed values (Table 1).

Table 1. Integral indicators of sustainable development

The range of values of the integral assessment	Boundaries	Interpretation of integral evaluation
Sustainable	0.85–1.00	High level of sustainable development
	0.65–0.85	Sustainable development
Quasistable	0.50–0.65	Development level close to sustainable
	0.25–0.5	Development level with signs of instability
Unsustainable	0.1–0.25	Unsustainable development
	0.00–0.1	Critical state of sustainability
Note – Compiled by the authors on the basis of the source [9]		

3. Research results

According to the presented methodology, calculations of the indicators for the period from 2008 to 2016 were carried out. The integral indices of each subsystem are obtained (Table 2).

Table 2. Integrated indicators of subsystems and the index of sustainable development of Almaty, 2008–2016

Year	I_{EcD}	I_{SocD}	I_{EnvD}	I_{CSD}	Level of sustainability of development
2008	0.22	0.39	0.62	0.38	Development level with signs of instability
2009	0.14	0.33	0.59	0.32	Development level with signs of instability
2010	0.24	0.34	0.58	0.35	Development level with signs of instability
2011	0.33	0.36	0.59	0.39	Development level with signs of instability
2012	0.43	0.48	0.58	0.48	Development level with signs of instability
2013	0.57	0.59	0.59	0.58	Development level close to sustainable
2014	0.72	0.55	0.33	0.57	Development level close to sustainable
2015	0.76	0.56	0.38	0.59	Development level close to sustainable
2016	0.89	0.65	0.33	0.67	Sustainable development
Note – Compiled by the authors					

Economy subsystem. The economy of the city shows rapid growth. Investments in fixed assets increased by 20 %, in housing construction – 33 %. During the period under review, the gross regional product of Almaty increased 3.5 times, the gross regional product per capita – 2.8 times, and industrial production – 2.1 times. The remaining indicators also show positive dynamics. The role of small and medium-sized businesses in the city's economy has increased. Thus, the share of its gross added value in GRP increased from 27.8 % to 33.7 %. The level of innovative activity of firms has grown from 6.4 % to 7.6 %, while the volume of innovative products has increased by 6.1 billion tenge, which exceeds the level of 2008 by 55 %. There is a reduction in the unemployment rate by 28 %. Due to the positive dynamics of economic indicators, the integral indicator of the economic subsystem showed an increase from 0.22 to 0.89.

Social subsystem. Due to an increase in the housing stock by 84 %, the provision of housing for a person by the population grew by 42 %. At the same time, the housing construction in water supply systems in 2016 was 100 %, with sewage systems – 97 %, with water supply systems – 70.2 %. The number of fixed Internet subscribers increased by 5.5 times, while the share of Internet users aged 16-74 grew from 23 % to 86.2 %. To date, there is a digitalization of society and the economy of the city. The number of hospital organizations and doctors has increased, but the number of hospital beds per 10,000 population has decreased. The number of students in colleges fell by 3 327, universities – by 64 460. The crime rate increased 3.5 times, the number of criminals increased 2.2 times. Despite the fact that in the presented subsystem fluctuations of indicators were observed during the considered period, the social subsystem of the city in 2008 showed development with signs of instability, in 2016 it became more stable.

Environment subsystem. To protect the environment of Almaty in 2016, 2.8 billion tenge was allocated, which is 39 % higher than in 2008. However, the amount of emissions in comparison with this year has significantly increased. So, emissions of solid pollutants increased by 2.3 times, liquid and gaseous pollutants – 3 times, pollutants into the atmosphere – 2.8 times. The pollution of the city during the period under review was very high. Pollutants were captured and rendered harmless 8 times more than in 2008, only 3.3 thousand tons were disposed of, which is 19 % of the 2008 level. Significant fluctuations in the indicators of the city's ecological development led to the instability of the development of this subsystem. If in 2008 its development was close to a stable state, then in 2016 there are signs of its instability.

4. Conclusion

During the period under study, the economic and social subsystems for the sustainable development of the city of Almaty showed positive dynamics, while the ecological subsystem was negative. However, the instability of one subsystem was compensated by the higher value of stability of others, therefore the Almaty city sustainability index is defined as increasing. In 2016, the development of Almaty became stable due to the high level of economic subsystem stability. The social subsystem was also stable. The development of the ecological subsystem showed the presence of signs of instability. The obtained results confirm the results of international rating. In the development of Almaty, much attention should be paid to environmental issues, as well as stabilization of the economic and social sustainability of urban development.

References

[1] Jianfeng L, Xiaotong Zh and Jianqing H 2015 *Urban Sustainable Development: Measuring Capacity Outweighs Measuring Development Level*
[2] Alkire S and Foster J 2011 Counting and Multidimensional Poverty Measurement *Journal of Public Economics* **95(7–8)** 476–87
[3] 2016 *Measurement of city prosperity Methodology and Metadata* (UN-Habitat)
[4] Lazaroiu G and Roscia M 2012 Definition Methodology for the Smart Cities Model *Energy* **47** 326–32
[5] Jaderi F, Ibrahim Z, Jaafarzadeh N, Abdullah R, Shamsudin M, Yavari A and Nabavi S 2014 Methodology for modeling of city sustainable development based on fuzzy logic: a practical case *Journal of Integrative Environmental Sciences* **11 (1)** 71–91
[6] Bobylev S, Kudryavtsev A and Solovyov S 2014 Sustainable development indicators for cities *Economy of region* **3** 101–10
[7] Denevisyuk D 2012 Sustainable development of the city: questions of theory and methodology of the evaluation *Regional issues of economy transformation* **2** 103–12
[8] Tsapieva D, Denevisyuk D and Agaragimov M 2007 Integral assessment of the city's sustainable development *Economics and management* **46 (7)** 64–70
[9] Golovanov E 2015 Methodical approach in assessment of sustainable development of regional economy *Modern management technologies* MMT03(2015)5104

Industrial pollution of the hydrosphere and problems of water treatment in Russia

E Ikanina

Physical Chemistry and Chemistry of Colloids Department, Institute of Chemical Engineering, Ural Federal University Named after the First President of Russia B. N. Yeltsin, 28 Mira str., Ekaterinburg, 620002, Russia

e-mail: e.v.ikanina@urfu.ru

Abstract. The statistical data of water use by various industries in Russia were presented and analyzed for the years 2010 to 2016. The problems of saving water resources and developing market for environmental services in Russia were considered. The dynamics of industrial water consumption and water treatment was estimated. The structure of water use and the wastewater disposal for different branches of economy is represented. The constructive proposals for industrial water treatment systems were made. It was shown that sorption method is one of the most promising and environment friendly. The composite sorbents were synthesized from reagents being produced in Russia. For comparison the use and effectiveness of the composite sorbent and imported ion-exchanger was investigated. The developing closed water systems based on the composite sorbents was suggested.

1. Introduction

Currently in the world there is no country, which would have not affected by environmental problems. Compared with other countries, especially with Europe, Russia has huge water resources [1]. Unfortunately, it is for this reason the problem of saving water resources has not been given proper attention for a long time. After long ignoring problem at the end of the 20th century Russia has become a consumer in the global market for environmental goods and services, but such a way of development is not in the interests of our country.

Firstly, this way contributes to the economic growth only of the exporting countries.

Secondly, many imported building materials and reagents which use in the water treatment cannot withstand severe climatic conditions. These can be successfully used only in the southern regions of our country. On the other hand, most of the companies are in the Urals and Siberia, so before start-up of the imported equipment it needs to be retrofitted (covered space, additional heating, etc.). This requires additional costs instead of resource and energy saving. In addition, after about two years, our country must to purchase spare parts, materials, etc. from a foreign supplier for a hefty price.

Thirdly, data about principle of the equipment action are commercial secrecy of the exporting countries. Our engineers often use imported equipment and do not understand well how it works.

This situation may have very undesirable consequences for economic development of Russia. Today the changing situation is important, a way out of it is new scientific developments in the field of ecology.

Content from this work may be used under the terms of the Creative Commons Attribution 3.0 licence. Any further distribution of this work must maintain attribution to the author(s) and the title of the work, journal citation and DOI.
Published under licence by IOP Publishing Ltd

Sorption method is one of the most promising among the known methods of water treatment. The sorption efficiency is comparable to the same value for complex and expensive methods as electrodialysis and vaporization. However, sorption systems consume little power. These are environment friendly and have simple working mechanism. The main problem is making effective sorption materials. The search for the optimal sorbent is frequently carried out according to the approach that «each physicochemical system needs its own sorbent» [2].

Modification of carbon matrices by micro- and nanodispersed metal hydroxides is upcoming direction for synthesis of effective sorbents. The authors [3] synthesized the composite sorbent based on ion exchangers and polyvalent metal hydroxides. All reagents used for synthesis are produced in Russia. To modify the ion exchanger in optimal conditions the authors investigated and modeled complexation in solutions containing ions of polyvalent metals [4].

A polyvalent metal (iron, tin, titanium, etc.) dispersed in the composite sorbent has a large specific surface area, due to the stabilizing action of the porous carrier, and, thus, high reactivity. The composite sorbent has pronounced physicochemical properties conforming to the operating requirements. Possible application areas of composite sorbents being synthesized are the drinking water preparation, the wastewater treatment and the elimination of emergencies.

The aim of the present study was analyzing statistical data of water use by various industries in Russia in the past seven years (from 2010 to 2016) and investigating the use, effectiveness of the composite sorbent for the water treatment in comparison with imported ion-exchangers.

2. Analysis of statistical data

In 2015 there was a decline in branches of economy by 3.4–5.4%. Manufacturing has suffered the most [5]. However, it practically did not affect the dynamics of water intake because the agricultural sector has increased water use.

In 2015 the total volume of water intake from natural sources was 68.6 km^3 (line 1, figure 1a). This is 3.1% less than in the previous year. The volume of fresh water used to the level of 2015 was 97.5% from the 2014 level (line 3, figure 1a). There are small decreases, but they are not significant [6].

The consumed volume of water for industry consists of fresh water (line 3, figure 1a) and water of consecutive and closed water systems (line 2, figure 1a). Saving of fresh water for production needs due to consecutive and closed water systems is on the average 78%.

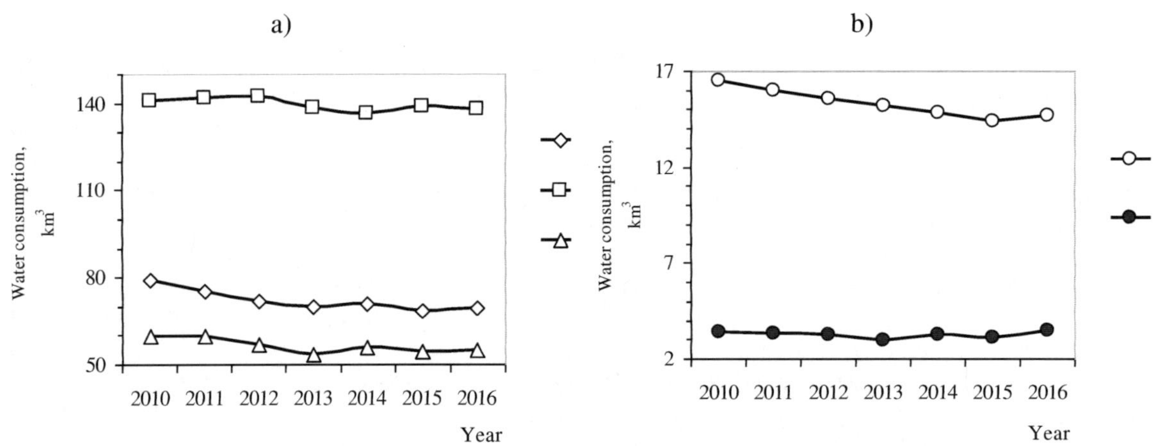

Figure 1. Dynamics of main indicators of water use in Russia from 2010 to 2016
a) intake, b) disposal
1 – water intake from natural sources, 2 – consumption in water recycling and reuse systems, 3 – using fresh water, 4 – total wastewater volume, 5 – wastewater discharge without treatment

A different situation is observed with the wastewater disposal. The volume of wastewater discharged into natural water bodies gradually decreased from 2010 to 2014 and began to grow in 2015 (line 4, figure 1b). In 2016 the volume of wastewater discharge without treatment is 13.5% above this value in 2013 (line 5, figure 1b).

The structure of water use and wastewater disposal for different branches of economy is shown in Figure 2. Pie charts were plotted on averaged data for the years 2014 to 2016 [6]. Agriculture and forestry are the second largest economic sector on water consumption and in last place on wastewater disposal. Water use of manufacturing is only 6% due to consecutive and closed water systems, but wastewater disposal is 18%. Portion of production and distribution of electricity, gas and water in the total water use and disposal is maximal. Mining and manufacturing has most toxic wastewater because of heavy metals.

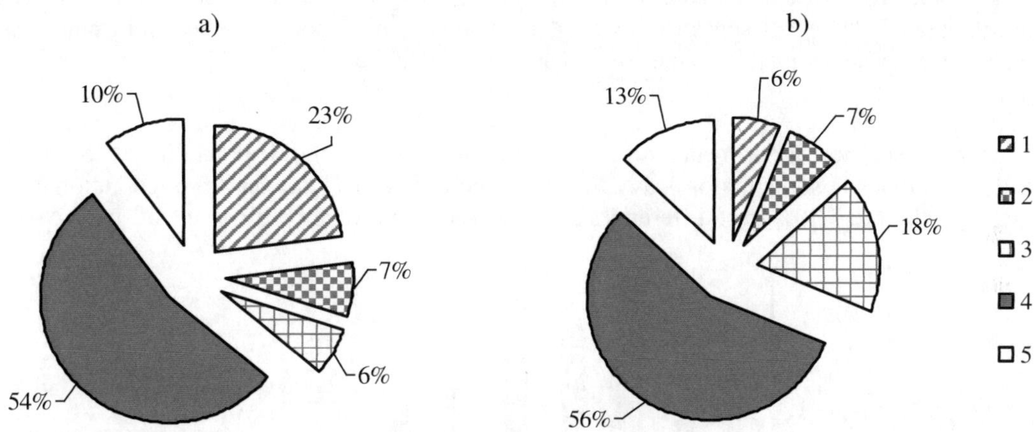

Figure 2. Structure of water use (a) and wastewater disposal (b) for branches of economy: 1 – agriculture and forestry, 2 – mining, 3 – manufacturing, 4 – production and distribution of electricity, gas and water, 5 – household and drinking water and other needs

The structure of wastewater disposal in 2016 are represented on Figure 3 [6]. It was shown that only 11.86% of wastewater are treated to normative values. The rest of 88.14% must be treated, 20.48% of wastewater are discharged into natural reservoirs without treatment. The wastewater treatment can be achieved by means of sorption technologies.

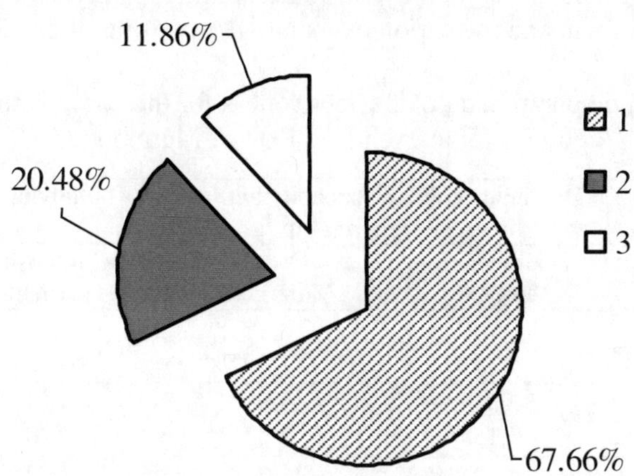

Figure 3. Structure of wastewater disposal in 2016
1 – insufficiently treated, 2 – without treatment, 3 – normative treated

3. Investigating the sorption materials

3.1. Experimental

The composite sorbents based on cation exchanger KU-2-8 and iron(III) hydroxide were made according to the methodology of stepwise synthesis [3]. For comparison determining the sorption capacity were carried out in parallel on the composite sorbent and the imported cation exchanger Dowex Marathon C which is often used in industry.

Heavy metal pollution of wastewater is one of the most important environmental problems throughout the world. In this regard, the investigated liquid phase was selected solutions of heavy metals. The certain volume of the test solution with known initial composition in the range pH from 5 to 14 was used for sorption processes. The pH value was adjusted by addition of ammonia solution and was monitored with the pH meter (pH 410 Akvilon). Experiments was carried out under dynamic conditions at 297±0.1 K in specially made glass column with stopcock. The equilibrium solutions were analyzed for heavy metals by the spectrophotometric method [7].

3.2. Sorption results

The results of sorption experiments are represented on Figure 4 and Table 1. It was found that the capacity of composite sorbents for heavy metals is about twice as large as imported cation exchanger Dowex Marathon C. The largest differences are observed in the alkaline pH range of the test solutions.

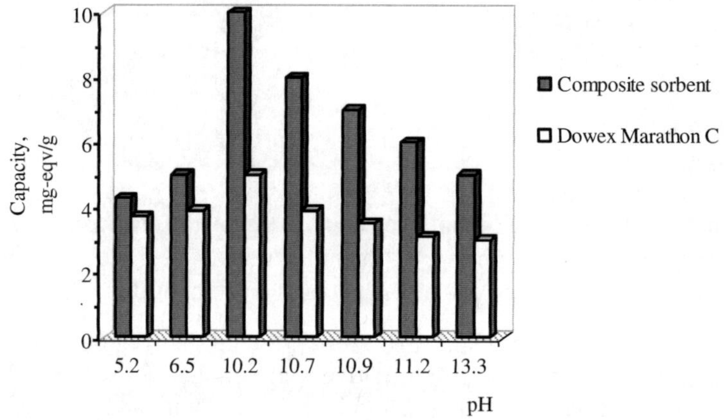

Figure 4. Dependence of the total dynamic exchange capacity for copper(II) of the composite sorbent and the cation exchange Dowex Marathon C on the pH

Table 1. Comparative data of effectiveness for the composite sorbent and the cation exchange Dowex Marathon C

Metal	pH solution	Dynamic exchange capacity before breakthrough, mg-eqv/g		Total dynamic exchange capacity, mg-eqv/g	
		Composite sorbent	Dowex Marathon C	Composite sorbent	Dowex Marathon C
Copper	5.20	2.10	0.18	4.27	3.87
	11.20	4.55	1.59	6.73	3.14
Zinc	5.53	2.01	1.55	4.24	3.82
	11.50	4.15	2.09	6.10	3.31
Nickel	5.66	3.02	1.74	4.67	3.08
	11.50	3.72	2.58	5.72	3.82
Cadmium	5.47	1.50	1.00	4.05	2.10
	11.80	2.80	1.72	5.18	2.40

Wastewater of many industries contain ammonia complexes of heavy metals. Removing complexes is very difficult [8]. The main method is thermal decomposition that requires a large energy consumption. The high capacity in the alkaline pH range allows using the composite sorbent to remove heavy metals from wastewater and to save energy. Due to the sorption degree is over 98% treated wastewater may be recycled into the production. Based on the experimental results closed water system in industries may be created.

4. Conclusions

Dynamics of the main indicators of water use in Russia from 2010 to 2016 hardly changed. There are small changes for branches of economy, but they are not significant. Saving of fresh water due to consecutive and closed water systems is about 78%. This value should be increased. Around 88% of the total wastewater must be treated.

Problems of saving of fresh water and wastewater treatment can be solved with the help of sorption technologies.

The composite sorbents were synthesized from cation exchanger KU-2-8 and iron(III) hydroxide which were made in Russia. The capacity of composite sorbents for heavy metals is about twice as large as imported cation exchanger Dowex Marathon C. Closed water systems based on the composite sorbents may be developed.

Acknowledgments

This research was supported by the scholarship of the President of the Russian Federation for young scientists and graduate students performing prospective research and development in modernization priority areas of the Russian economy, SP-622.2015.1

References

[1] 2009 *The Water Strategy of the Russian Federation for the Period up to 2020* (Approved by Decree № 1235-r of the Government of the RF)
[2] Melikhov I V, Berdonosova D G and Sigeikin G I 2002 *Russ. Chem. Rev.* **71** 159–79
[3] Ikanina E V, Kalyaeva M I and Markov V F 2017 *AIP Conf. Proc.* (Ekaterinburg) vol 1886 (College Park: American Institute of Physic) 020066
[4] Ikanina E V and Markov V F 2017 *Theor. Found. Chem. Eng.* **51** 45–51
[5] Dzagoeva M R, Kaytmazov V A and Ikoev S K 2015 *Fundamental research* **2-26** 5860–3
[6] The state report *On the state and use of water resources of the Russian Federation in 2016*
[7] Sharlo G 1969 *Methods of Analytical Chemistry* (Moscow: Chemistry) Part 2 881–9
[8] Vinogradov S S 1998 *Environmentally Friendly Galvanic Production* (Moscow: Globus)

Published under licence in *IOP Conference Series: Earth and Environmental Science* by IOP Publishing Ltd.

On Passenger Flow Estimation for new Urban Railways

A Misharin[1], D Namiot[2] and O Pokusaev[3]

[1] RUT (MIIT), 9b9 Obrazcova str., Moscow, 127994, Russia
[2] Lomonosov Moscow State University, Leninskie Gory Moscow, 119991, Russia
[3] Center for High-Speed Digital Transport RUT (MIIT), 9b9 Obrazcova str., Moscow, 127994, Russia

e-mail: dnamiot@gmail.com

Abstract. In this paper, we discuss the questions, associated with the forecast for passenger traffic for urban railways. The aim of the study is to select and verify the model for predicting passenger traffic of new urban railways. The article is based on the practical tasks implemented during the project phase for new urban railways in Moscow, Russia. We are considering data sources for building the forecast, as well as practical models that can be used to obtain numerical estimates. Among the discussed data sources, we target migration data that can be collected with the help of telecommunications operators, and information on the use of public transport, obtained from the validation of transport cards. Also, in this paper, we investigate the metrics for traffic along the new city rail line, which can be determined on the basis of the projected passenger traffic. The result of the work was the constructed model of the transport behavior of passengers, taking into account the availability of new urban railways and a set of metrics for assessing the functioning of this transport tool.
Keywords: Smart City; smart mobility; urban railways.

1. Introduction

It is a well-known fact that mobility (smart mobility) is one of the key elements for Smart Cities. Urban railways recently attracted much attention, because they are almost the only transport in the city, which can move without traffic jams. This article discusses the issues of modeling of passenger flow for urban railways. City railways, similar to other transport projects in cities, are quite expensive enterprises, which require comprehensive calculations and justifications. An expected traffic is, of course, one of the key parameters. This traffic is the basis for calculating social and economic effects from the railroad, many of which are based on the expected traffic [1]. For example, the predicted traffic allows us to estimate the revenue from the future passenger transportation. Also, it allows us to estimate saving time by passengers from the introduction of new routes, it lets us estimate a decrease in transport fatigue, reachability of new areas, etc. In this connection, we could mention once again the paper [1], where we, in fact, considered various tasks assessment of social and economic effects.

The objectives of this study were the selection and verification of a model suitable for estimating (predicting) the traffic of new urban rail lines, selecting initial data for modeling, selecting metrics for assessing the functioning of the constructed transport lines. This is the first such kind of research for Moscow.

The rest of the paper is organized as follows. In Section 2, we discuss research methods. In Section 3, we provide our analysis. And Section 4 is devoted to the conclusion.

Content from this work may be used under the terms of the Creative Commons Attribution 3.0 licence. Any further distribution of this work must maintain attribution to the author(s) and the title of the work, journal citation and DOI.
Published under licence by IOP Publishing Ltd

2. On passenger flow models

This paper based on some work performed during early stages of Moscow radial urban railways project (so-called Moscow diameters). The base element for passenger flows prediction is data on migration within the city. Naturally, passengers are traveling residents or visitors, some of whom (and it is what will be evaluated in our model) could take advantage of a new mode of transport (a new railway line in this case).

Two basic approaches to assessing migration data could be mentioned here. As per the US, for example, the good sources of migration data are surveys (census), which the statistical service collects in the US. Answers to polls are required by law. These surveys include data on migration too [2]. As a good example of such data, we could mention paper [3] with collected statistics for San Francisco. The paper [4] presents detailed information on surveys in the Bay Area (California). At the same time, the accuracy of the information is quite high. Information is collected (monitored) at the level of city blocks (there are about 4,000 blocks in San Francisco). Accordingly, the correspondence matrix (it is a Russian terminology, US scientists use origin-destination or OD matrix abbreviation) marks the movement from block to block. It is illustrated in Figure 1.

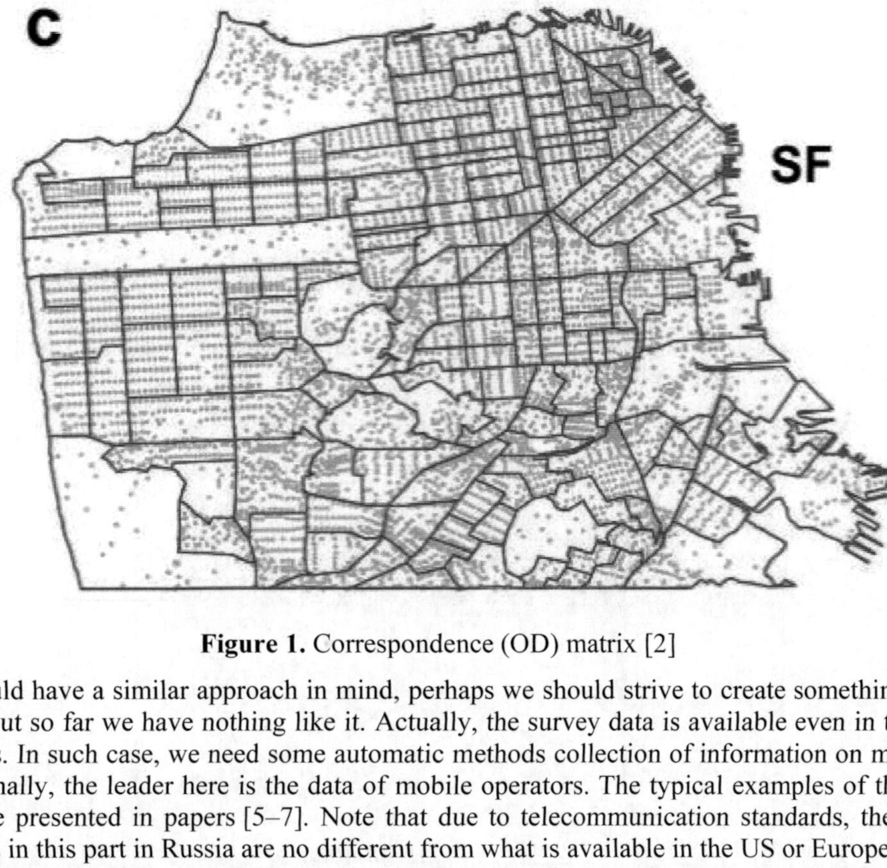

Figure 1. Correspondence (OD) matrix [2]

We should have a similar approach in mind, perhaps we should strive to create something like this in Russia, but so far we have nothing like it. Actually, the survey data is available even in the US not in all places. In such case, we need some automatic methods collection of information on movements. Unconditionally, the leader here is the data of mobile operators. The typical examples of this kind of research are presented in papers [5–7]. Note that due to telecommunication standards, the technical possibilities in this part in Russia are no different from what is available in the US or Europe. Data that are collected by telecommunication operators, technically, can describe different aspects of behavior for mobile subscribers that could be used for creating the same correspondence (or OD) matrix.

One of the first deployment models for telecom data is linked with the analysis of user activity, measured in Erlangs (1 Erlang is the use by a subscriber of a network during 1 hour). With this model it possible to determine areas of activity for users in the city (it is so-called social dynamics). The idea of use is very simple – if we register, for example, calls from a specific phone from the same place in the morning and in the evening, and then in the daytime from another, we can tell how the route looks in the morning from home to work for this phone (for its owner). For example, in [2] algorithm analyzed places for phone calls and it was assumed that time from 21:00 to 6:00 corresponds to a call from home, and time 9:00 to 17:00 corresponds to a call from work. It, accordingly, gave one route. In the paper [8], authors use a slightly different approach. For example, their algorithm assumes that the location from which the user leaves in the morning and to which he returns at night is his home. It is also assumed that location of the longest repeating stay on weekdays is a workplace of this mobile subscriber. This algorithm describes the construction of a general model of mobility in the city of Boston, where according to the operators' data house-work pairs have been restored for 75 % of subscribers. So, such model can predict user moves from a 10-minute interval.

Other ways of collecting parameters (for example, Location update, Handover [9]) allow come, as a result, to the same matrix of correspondence (starting and ending points in mobile phone movements). Note, that the end-point for the particular route can be defined empirically, as a certain time interval during which the mobile device remained in place. Basically, it is a complete analogy with the concept of a web session. A visit to a website by a specific user is also determined (completed) by time analysis for inactivity of the user [10].

The next important thing is to get the metrics for the passenger flow on the railway. The idea was borrowed from [2] – it is the distribution of travels by duration (Fig. 2).

As the next characteristics, we used the centrality [12]. The idea is that the structural properties of the network determine the flow through the network. The centrality for node v could be defined as:

$$b_c(v) = \sum_{s,t\in V, s\neq t} \frac{\sigma(s \to t|v)}{\sigma(s \to t)}$$

where $\sigma(s->t/v)$ is a number of shortest paths from station s to station t, passed through v, and $\sigma(s->t)$ is a total amount of shortest paths from s to t.

And according to the research, the traffic passing through the station correlates with its centrality. The more shortest paths pass through the station, the more passengers travel through it. That makes it possible to predict the volumes of passengers, since the characteristics of the nodes can be calculated statically (based only on the network topology).

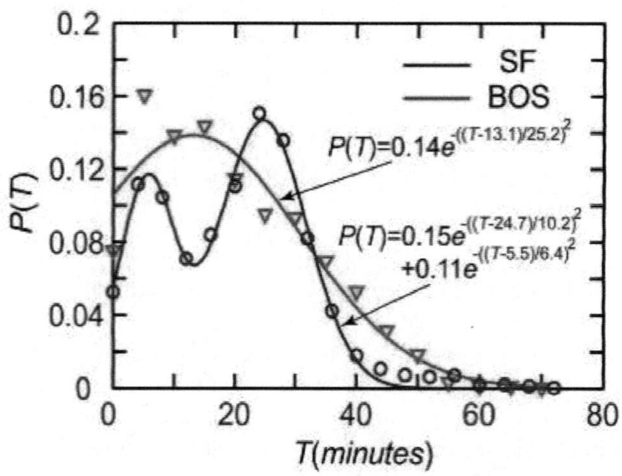

Figure 2. Distribution of travels by duration [2]

The centrality metric has also been used to analyze the resilience of a network [2]. By removing key nodes or edges between key nodes in the network (in the graph), it can be observed that the flow patterns through different paths will be highly affected. This remove operation will simulate the effect of disruption in real physical transportation networks (for railways it can simulate train breakdowns, for example). So, it could help to city administration (to transport department) identify whether transport system incorporates sufficient design robustness as well as justify its updates to overall resilience.

3. On proposed prediction model

Our model is based on migration data obtained from telecom operators and information about smart cards validation in public transport. In our case, it was possible to restore origin-destination movement for mobile devices in a square with a side of 250 meters.

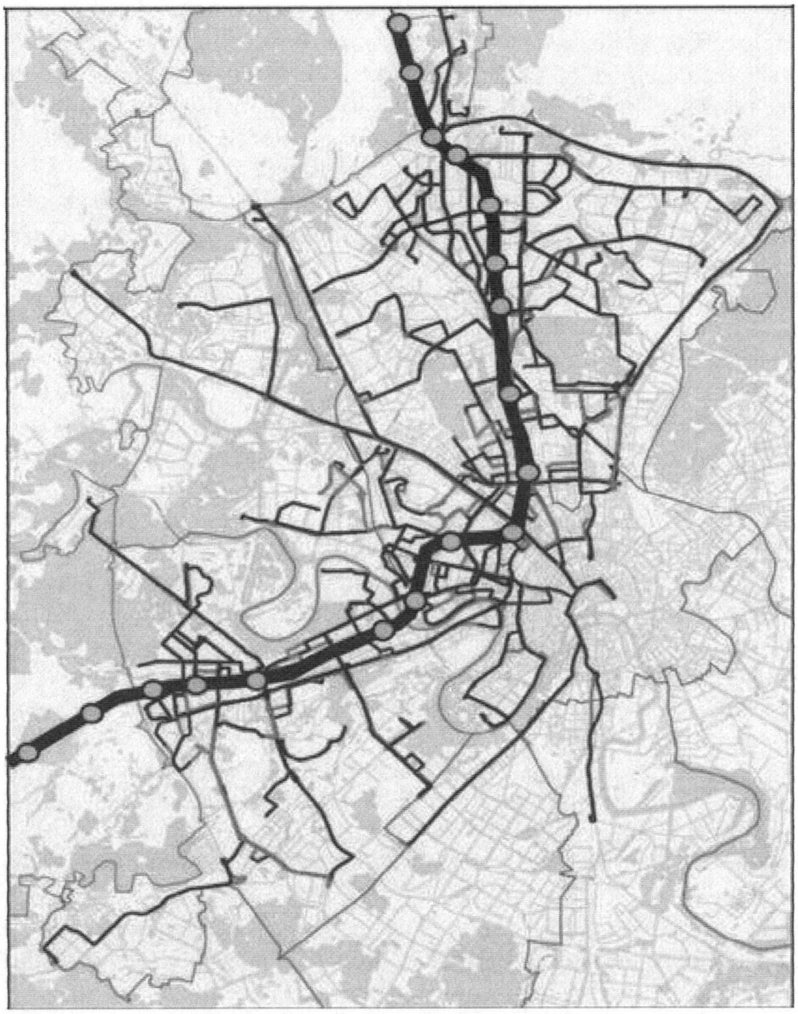

Figure 3. Public transport and railroad

After obtaining (restoring) the information about the movements (migration data), we must allocate (share, distribute) these trips to the different modes of transport.

As the first step (it's the simplest one), we can separate pedestrians. This, as a rule, is based on a simple assumption about the length of the route. If this length (and routes begin in a block and end in some other block – see Fig. 1) is less than some boundary value (for example, 1 km), it can be assumed that this is a walking route.

The remaining "migrants" are either passengers for public transport or take advantage of cars. The question about the proportion of drivers (motorists) in this flow is, of course, debatable. The above-mentioned papers (e.g., [2]) again used data from polls (Boston, San Francisco), where residents were directly asked for the ways they are moving around the city. For Moscow, such data is not available but we can use the advantage of smart cards. The most accurate way to determine the number of people using cars is to take all the transport migrants and remove (deduct) the public transport migrants. And for public transport passengers, we can use data from smart cards (travel card Troika, in Moscow, for example) validation. Transport cards in Moscow are using for landing (entry) only. It is an analogue of the check-in (the presence) in social networks [11].

Naturally, some of the users of cars will switch to the railway. This is noted in all works on urban transport (see [1], for example). In reality, the city railway is practically the only transport that is not subject to traffic jams. It is the reason, why drivers could switch to urban railways. The percentage of switched drivers is debatable (see review [1] again). The historical data is 5–20 %.

After receiving a real number of public transport passengers (we took into account both pedestrians and drivers), we need to distribute them according to the types of public transport. We assume that each passenger will choose the shortest route. Accordingly, for the distribution of passengers, we need a geo-information system with all stops of public transport (see Fig. 3).

On this geo-information system, we should place a new railway line too (bold line in Fig. 3). On this geo-information system, we should place our squares for which the movement routes were determined. In the process of determining the shortest paths for passengers, those who will use the railway will be identified (their shortest path implies moving by rail). Naturally, in this case, the used railway stations will be accurately determined too.

We can also calculate the movement and without a railway. After that, we can calculate the difference in time that will be spent by passengers on trips without taking into account the railway and with it. This difference presents a time that passengers will save on a daily basis in case of putting the road into operation. Multiplying this time by the average hourly wages, we can get a direct economic effect, expressed in money terms.

4. Conclusion

In this article, we have described a complete diagram of the representation of the passenger flow for a new urban rail line. The model is based on migration data, which was estimated (collected) with the help of mobile operators. Data on movements from mobile operators are compared with data on the use of travel documents in public transport. In the work, we managed to build a mashup for telecom data and data on the validation of travel documents. This allows, in the end, to identify both users of public transport, and those who use the car to travel around the city. To identify those passengers of public transport who will use the new railway line, a geo-information system with stops and public transport lines is used. This system allows us to calculate the shortest travel paths for passengers. We define the shortest path as the path which incurs the least amount of travel time between two stations. By the criterion for choosing the shortest path, the possible passengers of the new railway line are identified. The predicted passenger traffic allows us to calculate such metrics of the new railway line, as the distribution of the duration of the routes and centrality for the stations. The proposed model has been verified on the existing lines of the city railways, when the results of the simulation were compared with the actually recorded passenger traffic.

References

[1] Namiot D et al 2018 On the assessment of socio-economic effects of the city railway *International Journal of Open Information Technologies* **6** (1) 92–103

[2] Wang J et al 2013Vulnerability analysis and passenger source prediction in urban rail transit networks *PloS one* **8** (11) e80178

[3] 2010 *US Census Bureau*

[4] 2010 *Bay Area Census*

[5] Steenbruggen J et al 2013 Mobile phone data from GSM networks for traffic parameter and urban spatial pattern assessment: a review of applications and opportunities *GeoJournal* **78** (2) 223–43

[6] Ratti C et al 2006 Mobile landscapes: using location data from cell phones for urban analysis

Environment and Planning B: Planning and Design **33** (5) 727–48

[7] Calabrese F et al 2011 Real-time urban monitoring using cell phones: A case study in Rome *IEEE Transactions on Intelligent Transportation Systems* **12**.1 141–51

[8] Mearian L 2016 Ford, MIT use Bostonians' cellphone location data for traffic planning *Computerworld*

[9] Calabrese F, Ferrari L and Blondel V 2015 Urban sensing using mobile phone network data: a survey of research *Acm computing surveys (csur)* **47**.2 25

[10] Namiot D and Sneps-Sneppe M 2014 On the analysis of statistics of mobile visitors *Automatic Control and Computer Sciences* **48**.3 150–8

[11] Namiot D and Sneps-Sneppe M 2011 Customized check-in procedures *Smart Spaces and Next Generation Wired/Wireless Networking* (Springer, Berlin, Heidelberg) 160–4

[12] Ramli MA et al 2014 A method to ascertain rapid transit systems' throughput distribution using network analysis *Procedia Computer Science* **29** 1621–30

Published under licence in Journal Title by IOP Publishing Ltd

Evaluation of the effectiveness of the adaptive traffic management system for its development and interaction with electric transport

I A Anisimov, A D Gorbunova, L N Burakova and O D Burakova

Tyumen Industrial University, 38 Volodarskogo str., Tyumen, 625000, Russia

E-mail: burakova.1992@mail.ru

Abstract. Transport infrastructure is one of the important directions of urban environment sustainable development. The increase in the area of cities necessitates the implementation of measures aimed at improving the quality of transport services. This causes the development of intelligent transport systems, which allow to reduce the load on the transport network. However, the development of these systems in the Russian Federation lags behind the world experience and for the regulation of the traffic light it necessary the operator's intervention and the connection of the peripheral device with the traffic control center. Therefore, the use of these systems is difficult in remote areas such as the intersections that connect the industrial and residential parts of the city. These intersections, as a rule, have unstable transport demand in time and directions, which causes "unjustified" downtime of vehicles during the red light in the main direction. This article presents the adaptive traffic management system, which is autonomous and allows changing the cycle of traffic light regulation in remote territories. In the future, the plan of the development of this system is aimed at developing a new type of detector which communicates with the electric car.

1. Introduction

At the present time, the peculiarity of the socioeconomic state of the country is the concentration of the population in the centers of the Russian Federation subjects [1]. This causes the emergence of a number of negative living conditions, which are associated with intense pollution of the atmosphere, the expansion of urban areas, therefore, by removing the centers of attraction of the population [2]. The solution of these problems is the concept of sustainable development of these settlements [3]. Its main provisions are a long-term plan for changing economic and production activities, according to which a scheme for city's engineering infrastructure is developed, in particular transport.

The increase in the area of cities causes a decrease in the quality of transport services for the population, for the increase of which it is necessary to carry out activities. This is due to the reconstruction of the existing road network, the decentralization of places of employment and recreation, as well as the improvement of the organization of road traffic vehicles, both in the urban environment and for interaction beyond its borders [4–6]. However, the first two types of measures require large capital costs and are difficult to reconstruct the existing part of the city.

Improving the organization of road traffic is the most common method of increasing the capacity of the road network, and, therefore, reducing the time spent on moving. This is due to the high proportion of equipped intersections with traffic lights and the ability to search for an optimal cycle

Content from this work may be used under the terms of the Creative Commons Attribution 3.0 licence. Any further distribution of this work must maintain attribution to the author(s) and the title of the work, journal citation and DOI.
Published under licence by IOP Publishing Ltd

of their regulation at minimum cost [7]. However, during the last decade, intelligent and adaptive transport systems that increase the capacity of the intersection in the absence or partial intervention of the operator by automatically searching for the optimal traffic signal regulation cycle have been widely applied [8, 9]. Its local project is automated traffic management systems that contribute to increasing the carrying capacity of intersections, reducing the time of vehicle delays and improving the ecological situation of the environment [10]. However, these devices have, from a technical point of view, some drawbacks that lie in the mandatory presence of the operator, in the constant interaction of peripheral equipment with an automated workplace and high capital costs. This leads to some limitations in the area of application of these systems, in particular, this is the lack of possibility of their installation in remote areas, as well as at intersections with low or unstable transport demand in directions and time.

At the present time, an increase in the ecology of the urban environment consists not only in discouraging the population of using personal transport, increasing adaptability of vehicles to the required standards, reducing the time for vehicle delays to red light, but also in creating industrial zones in remote areas that adjoin the boundaries of a locality [11–15]. This causes the development of a new transport infrastructure, which is the connecting element of the places of employment and rest. Increased safety in these areas is achieved by their equipment with traffic lights, which is explained by low capital costs [16]. Therefore, we can often observe the intersection of the road network with entrances and exits from enterprises with a lack of transport demand from a secondary direction, which according to GOST R 52289-2004 (with changes from 2015) should be equipped with traffic lights. This leads to the formation of ineffective delay of vehicles during red light in the main direction, in the absence of transport demand for a secondary. The solution to this problem can be an adaptive traffic management system that will reduce the operating time of the red light in the main direction and increase the capacity intersection. Therefore, the goal of the work is the development of an adaptive traffic management system with low capital costs and an assessment of the possibility of its installation in remote areas, namely at intersections with unstable transport demand in time and directions.

2. Methodology

In Russia, the development of intelligent transport systems lags behind world experience, which is due to the uneven design of infrastructures by regions and the historical and architectural features of the urban environment [17–20]. Currently, as a rule, first and second generation systems that require operator intervention in the management process are applied in the Russian Federation. This causes the lack of autonomy of the system and the possibility of its installation in remote areas, in particular at entrances and exits from enterprises or remote administrative facilities. The solution to this problem is the use of an adaptive traffic management system that includes a detector and a controller with a program for autonomous control of the traffic light. At the present time in Russia, such systems are used to pass a pedestrian on regulated transitions and to provide an advantage when turning the vehicle. The efficiency of these systems is estimated with the help of indicators: the number and severity of the road accident, the capacity of the road network, transport delays, the number of vehicle stops, the length of the queues ahead of the intersections, the time of the trip, the speed of the message. However, as a rule, the leading indicator is the average delay of vehicles, which is due to the ability to analyze the nature of the operation of intersections and their capacity. Calculation and forecast of this value for intersections with unstable transport demand in time and directions is difficult, therefore earlier the authors of this work proposed an indicator that estimates the proportion of inefficient phases of traffic lights. It is defined as the ratio of the number of green light for which there is no transport demand to the total number of green light:

$$\eta_{t.r.} = N_{i.r.p} \Big/ N_{r.p} \quad (1)$$

IOP Conf. Series: Earth and Environmental Science **177** (2018) 012013 doi:10.1088/1755-1315/177/1/012013

where $N_{i.r.p}$ – number of green light with no transport demand; $N_{r.p}$ – total number of green light in this direction.

It allows you to analyze the efficiency of a traffic light and evaluate the rationality of its equipment with an adaptive traffic management system.

The paper [21] is defined the area of application of an adaptive pedestrian management system. The main indicator of the assessment is the delay of vehicles and pedestrians, as well as the intensity of their movement. The ratio of the intensity of pedestrian traffic to the intensity of vehicles at which it is recommended to install a pedestrian ringing device is 0.42. Based on the analysis of this work and according to GOST R 52289-2004 (with changes from 2015), where the traffic intensity of cars is indicated, at which a traffic light object with a rigid regulating mode is installed, the study was conducted for an adaptive traffic management system. This value is 0.5 for the application of a rigid mode of traffic light regulation.

The number of phases in which there is no transport demand depends on the traffic intensity and impacts on road capacity, that is, the greater the traffic intensity of vehicles, the less the number of phases not involved and the less the indicator of ineffective phases of traffic lights. Therefore, the value of 0.5 was proposed as the lower limit of the developed indicator. When it is less than 0.5, then the operation of a traffic light with a rigid control mode is effective. This value is the basis for assessing the effectiveness of traffic signal regulation at intersections with unstable transport demand in time and direction. The scope of the adaptive traffic management system is limited by the values of the proportion of ineffective phases of the traffic light equal to 0.5 or more.

Increasing the efficiency of intersections with unstable transport demand in time and directions can be achieved by increasing the capacity, reducing the time of delays of vehicles along the main road (with the prevailing transport demand during the day), and, consequently, by reducing the inefficient phases of the green light from the secondary direction. This necessitates the need to equip crossroads with a system that will include a red light when transport demand occurs from a secondary direction. Thus, the purpose of the experiment is to identify the places of application of the developed adaptive traffic management system and to assess the effectiveness of its application.

2.1. Research of application of adaptive traffic management system

The main criterion for determining the scope of an adaptive traffic management system is the proportion of ineffective phases of traffic lights. Initially, researchers at each of the proposed intersections in the morning, afternoon and evening time were counted the number of green light, in which there was no transport demand, and their total number in the direction in question. Further, for each intersection, the proportion of inefficient phases of the traffic light was calculated, which made it possible to determine in Tyumen the necessary locations for the installation of an adaptive traffic management system.

2.2. Assessment of the effectiveness of the adaptive traffic management system

The experimental study was carried out at the intersection, which was analyzed to assess the effectiveness of the traffic light object. As a result of the analysis of the data of the previous stage, the authors found that the traffic signal regulation of this intersection is ineffective. Therefore, it was chosen as a test site. The intersection scheme, which was selected for testing, is shown in Figure 1.

This intersection was equipped with an adaptive traffic management system, which was called the "Calling Automobile Device". The system includes a video detector and a controller that allows sending control signals to turn on the red light to the main directions 1 and 2 in case of a transport demand from a secondary 3–6 using the specified program. To assess the effectiveness of the application of this system, some indicators were revealed before and after the equipment of the traffic light of the developed system. This was the capacity in the main direction, the time of vehicle delays and the decrease in fuel consumption, spent in idle vehicles on an ineffective red light, and, consequently, emissions of harmful substances and carbon dioxide into the atmosphere.

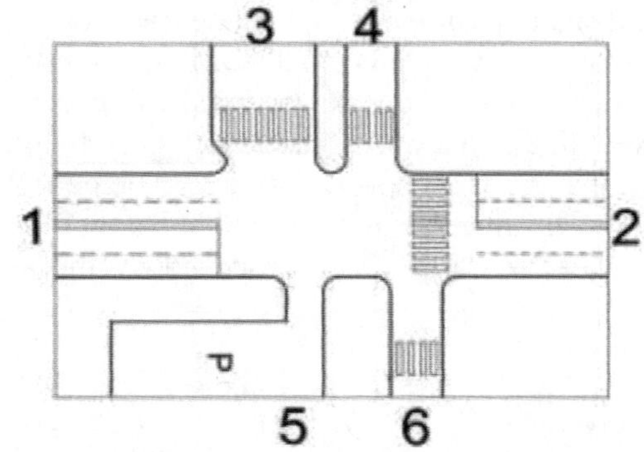

Figure 1. The intersection scheme, which is selected for testing

3. Results

3.1. Research of application of adaptive traffic management system

Tyumen was chosen for research, since it is a historically established settlement. The intensive development of its transport infrastructure, which provides an increase in the level of road safety, leads to a large number of intersections equipped with traffic lights. The city's territory can be divided into several zones: administrative and residential, and industrial zones, as shown in Figure 2

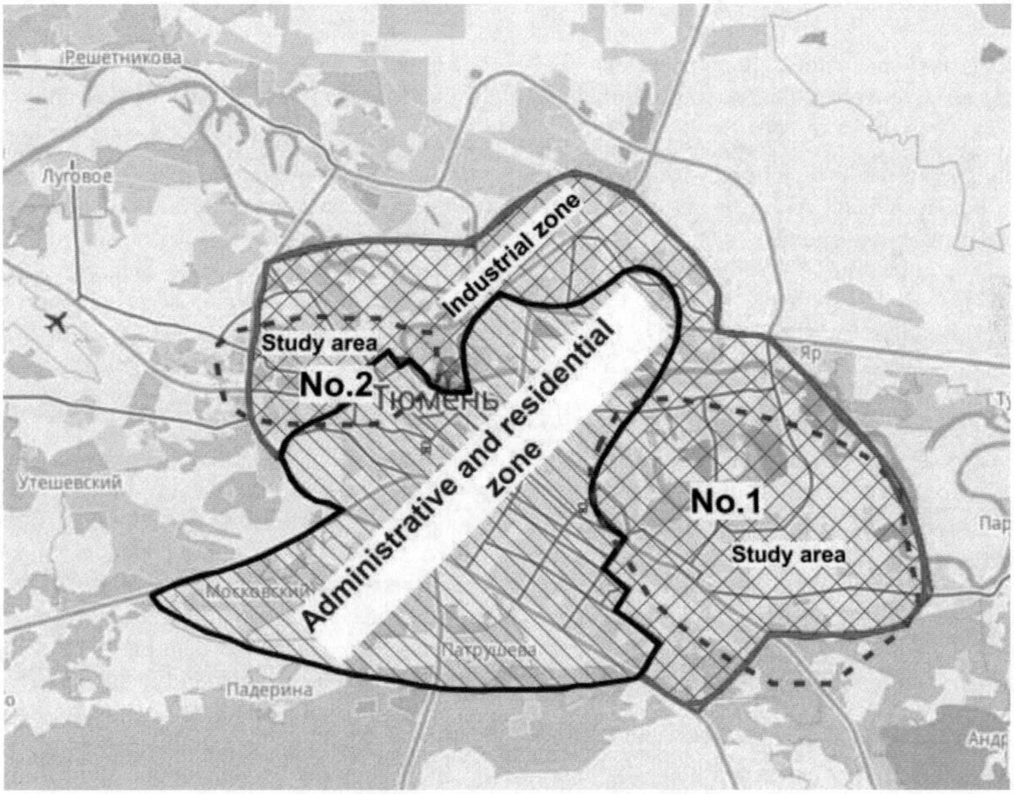

Figure 2. The zoning of Tyumen

The central part of the city is an administrative and residential zone and the semi-circle around it on the boundaries of the city territory are industrial sites: Antipinsky refinery, UMMC Steel, reinforced concrete products plant ZHBI-5. These large enterprises are the centers of industrial zones, in which there are also some medium and smaller company. When the transport infrastructure that connects the places of employment and recreation was created, intersections were equipped with traffic lights to ensure road safety. This led to the formation of intersections with unstable transport demand in terms of time and directions. However, this situation is typical not only for industrial zones, but it is also observed for municipal institutions and large sports facilities. The expected intersections that need to be equipped with an adaptive traffic management system were analyzed using the indicator of the proportion of ineffective phases of the traffic light, proposed by the authors, which is reflected in Figures 3 and 4.

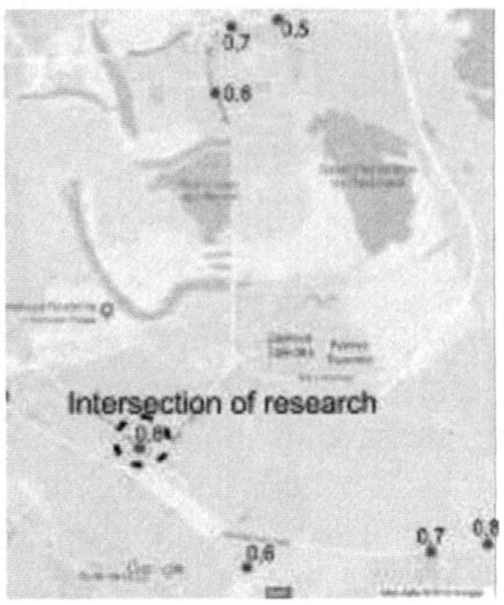

Figure 3. Indicator of the proportion of ineffective phases of the traffic light in the study area No. 1

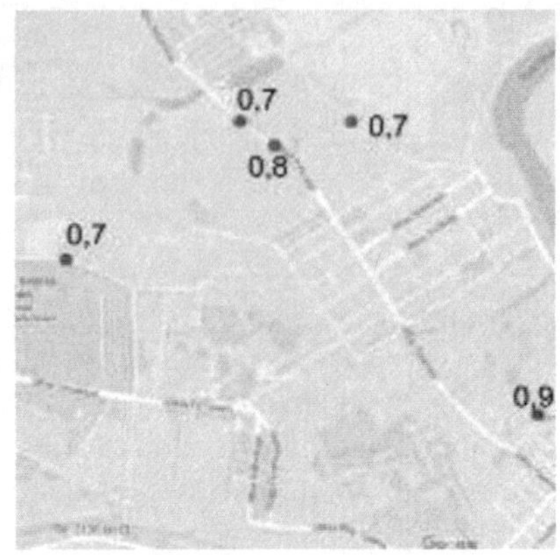

Figure 4. Indicator of the proportion of ineffective phases of the traffic light in the study area No. 2

Only part of the intersections with unstable transport demand in time and directions were analyzed during the experiment. Also, the authors identified the most characteristic places of application of the adaptive traffic management system. These are departures from adjacent territories, in particular hospitals, administrative buildings; crossing, which connect the major sports facilities, industrial enterprises with the city's road network. One of the crossroads analyzed, which is the section road that connects the large industrial enterprise (reinforced concrete products plant ZHBI-5) with the city's road network, was chosen to assess the effectiveness of the adaptive traffic management system.

3.2. Assessment of the effectiveness of the adaptive traffic management system

The average traffic intensity in the main directions 1 and 2 at this intersection is 1755 car/hour, while the traffic light regulation of this intersection is performed within 15 hours. As a result of the analysis of previous studies, the authors found that the proportion of ineffective phases of traffic signal regulation is 0.89 and 0.86 respectively for directions 3, 5 and 4, 6. Previously conducted theoretical studies and obtained data made it possible to establish that the operating time of the traffic light authorization signal, at which there is no transport demand, is 157 minutes during the day in directions 3, 4, while it is equal to 150 minutes in directions 5, 6. Therefore, data on its road capacity during the operation of the traffic light were obtained in the course of experimental studies before and after improving the organization of traffic at this intersection. The total road capacity was 67,350 cars within 15 hours of operation before the installation of the "Calling Automobile System" system. However, this value increased by 9,082 cars after improving the organization of traffic at this intersection, as shown in figure 5.

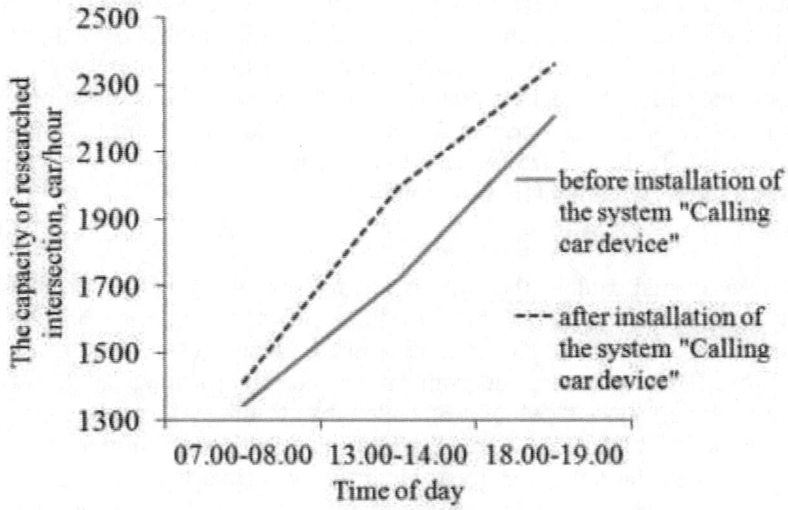

Figure 5. Change in the road capacity when equipping the traffic light with the "Calling Automobile Device" system

As a result of the equipment of the traffic light facility with the "Calling Automobile Device" system increases in the road capacity in the main direction by an average of 13 % and a decrease in the average time of vehicle delays.

The amount of fuel that is consumed by vehicles when idle for the red light was calculated on the basis of experimental and theoretical data on the average delay of vehicles and on fuel consumption by cars and trucks in idle mode during the year. The results of the calculation are presented in Table 1.

Table 1. The results of calculating the amount of fuel that cars consume when idle on the red light

Indicator name	Indicator value
The fuel consumption of cars and trucks during idling, l/h	1.2
The proportion of idle time of vehicles in the absence of the "Calling Automobile System" system (within 15 hours of traffic light)	0.34
The fuel that cars consume when idle for the red light, l/day (within 15 hours of traffic light)	3.705

When the "Calling Automobile Device" system is put into operation, the road capacity is increased, the delay time of vehicles to the red light is reduced, the fuel consumption of cars is minimized by 3.705 liters (2.816 kg), and carbon dioxide emissions are reduced by 8.828 kg during the day. This causes an increase in the mobility of the population and an improvement in the environmental situation.

4. Discussions

Currently, the application of this system is possible at 7 % of intersections in the urban environment, but new approaches to city planning and security on suburban routes lead to an increase in this value. The lack of transport demand from the secondary direction at the intersection, which is equipped with the "Calling vehicle system" system, increases the capacity of intersection in the main direction and reduces the time of vehicle delays, which causes an increase in population mobility and an improvement in the environmental situation. The cost of this system is low and amounts to 242.000 rubles. Its application is possible in remote areas, since it has a simple principle of operation and complete autonomy. Currently, the "Calling Automobile Device" system works by directs reading of vehicles that have arrived at the detection zone. However, in the future, the authors plan to develop a new type of detection, which is due to the development of transport.

The increase in the share of electric vehicles, which is one of the directions of the transport development strategy, leads to new types of detection that allow reducing the time and error of its readout. Therefore, in the future, the authors plan to create a new type of system that connects with an electric vehicle and includes a red light in the main direction.

5. Conclusion

In the course of experimental studies, the authors identified characteristic intersections that must be equipped with the developed system: departures from adjacent territories, in particular hospitals, administrative buildings; crossing, which connect the major sports facilities, industrial enterprises with the city's road network. The obtained results confirm the necessity to create an industrial sample of the adaptive traffic management system, which was called the "Calling Automobile Device", and lead to the need to evaluate the effectiveness of its implementation in operation. It represents a video detector and a controller that, with the help of a given algorithm, controls the traffic signal regulation and allows to increase the intersections capacity with unstable transport demand by time and direction by 13%, and also to reduce carbon dioxide emissions by 8.828 kg / day, which contributes to the improvement of the ecological situation. Also, the need for further development of this system was identified with the aim of adapting it to work with electric vehicles.

References

[1] Petrov A and Petrova D 2018 Economics of high-rise construction: The feasibility of skyscrapers building in the Russian cities *E3S Web of Conf. Rev.* **33** 03056
[2] Petrov A and Petrova D 2017 Atmospheric pollution in cities of Russia: Statistics, causes and characteristics *IOP Conf. Ser.: Earth Environ. Sci. Rev.* **72** 012007
[3] Anisimov I, Magaril E, Magaril R, Chikishev E, Chainikov D, Gavaev A, ErtmanS, Ertman Yu and Ivanov A 2016 Improving vehicle adaptability to the operating conditions of "smart" cities in the northern regions *E3S Web of Conf. Rev.* **6** 02003
[4] Ertman S, Ertman J and Zakharov D 2016 Adaptation of urban roads to changing of transport demand *E3S Web of Conf. Rev.* **6** 01013
[5] Panuccio P, Amodeo L, D'agostino P, Lamari D and Scattarreggia T 2015 Urban Regeneration and smart sity according to EU strategies: an urban distribution center in city ogistics *WIT Transactions on The Built Environment* vol 146 (UK: WIT Press) 313–24
[6] Jacyna M and Merkisz J 2015 Estimations in real urban traffic conditions s a deteminant of shaping sustainable urban development *WIT Transactions on The Built Environment* vol 146 (UK: WIT Press) 219–31
[7] Kolesov V, Danilov O and Petrov A 2017 Specific features of goal setting in road traffic safety *IOP Conf. Ser.: Earth Environ. Sci. Rev.* **90** 012059
[8] Lee J, Abdulhai B, Shalaby A and Chung E 2005 Real-time optimization for adpative traffic signal control using genetic algoritms *Journal of Intelligent Trasportation System* vol 9 (UK: Taylor and Francis Group) 111–22
[9] Wong S and Lo H 2004 Advanced algorithms in traffic signal control: editorial *Journal of Intelligent Trasportation System* vol 8 (UK: Taylor and Francis Group) 61–2
[10] Ceylan H and Bell M 2004 Reverse capacity for a road network under optimized fixed time traffic signal control *Journal of Intelligent Trasportation System* vol 8(UK: Taylor and Francis Group) 87–99
[11] Chainikov D, Chikishev E, Anisimov I and Gavaev A 2016 Influence of ambient temperature on the CO_2 emitted with exhaust gases of gasoline vehicles *IOP Conf. Ser.: Mater. Sci. Eng. Rev.* **142** 012109

[12] Chikishev E, Ivanov A, Anisimov I and Chainikov D 2016 Prospects of and problems in using natural gas for motor transport in Russia *IOP Conf. Ser.: Mater. Sci. Eng. Rev.* **142** 012110
[13] Chikishev E, Chainikov D and Anisimov I 2017 Increasing the use of natural gas on motor transport by an efficient location of the NGV RS *IOP Conf. Ser.: Earth Environ. Sci. Rev.* **50** 012004
[14] Chikishev E, Chikisheva A, Anisimov I and Chainikov D 2017 Natural gas use on minibuses, engaged in the carriage of passengers and baggage on the regular routes, as a measure for decrease in harmful environment effects *IOP Conf. Ser.: Earth Environ. Sci. Rev.* **50** 012008
[15] Ertman J, Ertman S, Anisimov I, Chainikov D and Chikishev Y 2015 Complex characteristic estimation of vehicle adaptability for low temperature operation conditions *Research Journal of Pharmaceutical, Biological and Chemical Sciences (Electronic Materials* vol 6*)* (India: RJPBCS) 1761–70
[16] Kolesov V and Petrov A 2017 Cybernetic Modeling in Tasks of Traffic Safety *Management Transportation Research Procedia (Electronic Materials* vol 20) (Netherlands: Elsivier) 305–10
[17] Russo F, Rindone C and Panuccio P 2014 The process of smart city definition at an EU level *WIT Transactions on Ecology and the Environment* vol 191 (UK: WIT Press) 979–89
[18] Langeland A 2015 Sustainable transport in cities: learnin from best and worst practice? *WIT Transactions on The Built Environment* vol 146 (UK: WIT Press) 93–103
[19] Galderisi A 2014 Adapting cities for a chenging climate: an integrated approach for sustainable urban development *WIT Transactions on Ecology and the Environment* vol 191 (UK: WIT Press) 549–60
[20] Cloke J and Layfield R 1996 The environmental impacts of traffic management schemes *WIT Transactions on The Built Environment* vol 26 (UK: WIT Press) 11
[21] Slobodchikova N 2010 *Improving the organization of traffic on the basis of the use of pedestrian ringing devices* (Irkutsk: Ikrut State Technical University)

Published under licence in *Journal Title* by IOP Publishing Ltd

IOP Conf. Series: Earth and Environmental Science **177** (2018) 012014 doi:10.1088/1755-1315/177/1/012014

Problems and prospects of sustainable low carbon development of transport in Russia

Yu Trofimenko[1], V Komkov[1] and V Donchenko[1, 2]

[1] Moscow automobile and road construction technical University (MADI TU), 64 Leningradsky prospect, Moscow, 125319, Russia
[2] Joint-stock company "Scientific and research institute of motor transport (NIIAT)", 24 Geroyev Panfilovtsev str., Moscow, 125480, Russia

e-mail: ywtrofimenko@mail.ru

Abstract. This article discusses the problems and prospects of sustainable low carbon transport development in Russia. Forecast scenario of Russian transport sector development was elaborated, which includes justification of the set of estimates of GHG emission by separate modes of transport and the transport industry as a whole, quantitative assessment of GHG emission from road, urban electric, rail, air, maritime, inland water transport, road building/maintenance and also justification of possible GHG emissions reduction targets when implementing the most effective measures for the period till 2030. Saving of gross GHG emission in 2030 at 2015 level or its reduction by 1.87–5.5 % is possible in case of implementation of a set of proposed measures. Package of comprehensive measures to improve energy efficiency and reduce GHG emission, first of all from motor transport based on balance of administrative, economic and mixed methods of the State regulation should be focused on introduction of carbon transport tax and norms of specific (per km of mileage) CO_2 emission and fuel consumption of vehicles, as well as the redistribution of the passenger and freight transport work from road and air transport to less energy-intensive transport modes (water, rail, urban electric cycling).

1. Introduction

Ensuring of sustainable low carbon development of transport sector has to be one of the main task of State transport policy. Such a policy has to foresee development of transport sector in direction of greenhouse gases (GHG) emission reduction, improving transport energy efficiency and reducing its negative impact on the environment.

Although the Transport Strategy of the Russian Federation adopted by the Government doesn`t directly use this term it provides for relative reduction of GHG emissions by separate modes of transport to the year 2030 in comparison with 2011. At the same time target values for gross and specific GHG emission from transport (as a whole and from its separate types) in this Document is not installed and is not linked to Russia's obligations in the framework of the implementation of Paris Agreement. This significantly reduces regulatory value of the Strategy as concern goal-setting, development and implementation of measures directed to improvement of energy efficiency and environmental safety both separate vehicles and the industry as a whole.

To solve this problem there was elaborated the transport sector development long-term forecast which includes: reasoning of the set of indices for GHG emission estimation from separate modes of

Content from this work may be used under the terms of the Creative Commons Attribution 3.0 licence. Any further distribution of this work must maintain attribution to the author(s) and the title of the work, journal citation and DOI.
Published under licence by IOP Publishing Ltd

transport and the transport industry as a whole; quantitative assessment of GHG emission from different modes of transport (road, urban electric, railway, air, maritime, internal water) and also from road building industry; establishment of targets for reducing of GNG emission when implementing the most effective measures.

2. Indicators of GHG emission by different modes of transport and methods of their evaluation

On basis of analysis of low carbon transport development indicators which are used for working out of international and national programmers there was defined the nomenclature of the most significant indicators used for assessment of effectiveness of different low carbon transport strategies (Table 1).

Table 1. The nomenclature of the most significant indicators of the development of transport industry

Recommended indicators	Comments
a) general transport indicators (key)	
Total gross GHG emission from transport, million tons CO_2 equivalent (CO_2e)/year	Allows to assess the cumulative effect of implementing of the policy directed to reducing of GHG emission in conjunction with overall economic and social processes taking place in society
Average specific GHG emission from transport in total per unit of transport work, g CO_2e/ton-km (passenger-km)	Allows to assess the cumulative effect from energy efficiency policy implementation in conjunction with general economic and social processes
Carbon intensity of transport in general (the ratio of GHG emission to GDP), ton CO_2e/bln rubles	Widely used in different countries
Total gross consumption of fuel and energy resources in transport sector, TJ/year	Allows to assess the cumulative effect of policy directed to reduction of carbon intensity of transport
Share of electricity in total energy consumption by transport, %	Allows to assess the cumulative effect of energy efficiency policy implementation
(b) indicators by mode of transport	
Motor transport	Allows to assess the effect of policy directed to reduction of GHG emissions, increasing energy efficiency of certain types of stationary and mobile sources of transport
Urban electric transport	
Rail transport	
Air transport	
Maritime and inland waterway transport	
Road Building/Management	
c) prospective indicators	
Average specific GHG emission during road construction (reconstruction), maintenance, renovation (rebuilding), ton CO_2e/km	Allows to assess the effect from the policy directed to improvement the energy efficiency and reduction of GHG emission from road construction and maintenance
Specific emissions of GHG by motor vehicles, g CO_2e/km	Allows to determine perspective normative requirements to specific GHG emission from single vehicles
Share of vehicles by mode of transport corresponding with certain classes of energy efficiency, %	Allows to assess the effectiveness of policy directed to renewal of rolling stock. Currently energy efficiency classes of cars in Russia is not installed

To quantify the values of the above indicators there was developed special methods based on the IPCC Guidelines which are used in the assessment of anthropogenic emissions from emission source and adsorption by GHG absorbers which are not controlled by the Montreal Protocol [1] for rail, air, sea and inland waterway transport as well as for road transport, urban electric transport and road construction/management.

Gross CO_2 emission from the mobile transport sources (in the absence of data on fuel and energy consumption by vehicle classes) is defined as:

$$E_{CO_2} = \sum_a \frac{AD_a \cdot EF_a}{10^6}, \text{Gg/year} \quad (1)$$

where: a – type of fuel (petrol, diesel, natural gas, liquefied petroleum gas, etc.); $AD_{(a)}$ – data on fuel type a consumption, TJ/year; $EF_{(a)}$ – specific CO_2 emission factor for fuel type a, kg/TJ.

Values of the specific CO_2 emission factor per unit of a type fuel burn mass are taken by recommendations of IPCC.

Table 2. Specific CO_2 emission factor for the level 1 [1, 2]

Type of fuel	kg/TJ	kg/ton of equivalent fuel	kg/ton, kg/th m3
Petrol	69 300	2 031	3 026
Diesel	74 100	2 172	3 149
LPG	63 100	1 849	2 903
CNG	54 400	1 594	1 840
LNG	54 400	1 594	2 710
DMT	73 300	2 115	2 096

If data on volumes of transport work for each mode of transport and respective specific emission factors per unit of transport work are available gross GHG emission should be calculated as (1), where: a – transport mode (road, urban, rail, air, water); $AD_{(a)}$ – the volume of transport work (passenger or freight) for transport mode a, million t-km or passenger-km/year; $EF_{(a)}$ – specific emission factor for CO_2 emission from transport mode a, kg/t-km or passenger-km.

If data on the fuel consumption by vehicle class are available emission of CO_2, CH_4 and N_2O may be calculated as:

$$E = \sum_{a,b,c} \frac{AD_{a,b} \cdot EF_{a,b}}{10^6}, \text{Gg/year} \quad (2)$$

where: E – emission of CO_2 , CH_4 or N_2O , t/year; a – type of fuel (diesel, petrol, natural gas or liquefied petroleum gas); b – type of vehicle; c – environmental class of vehicle; AD – energy content of fuel consumed, TJ/year; EF – specific emission factor for CO_2 , CH_4 or N_2O, kg/TJ. Normalization of GHG emission volume to CO2 equivalent (CO2e) is carried out in accordance with formula:

$$E_{CO2e} = E_{CO_2} \cdot 1 + E_{CH_4} \cdot 25 + E_{N2O} \cdot 298 \text{ , Gg} \quad (3)$$

Volumes of different fuels (electricity) consumption, for example, by motor fleet in formula (3) can be identified based on the number of passenger cars, light commercial vehicles, buses and trucks with different types of engines and used fuels, their specific fuel consumption (g/km) and the average weighted annual mileage. The formula for estimating emissions of CO_2, CH_4 and N_2O is:

$$E = \left(\sum_{a,b,c,d} \left(AD_{a,b,c,d} \cdot EF_{a,b,c,d} \right) + \sum_{a,b,c,d} C_{a,b,c,d} \right) / 10^9, \text{Gg/year} \quad (4)$$

where: E – emission of CO_2, CH_4 or N_2O, Gg/year; a – type of fuel/energy (diesel, gasoline, natural gas, liquefied petroleum gas, electricity); b – type of vehicle; c – environmental class; d – operating conditions (urban or rural roads); $AD_{a, (b), (c), (d)}$ – annual mileage on the road type (d) of type b vehicles with environmental class c utilizing type a fuel/energy , km; $EF_{(a), (b), (c), (d)}$ – emission factor for CO_2, CH_4 or N_2O for type b vehicles with c environmental class utilizing a type fuel/energy on road type d, g/km; $C_{a, (b), (c), (d)}$ – emissions during start-up and warming up of engines ("cold start") g/year.

The volume and structure of fuel consumption, for example for road transport, can be defined by actual (on the base of special investigations) or statistical data from organizations and people who operate motor vehicles.

IOP Conf. Series: Earth and Environmental Science **177** (2018) 012014 doi:10.1088/1755-1315/177/1/012014

Collection and analysis of actual and/or average statistical data on fuel consumption may carry out with the use of passport data for each type, subtype and model of vehicle or with the use of determined norms of fuel consumption for each vehicle`s type and subtype and conditions of traffic.

In the latter case norms of specific fuel consumption from the official document "Norms of fuels and lubricants consumption for road transport" [3] may be used.

For electric vehicles (with gross weight less than 3.5 tons) weighted estimation of specific energy consumption per km of mileage accounted for 0.15 kWh/km or 0.54 MJ/km. Specific energy consumption of tramcars and trolleybuses per kilometer of run should be based on technological standards of energy consumption for electric traction of most common models of tramcars and trolleybuses and estimation of their nominal weight.

According to MADI estimates specific number of motor transport enterprisers (service stations) in the Russian Federation (depending on the region and the number of vehicle fleet) is 0.6–0.9 per 1000 units of rolling stock, i.e. from 32 to 48 thousand enterprisers/stations [4]. For rough estimation of gross CO_2 emission from motor transport stationary objects (road transport enterprises, garages) the next formula may be used:

$$E_{CO_2} = \left[\sum_{k=1}^{h} (Y_k \cdot Z_k \cdot N_k \cdot L_k) \right] / 10^6 \tag{5}$$

where: E_{CO2} – gross emissions of CO_2, t/year; h – the number of considered types of motor vehicles; Y_k – per-unit energy consumption during k-type motor vehicle`s heating, maneuvering, maintenance and repairs on the territory of transport enterprisers, kWh/km; N_k – total number of k-type vehicles; L_k – the total mileage of k-type motor vehicles, km/year; Z_k – specific CO_2 emission factor values per unit of electricity consumed by motor transport stationary object, g/kWh.

Average weighted specific energy consumption (per km of motor vehicle run) for transport enterprises for different types (categories in accordance with international classification) of passenger and cargo vehicles are (in kWh/km [5]): M_1, N_1 – 0.174; M_2, N_2 – 0.148; M_3, N_3 – 0.110. Specific values of GHG emission per unit of electricity consumed taking into account also losses of electric power in electric networks for stationary emission sources of motor transport as well as for mobile sources of urban electric transport are: CO_2 – 435 g/kWh; CH_4 – 0.00723 g/kWh; N_2O – 0.0033 g/kWh [6].

The calculation of GHG emission from road construction/maintenance is made according to formula:

$$E_i = \sum_a \sum_k (FC_{ak} \cdot EF_{aTCE} \cdot EF_{aNCV} \cdot EF_{i,a}), \text{ t/year} \tag{6}$$

where: $E_{(i)}$ – greenhouse gas i emission, t/year; FC_{ak} – total mass of energy/fuel type a consumed on the roads of k technical category, t/year; $EF_{(a)TCE}$ – conversion factor in tons of coal equivalent by kind of energy resource a, tce/t; $EF_{(a)NCV}$ – conversion factor in energy units by type of energy source a, TJ/tce; $EF_{i\,(a)}$ – greenhouse gas i emissions coefficient by type of energy resource a, t/TJ.

Gross CO_2 emission from road construction/maintenance can also be defined if there are values of specific CO_2 emission for different types and categories of roads at all phases of road life cycle (Table 3) and proper roads length (ROSSTAT data) according to the formula:

$$E_{CO_2} = \sum_y \sum_k (EF_{yCO_2} \cdot L_k), \text{ t/year} \tag{7}$$

where: E_{CO2} – CO_2 emission, tons/year; EF_{yCO2} – CO_2 specific emission on k technical category roads at y phase of road life cycle, t/km ; L_k – length of k technical category road, km.

Indicators of CO_2 specific emission on the roads of different categories at different phases of their life cycle shown in table 3 take into account the specific features of road building/management activity

IOP Conf. Series: Earth and Environmental Science **177** (2018) 012014 doi:10.1088/1755-1315/177/1/012014

related to technologies of asphalt cement-concrete mixtures preparation, execution of works on construction, reconstruction, rebuilding, repair and maintenance of different categories of roads [7].

Table 3. Values of the specific CO2 emission on the different types of roads at different phases of their life cycle, t CO2/km (FAU "ROSDORNII", MADI)

Type of Roads	Life Cycle Phase	Road`s Technical Category				
		(I)	(II)	(III)	(IV)	(V)
Federal Roads	Maintenance	43.73	25.00	16.22	14.09	11.65
	Repair	533.28	271.70	195.84	190.16	52.05
	Rebuilding	1556.92	713.82	544.34	526.10	175.33
	Construction	2958.14	1356.26	1034.25	999.6	333.13
Regional and Inter-municipal Roads	Maintenance	10.93	6.25	4.06	3.52	2.91
	Repair	133.32	67.93	48.96	47.54	13.01
	Rebuilding	389.23	178.46	136.09	131.53	43.83
	Construction	739.54	339.07	258.56	249.9	83.28
Local (with hard or transitional pavement).	Maintenance	2.19	1.25	0.81	0.71	0.58
	Repair	26.66	13.59	9.79	9.51	2.60
	Rebuilding	77.85	35.69	27.22	26.31	8.77
	Construction	147.91	67.81	51.71	49.98	16.66

Table 4 shows the results of evaluation of the absolute and specific values of transport GHG emission obtained using the above formulas.

From this Table it may be seen that the gross transport GHG emission in 2015 was 286.01 million tons CO_2e/year, 73.03 % of which were from mobile sources and 26.97 % – from stationary (due to the lack of statistical data in this calculation there was not taken into account GHG emission from stationary objects of urban electric transport). It also may be seen that share of road and urban electric transport in total gross GHG emissions is 79.1 %, railway – 11.8 %, road building/maintenance and air transport (domestic) – 4.4 % and 3.9 % accordingly.

Table 4. Absolute and specific values of GHG emission by mode of transport, 2015

Mode of Transport	Total Gross GHG Emissions by Mode of Transport, Mio t CO_2e/yr	Specific Total Transport GHG Emission per Unit of Transport Work (distance), g CO2e/modified t km (t CO_2e/km*)	Volume of GHG Emission from Mobile Sources by Mode of Transport, Mio t CO_2e/yr	Volume of GHG Emission from Stationary Sources by Mode of Transport, Mio t CO_2e/yr
Motor transport	219.16	940.64	176.16	42.99
Urban Electric Transport	7.20	No data	7.20	No data
Rail Transport	33.80	14.46	12.70	21.10
Air Transport	11.21	2076.14	10.97	0.24
Maritime and Inland Water Transport	1.96	18.50	1.84	0.12
Road Building/Maintenance	12.67	12.12	12.67*	
TOTAL	286.01		**208.89**	**77.13**

* – total GHG emission in road building/maintenance is assigned to stationary sources

In the future growth of gross GHG emission is possible due to increase in transport work done by separate modes of transport as well as due to more complete and accurate accounting of direct gross GHG emission from transport infrastructure objects in case of the development of regional and sectoral GHG emission surveys. For the correct formation of transport policy, it is important to monitor not only gross but also specific indicators of GHG emission (in accordance with Task 6 of the Transport Strategy of the Russian Federation to 2030 adopted by the Government).

3. The forecast methods for gross and specific GHG emissions by different modes of transport

Probable scenarios of greenhouse gas emissions by transport of the Russian Federation for the future will be largely determined by the macroeconomic situation in Russia and in the world, rates of GDP growth, policies and measures in the fields of energy sector, industries and transport development, as well as results of implementing policies and measures aimed to reduction and limitation of greenhouse gas emissions and intensification of their outflows.

Two scenarios of changes in greenhouse gas emissions from transport may be considered – inertial (BaU) and basic (innovative). As a base of BaU scenario it was considered the "Basic plus" scenario of long-term socio-economic development project developed by the Russian Ministry of Economy and Development. The assumption was accepted that only measures already adopted in the framework of the State policies will be used to promote low carbon development, including energy efficiency measures, the development of nuclear energy, the use of non-conventional energy sources.

An innovative scenario is characterized by a significant strengthening of requirements to environmental safety and energy efficiency of transport development that in turn assumes changing the structure of used energy resources. This Scenario provides for more sophisticated model of transport management by the State and business and at the same time increase in the expenditure on transport infrastructure development, implementation of high-tech projects and human capacity development.

In both scenarios it is supposed the development and implementation of the State and market regulation mechanisms which foresee implementation of the regulatory, organizational, technical and economic measures grouped into three directions:

1) improving energy efficiency of vehicles and transportation technologies based on the use of traditional types of motor fuel;
2) widening the use of various sources of energy with reduce GHG emission for vehicles of all modes of transport;
3) mobility management – reducing the excessive, irrational, unreasonable movement of passengers and freight, restraint of hypermobility of population by development of transport systems, implementation of ITS and information/telecommunication technologies and so on.

Measures of the first direction include: development and introduction of new energy efficient and environmentally friendly vehicles and transportation technologies; renewal and formation of optimal structure of rolling stock of different modes of transport; implementation of new emission standards (e.g. standardization of specific emissions of CO_2), implementation of non-motorized modes of transport and movement; maintain technical conditions of rolling stock and transport infrastructure objects; encouraging of consumers to choice low carbon vehicles, etc.

Measures of the second directions include: development and implementation of energy-efficient vehicles, alternative fuels and vehicles capable to work on them; electrification of road transport; refueling infrastructure for alternative fuels and energy.

Measures of the third direction include: demand management for transport services; management of priorities in development of various modes of transport; formation of a rational structure of road and public transport networks in cities and agglomerations; low-carbon organization of transportation processes with mutual complementing (but not competition) of different modes of transport (digital transportation and logistics technologies, intelligent transport systems (ITS), etc.); formation of intelligent toll systems (tolls in accordance with distance travelled, vehicles weight, their level of energy efficiency and environmental safety), effective parking policy etc.

The implementation of the above-mentioned measures relates, above all, motor and urban electric transport, the share of greenhouse gas emission from which is almost 80 % of emission from all modes of transport.

Gross GHG emission from the mobile sources of motor transport up to 2030 is defined for prognostic number and structure of the fleet of passenger cars, light commercial vehicles, trucks and buses by type of their engine (type of fuel used), the annual runs and specific fuel consumption per km

(CO_2 emission). Prognostic number and structure of motor vehicles fleet by type of engine, annual runs were defined in accordance with [8] and are shown in Table 5.

Table 5. Forecast of the number of passenger cars, light commercial vehicles, trucks and buses in the Russian Federation to 2030, thousand

Years	Cars	Light commercial vehicles	Trucks	Buses	**Total**
2015*	44253	3054	3176	873	**51355**
2016*	44699	3033	3179	896	**51807**
2017*	46887	3188	3246	896	**54216**
2020	48960/54720	3325.4**	3840**	890**	**56250/62010**
2025	54900/60690	3509.5**	3953**	910**	**62510/68300**
2030	58400/65700	3818.0**	4130**	930**	**66330/73630**

* – GIBDD (Traffic Police) data;
** – for both scenarios: numerator/denominator – BaU/innovative scenario

From this Table it`s possible to see that the Russian motor fleet to 2030 will increased by 29.0–43.4 % as compared to 2015 reaching a size of 66.33–73.63 million units respectively (by BaU and innovative scenarios). The number of cars may amount to 58.4–65.7 million units (88 %), light commercial vehicles – 3.8 million units (5.2–5.8 %), trucks – 4.13 million units (5.6–6.2 %) and buses – 0.9 million units (1.3–1.4 %).

When forecast the structure and size of motor fleet by type of engine as the key condition there was defined the need to achieve the targets set out in the Transport Strategy of the Russian Federation to 2030: "the share of the vehicle fleet with hybrid, electric engines and engines used alternative fuels in the total number of vehicles in 2020 will reach 26 % for BaU scenario and 29 % for innovative scenario, and in 2030 – respectively 49 % and 54 %. Table 6 shows the results of the prediction of vehicle fleet structure by type of engine and fuel type which based on projections made abroad and in Russia (MADI and other organizations).

The proportion of electric vehicles and hybrids in car fleet for period concerned is characterized by the following values: 2015 – 0.026 %; 2020 – 7.7–8.5 %; 2025 – 10.8–24.0 %; 2030 – 23.0–48.0 % respectively for considered scenarios. The share of electric vehicles and hybrids in LDV fleet in 2030 could be 12.0–25.0 %, in trucks fleet – 3.0–10.0 % and in bus fleet – 3.0–4.0 %.

Table 6. Prediction of motor fleet structure by type of engine, % (numerator/denominator – BaU/innovative scenarios)

Type of Vehicle	Type Engine	2015	2020	2025	2030
Cars	Petrol	92.959	64.2/61.0	53.1/49.0	41.0/34.0
	Diesel	4.893	9.8/10.0	9.9/9.5	10.0/12.0
	Gas-engine	2.122	18.3/20.5	26.2/17.5	26.0/6.0
	Hybrids	0.026	6.7/7.0	8.8/20.0	20.0/40.0
	Electric, FC*, hydrogen	0.000284	1.0/1.5	2.0/4.0	3.0/8.0
	TOTAL	100.0	100.0	100.0	100.0
Light Commercial Vehicles	Petrol	61.166	40.0/37.6	31.0/26.6	21.0/17.4
	Diesel	32.935	34.0/33.4	32.0/ 31.9	30.0/28.6
	Gas-engine	0.5899	21.0/22.2	29.6/29.0	37.0/29.0
	Hybrids	0.0	4.7/6.0	6.4/10.0	8.0/20.0
	Electric, FC*, hydrogen	0.0	0.3/0.8	1.0/2.5	4.0/5.0
	TOTAL	100.0	100.0	100.0	100.0
Trucks	Petrol	8.463	8.6/9.0	7.6/7.0	1.0/2.0
	Diesel	85.4572	82.0/82.4	81.0/60.5	60.0/44.0
	Gas-engine	5.965	8.4/8.4	10.0/28.5	36.0/44.0

Type of Vehicle	Type Engine	2015	2020	2025	2030
	Hybrids	0.0	0.6/ 0.6	1.4/4.0	3.0/10.0
	TOTAL	100.0	100.0	100.0	100.0
Buses	Petrol	34.617	14.0/15.0	10.0/13.0	9.0/10.0
	Diesel	57.51	69.0/70.0	53.0/49.0	42.0/36.0
	Gas-engine	7.87	15.0/15.0	34.7 35.0	46.0/50.0
	Hybrids	0.0	1.0/1.0	2.3/3.0	3.0/4.0
	TOTAL	100.0	100.0	100.0	100.0

* – FC (fuel cells)

With taking into account target values fixed in the Transport Strategy to 2030 the share of trucks used natural gas as a fuel could reach 36–44 % of trucks fleet and buses – 46–50 %. Insufficient development of gas filling station network and gas-fueled vehicles service infrastructure is one from the key factors restrained the expansion of natural gas use as motor fuel for transport. This has been taken into account in the BaU scenario.

Analysis of data presented in Tables 5 and 6 showed that the total number of Russian vehicle fleet for the period up to 2030 in accordance with the type of engine may undergo significant changes as compared to 2015: the number of petrol vehicles in the Russian Federation in 2030 may be reduced almost in 2 times and will constitute about 23.1–24.7 million units; the number of vehicles with diesel engine on the contrary will increase by one third and can reach 9.47–10.8 million units; the number of vehicles with gas-driven engine, hybrids in 2030 may rise dozens of times and the number of electric vehicles – thousand times. This dynamics of the Russian motor fleet change by type of engine and fuel type can only be achieved in case of urgent adoption of a complex of regulatory, organizational, engineering and technical measures directed to reducing GHG emission.

Gross GHG emission from motor fleet for forecasted period was estimated by multiplying the volumes of consumption of different fuels (electricity) on specific CO_2 emission indices (kg CO_2/tce) (Table 7).

Table 7. Specific GHG emission by mode of transport in 2015, t CO_2/tce

Mode of Transport	Specific emission
Motor and Urban Electric Transport	1.62
Rail Transport	1.53
Maritime and Inland Water Transport	2.01
Air Transport	2.02

Consumption of different types of fuels (electricity) by motor transport for a given forecast period was defined by simplified method based on the number of cars, light commercial vehicles, trucks and buses with different types of engines (Tables 5 and 6), their specific fuel consumption (g/km) and the average weighted annual runs of cars, light commercial vehicles, trucks and buses, which were established for basic year taking into account fuel consumption balance and results of calculations with the use of COPERT programme [9] or [10] (NIIAT) and for forecast period – taking into account trends of increasing of vehicle`s fuel efficiency (2 % per year). Average values of fuel consumption per km of mileage by different types of vehicles in 2015 were accepted by EMEP/EEA data [11]. For the next period values of specific fuel consumption were received on the basis of analysis and synthesizing of transport development trends and expert judgements.

4. The results of transport greenhouse gas emissions forecast for the period until 2030

As the main parameters for forecast of gross GHG emission by separate modes of transport and road construction/maintenance there have been used:

1) transport work for passenger and freight transportation by all modes of transport which depends on transport services market development, volumes of freight and passenger transportation;

IOP Conf. Series: Earth and Environmental Science **177** (2018) 012014 doi:10.1088/1755-1315/177/1/012014

2) development of transport infrastructure (projected length of public roads set by the State programs);
3) development of vehicle fleet (forecasted fleet number by the type of engine and fuel used is estimated taking into account fleet renewal and introduction of major production capacities, balance of supply and disposal of vehicles);
4) GHG emission factors (forecasted values of specific GHG emission (per unit of transport work, per unit of mileage, per unit of road length, per unit of consumed fuel/energy) for all modes of transport are established on the basis of the target values of the Transport Strategy, analysis of trends of transport technics and technology development, tightening of normative values of specific emissions, as well as expert judgements).

Figure 1 shows the forecasted values of gross GHG emission from transport as a whole and from separate transport modes for the period up to 2030 in conditions of two scenarios realization.

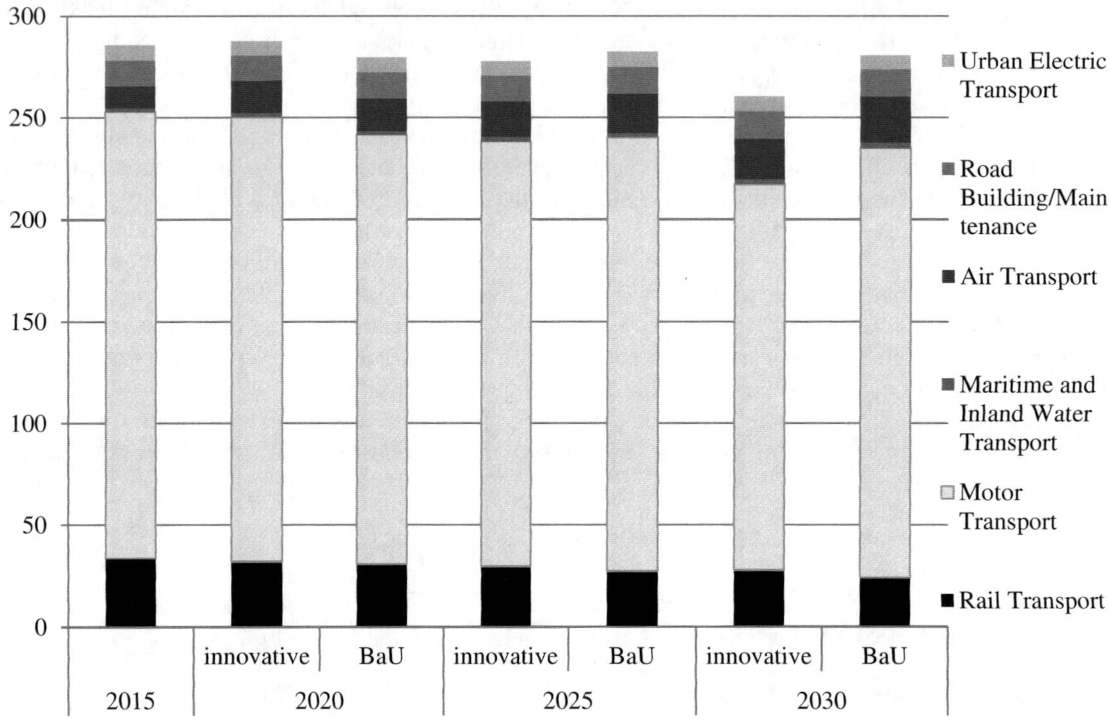

Figure 1. Total GHG gross emission by transport in the Russian Federation in accordance with two scenarios

In comparison with 2015 gross transport GHG emission for the period up to 2030 will decline in both scenarios under consideration. At that if for BaU scenario gross GHG emission reduction will be relatively small (about 1.4 percent or 4.0 million tons of CO_2e.) and sufficiently smooth (from 286.01 to 281.98 million tons CO_2e), then in innovative scenario in 2030 transport gross GHG emission reduction will be substantial – in 6.1 times more than for the BaU scenario (24.77 million tons of CO_2e). The most significant reduction in gross GHG emission in this scenario (from 277.81 to 261.25 million tons CO_2e i.e. 16.6 million tons CO_2e) is expected in the period after 2025 mainly on account of intensive replacement of benzene and diesel car by electric cars and hybrids as envisaged in the updated version of the Transport Strategy of the Russian Federation.

Maximum contribution to total gross GHG emission from transport in 2030 will make motor transport – 74.8–72.5 % from all emission what is in line with assessments by other authors and confirms the earlier findings on the importance of priority introduction of measures to reduce GHG emissions just by motor transport.

The railways share in total emissions from transport can constitute 8.6–10.7 %, road building/maintenance – 4.8–5.2 %, air transport – 7.8–8.4 %, urban electric transport – 2.64–2.48 %, maritime and inland water transport – 1.12–0.95 %.

Table 8 shows the forecasted values of major multimodal indicators of GHG emission and energy consumption of transport in the Russian Federation for the period until 2030 obtained on the basis of the fulfilled projections of GHG emission and energy consumption, population projections, changes in GDP and other indicators of socio-economic development.

Table 8. Prognostic values of the main multimodal indicators of GHG emission and energy consumption of transport in the Russian Federation for the period until 2030

Indicators	2015	2020		2025		2030	
		Target	BaU	Target	BaU	Target	BaU
Total gross GHG emission from transport, Mio t CO_2e/yr	286.01	288.28	279.93	277.81	282.41	261.25	281.98
Average specific transport GHG emission in total per unit of transport work, g CO_2e/t-km (pass-km)	106.66	89.68	89.75	77.34	87.97	64.85	84.38
Carbon intensity of transport in general (ratio of GHG emission from transport to GDP), t CO_2e/billion rub.	3429.9	3457.1	3357.0	3248.6	3386.7	3050.3	3381.6
Transport GHG emission per capita, t CO_2e per capita	1.95	1.97	1.91	1.85	1.93	1.74	1.93
Total gross consumption of fuel/energy in transport sector, TJ/year·103	4846.03	4869.17	4709.73	4105.9	4764.57	3836.55	4730.66
Average specific power-intensity of freight transportation in general, MJ/t-km	1.81	1.51	1.51	1.17	1.49	0.98	1.42
Average specific power-intensity of passenger transportation in general, MJ/p-km	9.14	8.06	7.95	6.00	7.55	4.59	6.70
Share of electricity in total energy consumption by transport, %	29.58	N.d.	N.d.	N.d.	N.d.	N.d.	N.d.
Volume of final consumption of different types of fuels and energy by transport in general, Mio tce/year	165.39	166.18	160.74	140.13	162.75	130.94	161.46

From projections given in the Table 8 it follows that realization of both scenarios in the transport sector will lead to decline in gross (from 286.01 million tons CO_2e to 281.98–261.25 million tons CO_2e), specific GHG emissions and energy use. Besides in case of realization of innovation (target) scenario expected reduction of GHG emission and energy consumption by 2030 will be approximately 10–20 % greater than when implementing BaU (conservative) scenario. And in terms of specific GHG emission in total per unit of freight transport work the reduction could reach 33 % as compared with BaU scenario, for specific energy intensity of passenger transportation – 37.4 %, for specific energy intensity of freight transport – 48.2 %.

If for transport sector it will also be set a target level for reduction of gross GHG emission in 2030 to 25–30 % compared to 1990 in accordance with the declared aim of Russia, to which we should strive within the framework of the Paris Agreement, then received forecasted values of gross GHG emission by 2030 must be reduced even at 126.08–136.47 million tons CO_2e when implementing the BaU scenario and to 115.7–105.34 million tons CO_2e when implementing innovative scenario. To reach the target level of gross GHG emission in 2030 is possible only in case if all Russian motor fleet (66–73.6 million units) will consist from hybrids and electric vehicles (50:50). In this case total GHG emission of transport in general can amount in 140.17 million tons CO_2e, i.e. at 5.34–15.74 million tons CO2e less than possible target level.

A full replacement of petrol and diesel vehicles in Russian motor fleet by electric vehicles and hybrids even to 2050 is unrealistic particularly for political and socio-economic considerations associated with dominating today and in the future in the economy raw materials sectors. In this regard, it is not acceptable for the transport sector setting the target level of gross reduction of GHG emission to 2030 as 70–75 % from the level of 1990.

Thus, the persistence of gross GHG emission in 2030 at 2015 level or it reduction at 1.87–5.5 % is possible in case of implementation of a set of proposed measures and fits into the targets and indicators contained in the updated edition of the Transport Strategy of the Russian Federation for rail, air and water transportation, as well as in case of realization of predictive values of gross and specific GHG emission for motor, urban electric transport and road building/maintenance.

Package of comprehensive measures to improve energy efficiency and reduce GHG emission, primarily for motor transport, based on balance of administrative, economic and mixed methods of the State regulation should be focused on introduction of carbon transport tax and norms of specific (per km of mileage) CO_2 emission and fuel consumption of the vehicle, as well as redistribution of passenger and freight transportation from motor and air transport to less energy-intensive transport modes (water, rail, urban electric, cycling) by reorganization of transportation process through implementation of multimodal transport links, development of digitalization of logistic processes and so on.

A new trend on electrification of motor transport must be backed up by actions in the energy sector to create additional generating capacity (estimated from 2.76 to 7.95 billion kWh (electric vehicle fleet from 1.879 to 5.415 million units) and from 48.69 to 54.05 billion kWh – with full substitution of petrol and diesel vehicles by electric vehicles and hybrids (common fleet from 33.2 to 36.8 million units) required for the implementation of this trend preferably with the use of renewable energy sources.

References

[1] Eggleston H, Buendia L, Miwa K, Ngara T and Tanabe K 2006 IPCC Guidelines for National Greenhouse Gas Inventories *Intergovernmental Panel on Climate Change (IPCC)* (Japan) NGGIP(2006)

[2] 2015 *Methodical recommendations on carrying out of voluntary greenhouse gas inventory within the entities of the Russian Federation* (Moscow: Decree of the Ministry of Natural Resources of Russia)

[3] 2015 *Norms of consumption of fuels and lubricants in motor transport* About introduction in force of methodical recommendations (Moscow: Ministry of Transport of Russia)

[4] Trofimenko Yu and Jakubowicz I 2016 *Engineering structures and environmental safety of service centers* (Moscow: Publishing Center Academy)

[5] Lukanin V and Trofimenko Yu 2003 *Industrial and transport ecology* (Moscow: Vysshaya Shkola)

[6] 2016 *Methodical recommendations for transport organizations engaged in transportation by motor and electrified urban transport (trams, trolleybuses) on conduction of GHG inventory* (Moscow: Ministry of Transport of Russia (draft))

[7] Trofimenko Yu, Grigoreva T and Evgenev G 2014 Energy-saving problems of road facilities in Russia *WIT Transactions on Ecology and the Environment***190** 535–42

[8] Trofimenko Yu, Komkov V and Grigoyeva T 2018 Forecast of the vehicle fleet size and structure in the Russian Federation by ecological classes, a type of power installations and a

IOP Conf. Series: Earth and Environmental Science **177** (2018) 012014 doi:10.1088/1755-1315/177/1/012014

fuel type for the period up to 2030 *Proc. int. Env. Cong. on ELPIT 2017* (Samara: Publishing House Scientific Centre) 311–26

[9] Gkatzoflias D, Kouridis C, Ntziachristos L and Samaras Z 2012 *Computer programme to calculate emissions from road transportCOPERT 4* (User manual vers. 9.0)

[10] Donchenko V, Kunin Yu, Ruzski A, Mekhonoshin V, Barishev L and Trofimenko Yu 2016 Estimated atmospheric emission from motor transport in Moscow based on transport model of the city *Transportation Research Procedia* 2649–58

[11] EMEP/EEA air pollutant emission inventory guidebook 2016 *Technical guidance to prepare national emission inventories*(Copenhagen: EEA Report)

Published under licence in *Journal Title* by IOP Publishing Ltd

Entropic analysis of dynamics of road safety system organization in the largest Russian cities

A Petrov and V Kolesov

Tyumen Industrial University, 72 Melnikaite str., Tyumen, 625000, Russian Federation

e-mail: ArtIgPetrov@yandex.ru

Abstract. The article presents the method of road safety system organization assessment in cities. The main idea of this method is the usage of entropic analysis in the solution of research problems. Information entropy is a measure of uncertainty and unpredictability of system functioning results. Therefore, it allows to quantitatively assess the quality of management of any city life aspect. In this article, by assessing of system entropy, we estimated dynamics of changes in road safety system organization of two biggest Russian cities – Moscow and St. Petersburg, in 2004–2017. It was established that the level of road safety system organization in Moscow and St. Petersburg has increased during the last 14 years. Comparison of entropy level of road safety provision systems of two Russian megalopolises allows to make a conclusion that level of organization of this system is slightly higher in Moscow than in St. Petersburg.

1. Introduction

Life safety is one of the most important components of life quality. Security issues has become especially significant in the largest cities. In 2017 8907 road accidents (5.25 % of all road accidents in Russia) occurred in Moscow [1]. The share of deaths in road accidents in Moscow was 494 deaths or, in other words, 2.58 % of deaths in road accidents in Russia [1]. Indicators of road accident rate in the second most important city of Russia – St. Petersburg, are less than in Moscow but are also very significant. In the last 12 years, i.e. in the period when federal target road safety programs [2, 3] were actively realized, the level of road accident rate has decreased in the country in general and in the biggest cities in particular. However, we can't draw far-reaching conclusions based on statistics of road safety first-level indicators (count of road accidents, count of deaths and victims in road accidents). To do this we need to know dynamics of changes in road safety second-level indicators (Human Risk and Transport Risk) [4] and understand how the organization of road accident prevention system changes over time [5]. For this it is necessary to study tendencies of significance change of different sections of the cause-and-effect chain "Population – Vehicle fleet – Road accident – Victims in road accident – Deaths in road accident" [6]. On the basis of this information about structural changes in the studied system, we can make a conclusion about dynamics of systems of traffic safety management organization.

Content from this work may be used under the terms of the Creative Commons Attribution 3.0 licence. Any further distribution of this work must maintain attribution to the author(s) and the title of the work, journal citation and DOI.
Published under licence by IOP Publishing Ltd

2. About possibility of using the entropic analysis in the assessment of system organization dynamics

The author of information theory C.E. Shannon [7, 8] introduced the concept of information uncertainty – entropy. According to C.E. Shannon, entropy characterizes missing information about the studied system. Information is the primary concept of our world, along with material, energy, space and time, and it can't be strictly defined [9]. At the same time, the usage of entropic analysis allows to assess the degree of informational organization of system processes. According to C.E. Shannon [7, 8] information is discrete.

The basis of information theory is based on the concept of four initial elements: transmitter, channel lines, receiver and information. The information, transferred by channel line from the transmitter to the receiver, can consistently change its quality. The degree of information loss in the process of passing through the channel is characterized by entropy.

Entropy is low in the highly organized systems and high in low-organized systems. It is important to note that the degree of human-technical systems organization changes over time. In qualitatively managed systems chaos level and entropy value are decreased. The understanding of entropy dynamics allows to find out whether the degree of managed system organization is raising. That is why entropic analysis is an appropriate tool for assessment of the quality of road safety provision processes in the largest Russian cities.

3. Aim, problems and research method

The aim of the research is studying the dynamics of entropy at the road safety provision systems in two largest Russian cities – Moscow and St. Petersburg (2004–2017).

Research problems:

1. Gathering of official statistics by blocks of the cause-and-effect chain "Population – Vehicle fleet – Road accident – Victims in road accident – Deaths in road accident" in Moscow and St. Petersburg.
2. Calculation of transitional coefficients K_{i} between blocks of chain of road accident rate formation process (fig.1). Describing of studied process as 3 subprocesses with specific coefficients K_{n} (transformation of population into count of vehicles in transport system), K_{dtp} (transformation of vehicle fleet into road accidents count), K_{pg} (transformation of road accidents count into count of deaths in road accidents).

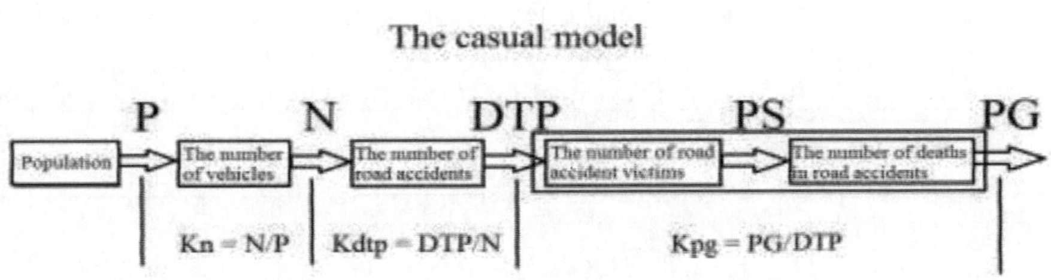

Figure 1. Cause-and-effect model and coefficients of transferring of road accident rate formation process elements

3. Identifying of the positive of the contribution Q relatively to weights of appropriate elements of examined transformational process within the chain "Population – <...> – Deaths in road accident".

$$Q = Q_n + Q_{dtp} + Q_{pg} = \ln(1/K_n) + \ln(1/K_{dtp}) + \ln(1/K_{pg}) = 1 \tag{1}$$

The physical meaning of the positive of the contribution Q of different elements of the chain "Population – <...> – Deaths in road accident" into the final result of road accident rate is the measure of information amount or derivative of examined process entropy.

4. Identifying the structure of weight coefficients w_i for assessing the positive of the contribution Q of different elements of the chain "Population – <...> – Deaths in road accident". Availability of calculated values w_n, w_{dtp}, w_{pg} of positive allows to solve the main problem of entropic analysis – assess the impact of different elements of the chain "Population – <...> – Deaths in road accident" on formation of final road accident rate. Above-stated researches were held for each year from period of 2004–2017.
5. Calculation of entropy H in road safety provision systems of two largest Russian cities – Moscow and St. Petersburg by classic C.E. Shannon's formula (2):

$$H = -\sum_{i=1}^{n} w_i \cdot \ln w_i \tag{2}$$

where
n – system elements count (in our case n = 3);
w_i – weight coefficients, satisfying the normalization condition, $\sum_{i=1}^{n} w_i = 1$.
6. Calculation of relative entropy (3):

$$H_n = H/H_{max} = H/\ln(n) \tag{3}$$

7. Analysis of dynamics of entropy H of road safety provision systems of Moscow and St. Petersburg over the period of 2004–2017.
8. Detection of specific (for researched cities) features of entropy H dynamics over time and formulating final comparative conclusions.

4. Research results

Table 1 shows initial data (2017) and the example of calculation of entropy value at the road safety provision system in Moscow.

Table 1. Initial data and elements of calculation of entropy value at the road safety provision system in Moscow (2017)

Numerical values of elements of a cause-and-effect chain of road accident rate formation process (Moscow, 2017)			
Population	Vehicle fleet [1]	Road accidents count [1]	Deaths count [1]
12380664	4704652	8907	494
Numerical values K_i			
K_n		K_{dtp}	K_{pg}
4704652/12380664 = 0.380		8907/4704652 = 0.00189	494/8907 = 0.05546
Numerical values of positives $Q_i = \ln(1/K_i)$			
0.967584		6.271178	2.892093
Numerical values of weight $w_i(Q_i)$			
0.0955		0.6190	0.2855
Numerical value ($\ln w_i$)			
-2.34854		-0.47962	-1.25361
Numerical values $(w_i \cdot \ln w_i)$			
-0.22433		-0.29693	-0.35788
Numerical value of entropy H = 0.8791			
Numerical value of relative entropy H_n = 0.8002			

Analogously performed calculations for the road safety provision systems in Moscow and St. Petersburg on the basis of the data of 2004–2017 allows forming the Table 2.

Table 2. Summary table of results of assessment of entropy at the road safety provision systems in Moscow and St. Petersburg (2004–2017)

Year	Calculated values of entropy H and relative entropy H_n in road safety provision systems in the largest Russian cities			
	Moscow		St. Petersburg	
	H	H_n	H	H_n
2004	0.9103	0.8286	0.9666	0.8798
2005	0.9521	0.8666	0.9666	0.8798
2006	0.9499	0.8647	0.9630	0.8766
2007	0.9429	0.8583	0.9546	0.8689
2008	0.9346	0.8507	0.9406	0.8562
2009	0.9209	0.8383	0.9274	0.8441
2010	0.9190	0.8365	0.9359	0.8519
2011	0.9109	0.8291	0.9268	0.8436
2012	0.9066	0.8252	0.9284	0.8451
2013	0.8909	0.8110	0.9233	0.8404
2014	0.8937	0.8135	0.9239	0.8410
2015	0.8824	0.8032	0.9172	0.8349
2016	0.8784	0.7996	0.9154	0.8332
2017	0.8791	0.8002	0.9145	0.8324

5. Discussion of results

Fig. 2 shows the dynamics of changes of relative entropy H_n at road safety provision systems in Moscow and St. Petersburg (2004–2017). Level of road safety provision systems organization in Moscow and St. Petersburg has gradually increased during 2004–2017. Relative entropy in Moscow decreased from 0.8666 to 0.8002 (by 7.7 %), in St. Petersburg – from 0.8798 to 0.8324 (by 5.4 %).

This fact allows us to conclude that actual level of road safety provision system organization and the rate of decline of the entropy of given system in Moscow is slightly higher than in St. Petersburg. In general progress in decreasing the road accident rate is better in Moscow.

6. Conclusion

Unfortunately, currently entropic analysis practically is not used in the analysis of the effectiveness of management decisions in the sphere of road safety assessment [5, 10]. People responsible for analytical work (for example, employees of Research Center of the State Road Safety Inspectorate of the Ministry of the Interior of Russia are used to assess progress using the method of simple comparison of actual first-level road safety indicators with the last year's ones [4].

That does not allow to make serious conclusions, excepting assessment of the simplest tendency. Usage of entropic analysis of road safety provision systems organization allows to estimate quality of professional activity of specialists who organize functioning of transport systems in cities [10, 11].

Analysis of long-term dynamics of road safety provision systems organization in the largest Russian cities – Moscow and St. Petersburg, showed that the level of these systems organization is although relatively slowly, but raising. Entropy, as the measure of disorganization of these systems, is decreasing in both cities. This fact means that dynamics of the condition of road safety provision systems organization in Russian megalopolises is positive and we should develop it [12].

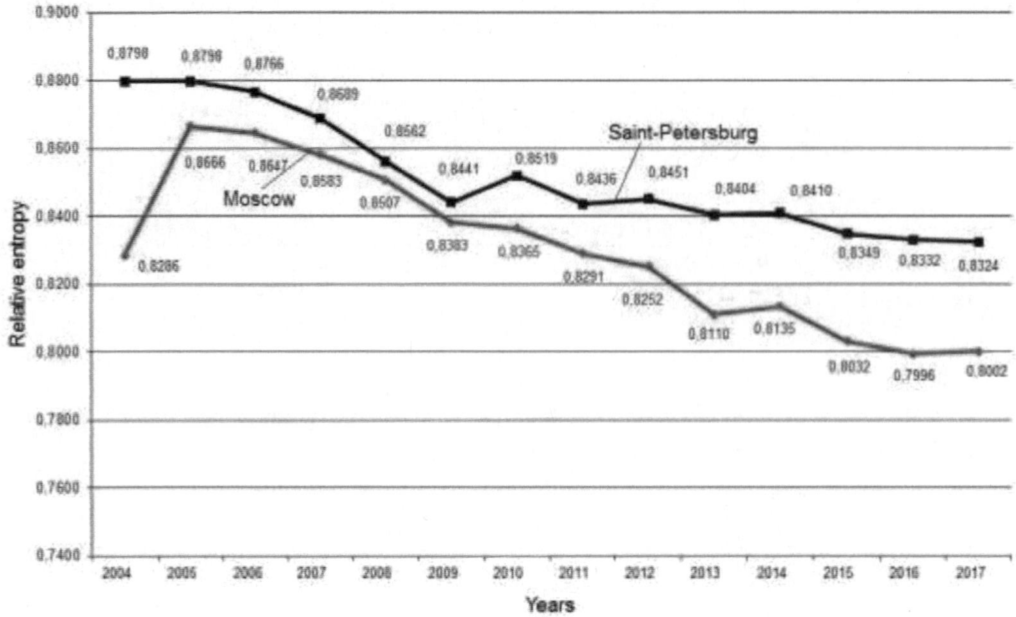

Figure 2. Changes of relative entropy H_n at the road safety provision systems in Moscow and St. Petersburg (2004–2017)

Reference

[1] Indicators of security status of traffic *State Inspection for Road Traffic Safety of the Ministry of Internal Affairs of Russia* (Official website)

[2] The resolution of the Government of the Russian Federation from 2/20/2006 № 100 (an edition from 11/22/2012) "About the Federal target program "Increase in Traffic Safety in 2006–2012"

[3] The resolution of the Government of the Russian Federation from 10/3/2013 № 864 (an edition from 12/13/2017) *About the federal target program "Increase in Traffic Safety in 2013–2020"*

[4] Kolesov V and Petrov A 2015 *Security concern and emergency situations* **3** 21–7

[5] Petrov A and Kolesov V 2018 *Problems of functioning of transport systems* (Tyumen: Industrial University of Tyumen Press) 368–75

[6] Kolesov V and Petrov A 2017 *Transportation Research Procedia* **20** 305–10

[7] Shannon C E 1948 *Bell System Technical Journal* **27** 379–423

[8] Shannon C E 1963 *Works on the Information theory and Cybernetics*

[9] Chumak O V 2011 *Entropy and fractals in the analysis of data*

[10] Kolesov V I, Danilov O F and Petrov A I 2017 *IOP Conf. Series: Earth and Environmental Science* **90** 012059

[11] Petrov A and Petrova D 2016 *IOP Conference Series: Materials Science and Engineering* **142** 012116

[12] Petrov A I 2016 *Economic and Social Changes-Facts Trends Forecast* **3**(45) 154–72

Published under licence in Journal Title by IOP Publishing Ltd

Methodology for generating a comprehensive assessment criterion for passenger traffic conditions in the space of roofed communication elements of intermodal transport hubs

D Vlasov and P Kozlov

Moscow State University of Civil Engineering, 26 Yaroslavl highway, Moscow, 630099, Russia

E-mail: poil_off@mail.ru

Abstract The paper considers an assessment methodology for pedestrian environment in the space of communication elements related to intermodal transport hubs, which planing is performed by construction of transfer unit. Pedestrian environment is proposed to be assessed using an integral criterion that comprises quantitative features of the main parameters for passenger traffic by analogy to the practice applied in foreign experience.

1. Introduction

Under condition of high-intensive passenger flows being generated in the major cities, and against the background of deficient area for development, and strict requirements for the distance of pedestrian approach when transferring between transport types, native town-planing school, by analogy to the global experience, considers a possibility of planing development for transport hubs using construction of transfer unit [1–3].

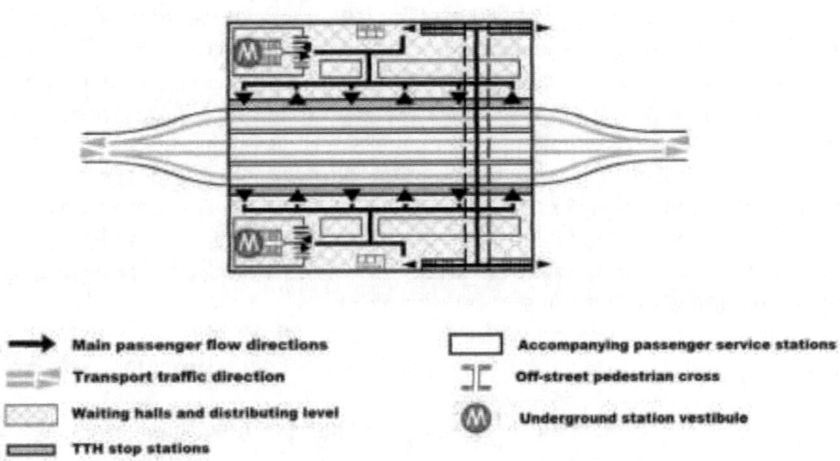

Figure 1. Transfer unit distribution level arrangment flow chart [4]

Content from this work may be used under the terms of the Creative Commons Attribution 3.0 licence. Any further distribution of this work must maintain attribution to the author(s) and the title of the work, journal citation and DOI.
Published under licence by IOP Publishing Ltd

Fundamental provisions that ensure an efficient function of a transport and transfer hub is a possibility to ensure passengers that use municipal and individual transport transfer with the maximum possible comfort, and in the shortest possible time. The maximum comfort for passengers may be provided by ensuring spatial and functional interconnection between any basic elements related to TTH within a common unit, comfort and transfer time in such facilities is ensured by a developed system of pedestrian links, the core of which is a distributed pedestrian level also called as communication area of transport hub. TTH communication area shall implement the main task when developing pedestrian link system related to the unit i.e. distributing passenger flows by directions and levels minimizing conflicts at crosses (Fig. 1).

Basic communication elements of TTH distribution level that ensure distribution of pedestrian flows, and those considered by the authors is a system of pedestrian galleries, stairs descents, and off-street pedestrian crosses.

The main issue related to such a way to develop TTH in our country is a complete absence both of transfer units, and mechanisms that allow for assessing their pedestrian environment from the point of passenger servicing quality.

2. Study overview

Passenger servicing quality in the space of roofed communication elements of transfer hub in general, and especially the distribution level is understood by the authors as a condition of pedestrian environment, under which the following traffic provisions shall be implemented for each individual participant:

- possibility to move along the entire distance of communication element section at a velocity that conforms to individual psychophysiological condition at this moment of time;
- possibility of unhindered overtaking of passengers that move at a less velocity;
- contact to other passengers with body parts or clothes elements is practically excluded when moving forward and overtake;
- possibility to select freely movement direction at any moment of time [5, 6].

In the paper, the authors propose assessing pedestrian environment of transfer units using an integral criterion (servicing level) that comprises quantitative features of main parameters for passenger traffic by analogy to Level off Service (LOS) a practice applied in the US. As the main element of assessing function of transport infrastructure facilities that embraces any phase of their life cycle [7–9].

LOS assesses various movement and access types being supported by a clear hierarchy, priority in which is given to comfort and safety of road traffic. The main advantage of such approach is a possibility to acquire as soon possible and assess the data required at any phase of lifecycle related to transfer hub at a minimum amount of assessment criteria. As the assessment criteria, the authors consider the main parameters for the passenger flow taking into account their mutual dependencies [10].

Since earlier passenger traffic in native transport facilities was assessed exceptionally from the point of safety and evacuation, no general conditions for comfortable transfer of passengers in TTH, and quantitative values for the passenger flow related parameters were proposed [11]. In this context, the authors, based on mass service theory, developed a theoretical model for passenger flow traffic in communication area of transport and transfer hub being expressed via a multi-channel mass service system, which efficiency factor is characterized by the quantitative value for estimated possibility of fault in passenger movement through the functional space of structural element related to communication area.

The model implements a probabilistic assessment format for functional efficiency of structural elements related to communication area [12].

Passenger flow mathematics model schematic as a service processing system comprising several service channels (Fig. 2).

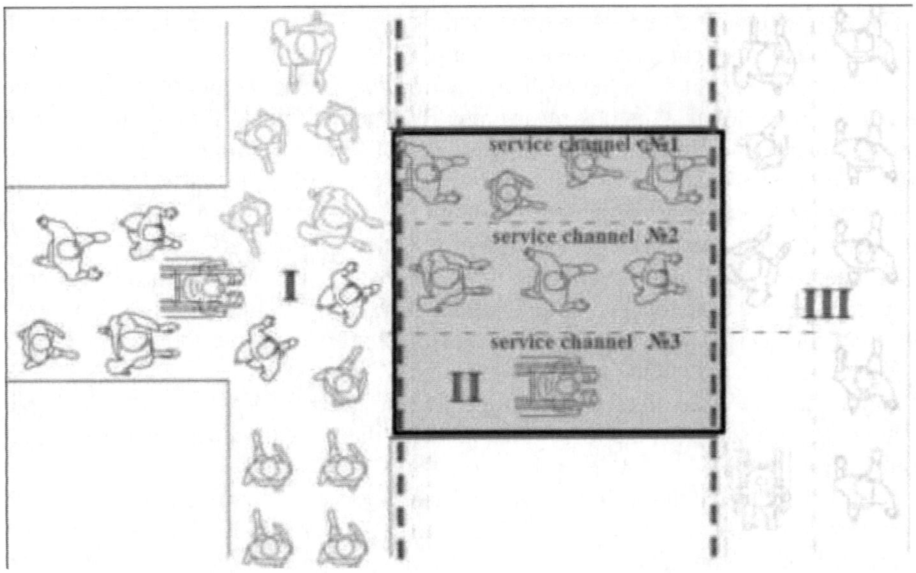

Figure 2. Passenger flow mathematics model schematic
Legend: I – incoming flow formation zone; II – incoming flow service zone; III – outgoing flow zone

Accordingly, as the main criterion for functional efficiency of transfer unit system it is expedient to utilize service failure probability indicator, or a possibility of such an event when any service channel available are occupied when a passenger approaches the functional space of structural element related to communication area in interest:

$$p(n) = \frac{(И_ч \cdot t_{co})^n \cdot \frac{1}{n!}}{\sum_{m=1}^{n} (И_ч \cdot t_{co})^m \cdot \frac{1}{m!}} . \qquad (1)$$

where:
p(n) – estimated fault probability for passenger's movement through the functional space of structural element related to communication area (zone II, Fig. 1), when the system has n service channels available;
t_{co} – average time for passenger movement
n – number of service channels
I_h – hourly amount of passenger transport in the transfer unit system.

Based on the simulation data, the authors produced theoretical conditions for passenger movement in the space of communication of area related to transport and transfer hub. In passenger flow related studies conducted earlier, a regularity of interconnection between passenger flow velocity V (m/min) and passenger flow density **D** (м^2/pass) – (**V=*f*(D)**) was proved, increase in passenger flow velocity reduces the space for movement, and a sequence results in reduce in velocity.

The authors in order to assess comfort of movement in passenger flow propose to use **M** (m^2/pass) parameter i.e. a free space per a passenger in a communication element, a value reciprocal to density **D** (pass/m^2). In our opinion, **M** parameter reflects more precisely traffic conditions from the point of passenger servicing quality, and fits better for estimations by analogy to LOS.

Taking into account that under the most favorable traffic conditions **D** and **M** indicators fails to impact the average passenger flow velocity, then it is proposed to be deemed a free movement velocity, length of separate sections of communication element will not impact significantly passenger floe parameters subject to conditions, where $t_{co} \leq$ 5 min, and the total maximum length of communication element will satisfy S≤150m condition, the lower threshold value for the average passenger flow velocity shall comply with V ≥ 30 m [13].

In the global experience, it is believed that the free movement velocity for horizontal paths, in apertures and when moving on stairs downward may fluctuate within 48 m/min and 155 m/min. Based on the estimation that a velocity below 48 m/min will be deemed a limited one, shuffling movement, and over 155 m/min will be already running movement. According to native and foreign studies, a velocity of passenger flow for horizontal paths within 90 and 155 m/min is deemed a movement of high activity degree, and observed at emergencies, or under conditions when the most pedestrians in the flow is limited by a strict time specific frameworks, which conforms to movement conditions for passengers at transfer. Ascend on staircase is characterized by velocity range between 12 and 80 m/min, where an interval between 50 and 80 m/m in conforms to the movement having a higher activity degree [14–16].

The authors propose to take a velocity for movement on horizontal paths at V ≥ 80 m/min, velocity of upward movement on stairs, as the lower threshold for theoretical values for the average passenger flow V ≥40 m/min.

To conduct theoretical estimation for M, averaged geometrical parameters for communication area sectoins were selected based on existing inter-station passages in Moscow undergound as the closest with respect to passenger flow level and a set of communication elements to pedestrian environment related to TTH transfer units.

Charging of a communication element by passenger flows is taken minimum posible one, which complies with the moment between first 15 minutes from the opening of undergound station, and according to the data submitted by this instutition, passenger flow velocity will be 110 pass/15 min.

From the point of practical application of estimation results obtained, the authors also propose considering such a significant parameter related to passenger flow as 'flow in section' P (pass*min/m) i.e. passenger floe in section that passes through a meter of width of communication element to be estimated by formula:

$$P=V/M \tag{2}$$

where:
V – average passenger flow velocity, m/min;
M – pedestrian space m^2/pass;

Table 1. Optimum functional conditions for structural elements of transfer unit communication area

Element description	Flow parameter values	Estimated value p(n)
Stairs upward	M>2.3 m^2/pers; P≤21.7 pass*min/m; V≥40 m/min;	0.022 ≤
Horizontal paths with double direction movement	M>2.7 m^2/pers; P≤33.0 pass*min/m; V≥80.0 m/min;	0.040≤
Horizontal paths with single direction movement	M>3.8 m^2/pers; P≤23.7 pass*min/m; V≥80.0 m/min;	0.034≤

Parameters of theoretical model ver validated and specified based on 4 inter-station passages of Moscow undergound determined by the outcomes of representative sample.

The outcomes of field investigations after statistical and quartile analysis were summarized in tables for each communication element.

Conducted quartile analysis for distributions determined quantitative values for **P** indicators in each equal part of 4 communication element groups.

When comparing the outcomes of quartile analysis to theoretical values for movement conditions, one can conclude that movement conditions obtained from theoretical estimations comply with the value range between the first and second quartiles, which, according to analysis outcomes, conform to the passenger movement conditions during off-peak time intervals having possibility to overtake freely other passengers, and select freely the velocity of movement on flow.

a) b) c) d)

Figure 3. (a, b, c, d) – Passenger servicing levels

The most favorable service level that conforms to the first quartile is characterized by a possibility to select freely by passenger the velocity and direction of movement in flow not considering movement of other passengers (Fig. 3a).

A favorable service level that conforms to the second quartile is characterized by a possibility to select freely by passenger the velocity and possible overtake considering movement of other passengers (Fig. 3b).

An optimum service level that conforms to the third quartile is characterized for passenger by a free selection of velocity (Fig. 3c).

A restricted service level that conforms to the fourth quartile is characterized for passenger by a limitation on selection of velocity and possibility to overtake, through a forward movement only by minor pace (Fig. 3d).

Thus, once can say that the borders to applying the service levels generated comprise any phase of TTH lifecycle, except for disposal.

When developing regulatory documents related to planning, engineering and reconstruction of transport and transfer units, the authors propose utilizing an assessment procedure for the required width of communication element i.e. it will allow for generating the maximum comfortable conditions at the planning phase with respect to passenger movement in the space of transfer unit considering prospective development of TTH adjacent area.

Table 2. Passenger service level parameters
for the main inner communication elements related to TTH transfer unit

	Stairs descent			Horizontal paths with double direction movement			Horizontal paths with single direction movement		
Service level	Passenger flow in section, pass*min/m	Free room, m^2/pass	Passenger flow velocity, m/s	Passenger flow in section, pass*min/m	Free room, m^2/pass	Passenger flow velocity, m/s	Passenger flow in section, pass*min/m	Free room, m^2/pass	Passenger flow velocity, m/s
The most favorable	≤19.4	>2.4	≥46.0	≤19.4	>5.4	≥94.0	≤17.5	>6.2	≥97.0
Favorable	≤31.5	>1.1	≥35.0	≤36.1	>2.3	≥74.0	≤25.8	>3.5	≥70.0
Optimum	≤41.5	>0.8	≥23.0	≤49.6	>1.4	≥60.0	≤36.8	>2.2	≥62.0
Restricted	>41.5	<0.8	<23.0	> 49.6	<1.4	<60.0	>36.8	<2.2	<62.0

To ensure comfortable conditions for passenger traffic in the space of roofed communication elements related to transport and transfer unit, the authors recommend applying characteristics of an optimum service level and higher.

IOP Conf. Series: Earth and Environmental Science **177** (2018) 012016 doi:10.1088/1755-1315/177/1/012016

3. Conclusion

Thus, generating a service level system in communication elements related to transport and transfer hubs, and quantitative values for the passenger flow parameters may be utilized when determining a sufficient width of communication elements at planning, engineering or reconstruction phase, further investigation of passenger flows in transport and transfer hubs will generate assumptions for developing native regulations that determine practices for assessing passenger servicing quality at any level of TTH native system.

References

[1] Vlasov D 2016 Regulatory support of sustainable intermodal transport transit hub system development *E3S Web of Conferences* **6** *International Conference on Sustainable Cities* 01012

[2] Vlasov D 2017 The chance of transport behavior paradigm in the largest cities of Russian Federation *IOP Conference Series: Earth and Environmental Science* **72** *International Conference on Sustainable Cities* 012017

[3] Vlasov D N 2013 Priority areas for development of agglomeration transport and transfer hub system *Architecture and construction* **3** 86–89

[4] Vlasov D 2013 *Scientific and methodological grounds for developing agglomeration systems of transport and transfer hubs* (by example of Moscow agglomeration) (Moscow)

[5] Zhvioglyadov V G 2005 Movement theory for transport and pedestrian flows *Publisher of Izv. vuzov. Sev.-Kavk. region magazine* 1082

[6] Kholschevnikov V V 1983 *Human flows in buildings, structures and in area of their units* (Moscow)

[7] Fruin J J 1971 *Pedestrian planning and design* (New York: Metropolitan Association of Urban Designers and Environmental Planners)

[8] Harney D 2002 Pedestrian modeling: current methods and future directions *Road & Transport Research* **11** (4) 2–12

[9] Ewing R 1996 *Pedestrian and transit friendly design* (Florida Atlantic University: Joint Center for Environment and Urban Problems)

[10] 2008 *Station Site and Access Planning Manual* (Washington DC)

[11] Predtechensky V M and Milinsky A I 1979 *Designing buildings considering arrangement of human flow arrangement* (Moscow: Stroyizdat)

[12] Vlasov D and Danilina N 2014 Scientific and methodological basis of development of the park-and-ride facilities in the intermodal transport hubs of Moscow agglomeration *Advanced Materials Research* **869–870** 201–4

[13] *Planning and Development of Urban and Rural Settlements* (Urban Planning. Code SP 42.13330.2011 Moscow)

[14] Sherbina E V, Danilina N E and Vlasov D N 2015 City planning issues for sustainable development *International Journal of Applied Engineering Research* **10** (22) 43131–38

[15] Bowman B L and Vecellio R L 1994 Pedestrian Walking Speeds and Conflicts at Urban Median Locations *Transportation Research Record* **1438** 67–73

[16] Rivasplata Ch R 2001 Intermodal transport centers: towards establishing criteria *20th South African Transport Conference: Meeting the Transport Challenges in Southern Africa: Conference Papers*

Betweenness centrality in urban networks: revealing the transportation backbone of the country from the demographic data

M V Goremyko[1], V V Makarov[1], A E Hramov[1], D V Kirsanov[1], V A Maksimenko[1], A V Ivanov[1], I A Yashkov[1] and S Boccaletti[2]

[1] REC 'Artificial Intelligence Systems and Neurotechnology', Yurii Gagarin State Technical University of Saratov, 77 Polytechnicheskaja str., 410054, Russia
[2] CNR-Institute of Complex Systems, Via Madonna del Piano 10, 50019 Sesto Fiorentino, Florence, Italy

E-mail: hramovae@gmail.com

Abstract. The theory of complex networks plays an important role in the modelling and analysis of processes in urban systems, for example, in studies of urban transport networks, the evolution of street networks, passenger flows in the city, etc. In this study we analyze a set of demographic data taken from official population census of Republic of Kazakhstan considering the complex network approach based on spatio-populational principles, namely, the estimation of the distance between cities and taking into account their populations. To determine the geopolitical importance of particular city we use very common network characteristics such as the degree of the node and the betweenness centrality. We show how the values of betweenness can reveal the main transport routes in the country and evaluate the wealth of transportation network.

1. Introduction

In recent years, the interest of researchers dealing with the problems of modeling various complex socio-economic systems is attracted by the network theory as an adequate mathematical tool capable of describing the structure and dynamics of complex processes in many real systems [1]. The formation of social, demographic, economic and other links inevitably leads to the formation of specific structures characterized by the presence of separate elements (or, as they are called in the theory of networks, nodes), between which various connections are established, which means the formation and development of various networks structures [2]. It should be noted that when building network models, communication between nodes can be very diverse and depend on the modeled object [3]. For example, links between network elements can be specified by existing material objects, for example, a network of railway or motorway connections, where links are determined by the presence of a corresponding path, power grids, etc [4]. However, links between network nodes can be defined in abstract space, such as an airline networks, when there are no physical connection between the nodes of the network (in this case between airports), and links are determined by the presence and intensity of air communication between two network vertices [5].

The theory of complex networks plays an increasingly important role in the modeling and analysis of processes in urban systems, for example, in studies of urban transport networks, the evolution of

Content from this work may be used under the terms of the Creative Commons Attribution 3.0 licence. Any further distribution of this work must maintain attribution to the author(s) and the title of the work, journal citation and DOI.
Published under licence by IOP Publishing Ltd

street networks, passenger flows in the city, and so on [6]. In fact, the city is a network of networks where different structures are "nested" in each other and closely intertwined, dynamically changing in the process of urban development [7]. At the same time, the majority of urban studies, which use the complex networks theory, are aimed at the analysis of particular cities or urban area, its economic and transport system. The use of network theory is also of great interest in the study of socio-economic indicators of cities, where a number of interesting results have also been achieved [8]. In particular, various approaches have been proposed to identify the links between cities that are of a socio-economic nature [9]. At the same time, cities are also connected with modern communications and telecommunications, it has been shown that different types of networks are formed between cities, both hierarchical in nature and distributed, and interaction between nodes can be both cooperative and competitive.

Analysis of the available literature allows us to conclude that the network models are promising for understanding the evolution of the urban structure of a particular territory (country) and to identify the features of urban organization, in particular, to analyze the growth or degradation processes (due to the extinction of settlements) of the city network, the network characteristics during the administrative reorganizing processes (for example, transfer of the capital or changing the administrative division).

2. Urban network construction

2.1. Overview of demographic data

Studies of real systems are often complicated by a limited set of statistical data available. This, in turn, encourages us to search for new approaches to extrapolate and to transform the existing data in more vivid representation. In this study we analyze a set of demographic data taken from official population census of Republic of Kazakhstan for three time points, 1989, 1999 and 2009. For each year, the data includes the population of each city, its name, geographical coordinates, and administrative affiliation.

We should note some basic features of the data under study. Firstly, the number of cities decreases during the considered 20 years from 292 to 139 in 2009. Secondly, this process is accompanied by the sharp growth of the average population of the city from 31,769 to 64,096. Such dynamics implies that the cease of the number of cities involved mostly the small towns together with overall outflow of people to regional centers. This also correlates with the relatively stable value of summary urban population of the country: from 9276510 in 1989 it decreased only to 8909333 in 2009.

2.2. The mathematical model

We consider approach to construct a mathematical model of the urban network, which is based on spatio-populational principles, namely, the estimation of the distance between cities and taking into account their populations. We are inspired by early works which attempted to estimate the interaction between cities in the gravity framework [10] together with the principles of spatial growth [11].

In this approach, in the first stage, for all pairs of cities i and j we calculate the weight of connection, w_{ij}, as

$$w_{ij} = \frac{lgP_i lgP_j}{R_{ij}}, for\ all\ i \neq j, \tag{1}$$

where P_i is the population of the i-th city, R_{ij} is the Euclidian distance between the i-th and j-th cities, $i, j = 1..N$, N – the total number of cities in the considered year. We also normalize the set of w_{ij} on the maximal value, so $\max_{i\in 1..N, j\in 1..N} w_{ij} = 1$ and all other values of w_{ij} are in range (0:1). Before proceeding to the construction and analysis of the characteristics of the network, let's consider the question of a specific choice of relation (1) for calculating the weights of links between nodes. The weight of the connection between the two nodes (cities) of the network is proportional to the logarithms of the populations of the corresponding cities and inversely proportional to the distance between the cities. At first glance, it would be more natural to evaluate the weight of connection as $w_{ij} = P_i P_j / R_{ij}$, i.e. in proportion to the product of the population of cities. Why there is a need for complication of the form of the function? To answer this question, we will consider two fragments of the network in different scale, both including a node corresponding to the city of Almaty (Fig. 1 *a, b*). The Fig. 1 *a* shows a fragment of the nearest cities of Almaty. To the left is a fragment of a map with a marked characteristic scale of 20 km and four cities, three of which are rather small ones having the population of less than 40 thousand people. The diameter of blue circles is proportional to the population, P_i, of corresponding city (the number is also indicated on the map under the city name). The Fig. 1 *b* shows another situation – four large cities, located at large distances from each other (the characteristic scale on the map is 200 km).

On the middle and right panels on Fig. 1 we show the corresponding parts of the network with the values of the weights (marked by the thickness of the lines, and also by the numbers next to the corresponding link) for the two equations for w_{ij}. It is clearly seen that in the first case (the middle row), we have a very large spread of weight values, although it can be assumed that in such a compact urban agglomeration the strength of communication between cities should not be very different. Taking into account that we want to isolate and carry out the study of the formation of the backbone of strongly connected clusters of cities, we introduce nonlinear functions (by adding logarithm in the equation) that grow more slowly than the linear function, as the influence of the population of cities effects on the weight of links, as can be seen from the right figures.

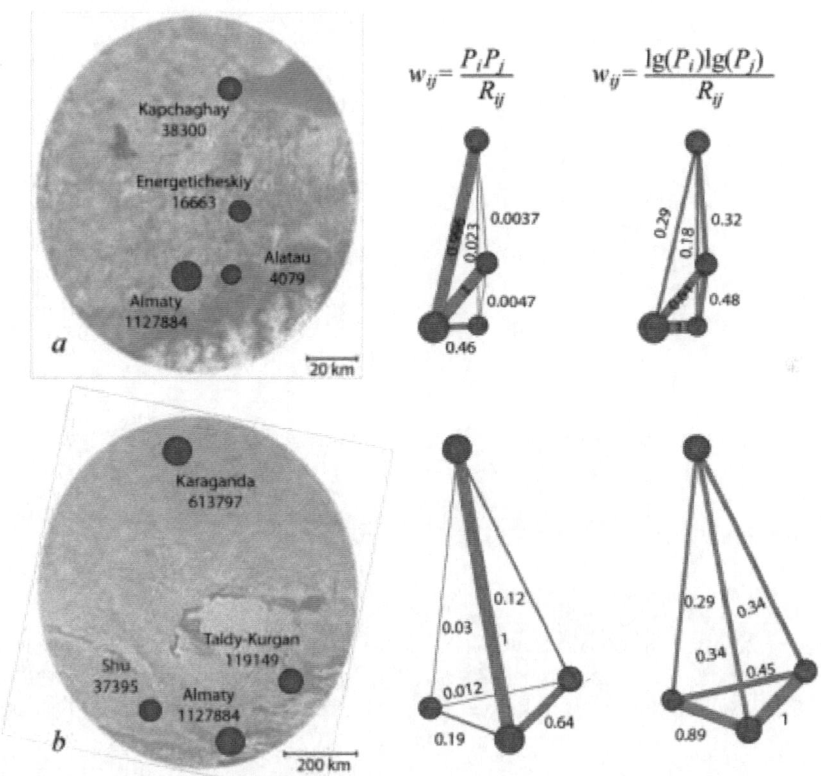

Figure 1. Fragments of a network of urban settlements in the vicinity of Almaty in 1989 explaining the choice of relation (1) for the weights between the nodes

A similar situation arises when we consider large cities located at a great distance from each other, as shown in Fig. 1 *b*. Similarly, when considering the linear connection (middle figure), we get a significant spread in the link weight between cities located at approximately the same distance with each other, while closely located cities are characterized by significantly less ties. The right illustration shows the same network fragment constructed using relation (1), where the links are more homogeneous, which seems to be more correct. Based on these estimates, we have designed the equation (1) to calculate the weights of the urban network.

However, to obtain more simple model, which can be easily studied using the complex network approach, we need to binarize the links between the cities. A large number of weights, w_{ij}, is characterized by negligibly small values, for example, for pairs of cities with small populations, located at a large distance, R_{ij}, from each other. This is illustrated is Fig. 2, which shows the weight distribution of the 1989, when the number of settlements is the largest (N = 292), and hence the number of links. It should be noted that the distributions obtained for other years are qualitatively similar to the distribution presented in Fig. 2. On this basis, it is convenient to enter some threshold value Q, which defines the threshold weight of the connection, which is necessary to connect the two nodes (cities) with each other. The choice of threshold value can be determined from the weight distribution. One can see, that the most of it fits the power law (solid line) nicely. This marks the fact, that the obtained network demonstrates the scale-free property [12], which is essential for the majority of real networks [13]. However, the first point, corresponding to the negligibly small weights does not fits the power law, hence this weights can be removed from the system. More specifically, we choose the value of Q at which one of the networks nodes becomes disconnected, and restore the removed link. This vale of Q accurately vanishes the last bin of distribution as seen from Fig. 2.

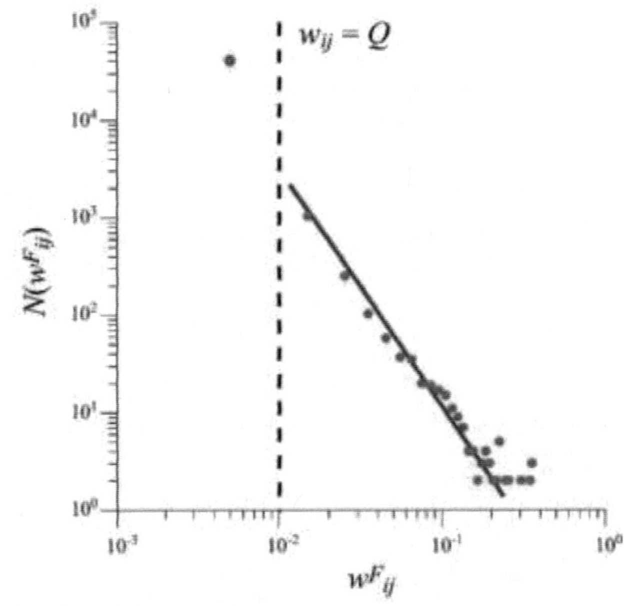

Figure 2. Distribution of weights (red dots) shown in double logarithmic scale. The distribution is based on the data of 1989 year. The vertical dashed line corresponds to a threshold value of Q, by which the weak links were cut off. A solid line corresponds to a power law approximation

The corresponding networks constructed on the basis of our model are shown in Fig. 3 a-c. One can clearly see from them that the model reflects the connections between nearby cities that make up urban agglomerations (clusters) in the north of Kazakhstan, in the south it is tied to the former capital of Almaty, to the west, in the vicinity of the city of Uralsk, etc. There is also the establishment of chains of links through nodes (cities) which are located between such strongly connected clusters, being in good agreement with the form of transport arteries (roads and railways) in Kazakhstan making the connection between large cities through small towns located along roads. In our opinion, this indicates the correctness of the constructed model, which correlates, among other things, with the network of real transport flows.

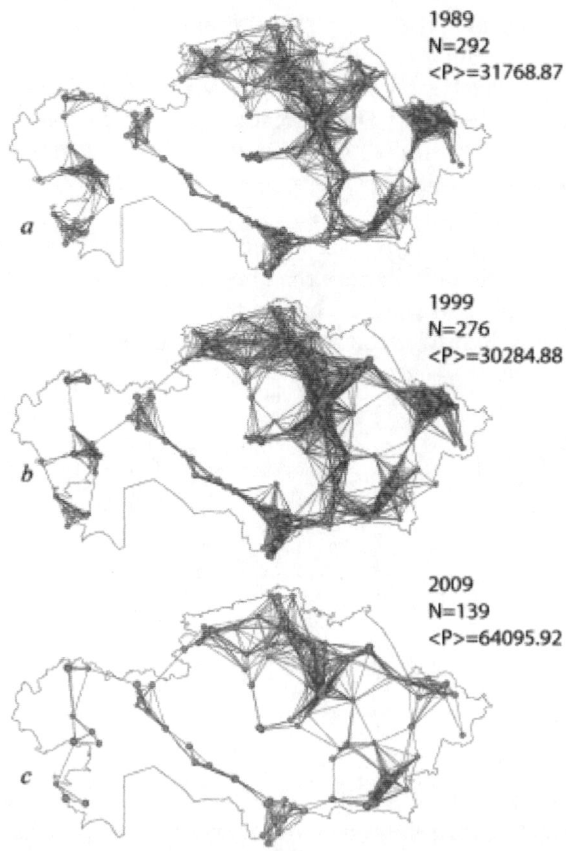

Figure 3. Spatial visualizations of the obtained networks

3. Betweenness centrality in the urban network

To determine the geopolitical importance of particular city we use very common network characteristics such as the degree of the node and the betweenness centrality. The power of node i is a simple sum of all its connections:

$$K_i = \sum_{j=1}^{N} w_{ij}, \tag{2}$$

which allows to determine whether the given node is a structural hub aggregating connections with other elements, or a peripheral node. Nevertheless, this characteristic is very simple and does not allow to evaluate the real value of that or another network node as a transport hub. To assess the significance of a node in terms of transport flows we use the betweenness centrality, which is the number of shortest paths, g_i, between any two other nodes that pass through the node i [14]:

$$g_i = \sum_{j \neq i \neq l} \sum_{l \neq i \neq j} \sigma_{jl}, \tag{3}$$

where

$$\sigma_{jl} = \begin{cases} 1, if\ the\ shortest\ path\ between\ nodes\ j\ and\ l\ passes\ through\ node\ i, \\ 0, if\ the\ shortest\ path\ between\ nodes\ j\ and\ l\ does\ not\ pass\ through\ node\ i. \end{cases}$$

This type of centrality has shown itself very useful in urban studies, especially while studying street networks [15] or transportation paths [16].

First, we investigate how the betweenness centrality of the node correlates with its degree, i.e. number of links, connected to the node. The corresponding correlation plots are shown in Fig. 3 for each network: (a) 1989, (b) 1999, (c) 2009. Easy to see, that quantities, which we investige are completely uncorrelated between each other. This occurs as very surprising result, because, intuitively, the nodes with high number of links should aggregate the shortest paths, which are going through them. This also is an essential property of scale-free networks, where hubs accumulate connections with other hubs [17]. On the countrary, in our urban network we observe the existence of small-degree nodes, which demonstrate very high value of betweenness. At the same time, we can also see nodes with high degree, which have almost zero shortest paths going through them.

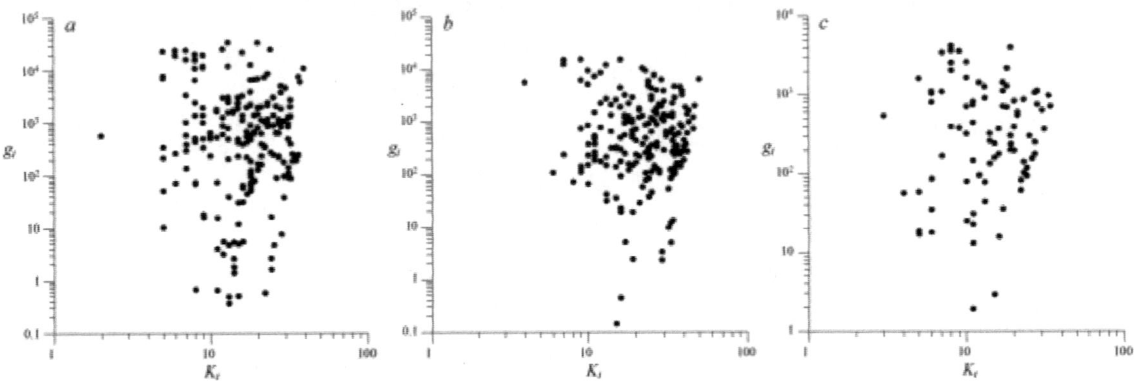

Figure 4. The correlation between the node degree, Ki, and its betweenness centrality, gi: (a) 1989; (b) 1999, (c) 2009. The plots are presented in double logarithmic scale

To understand the nature of the observed phenomena we present the spatial visualizations of urban networks depicting the betweenness by size of each node. The visualizations are presented in Fig. 5.

In 1989 nodes with high betweenness were located in the south of Kazakhstan and then spreaded to the northwest and north, roughly coinciding with the direction of the main transport routes of Kazakhstan connecting northwestern part of Kazakhstan, Russia and going up to the North, where a large number of cities are also densely concentrated. Moreover, the capital itself, Almaty (the orange dot in Figure 5), does not demonstrate a significant value of betweenness. Characteristics in 1999 and 2009. undergo significant changes related to a restructurization of urban network after the acquisition of independence by Kazakhstan. The changes that have taken place in this regard, first of all, the transfer of the capital and the rapid outflow of the population to big cities with the extinction of small settlements. It can be seen that the changes in the distribution of cities by number across the territory led to the fact that the nodes with the largest betweenness are now located in the north and central Kazakhstan. The southern region with the center in Alma-Ata becomes the periphery of the network and there are very few shortest paths going through it, basically, only those that are associated with this local fragment of the network. It is also pronounced that the role of Astana (the capital) is growing – the number of shortest paths passing through it has almost tripled since 1989. In general, if we consider the transfer of the capital from the point of view of optimizing communications within the urban network, then this was justified step and the place of the new capital was also chosen quite successfully, as indicated by the network analysis of the model.

The observed picture also gives us the answer, why the node betweenness is not correlated with its degree. The spatially-distributed urban network present itself the number of interconnected clusters of cities, which location is determined by complex historical process. This clusters of urbanization have a large number of link inside them, reflecting the strong communication between cities in the same region. At the same time, this dense clusters are interconnected by chain-like node structures, which mostly repeat the form of the main motorways of the country. Noteworthy, that the long chain of nodes connecting the south and west regions of the country lies alsmost at the location of the Great Silk Way. This chains of nodes are formed by the cities with rather small populations and, besides, we can observe the process of extinction of this nodes with time. Such dynamics can be considered as a marker of natural or political restructurization of transport paths inside the country, or degrading of transport communication between the regions.

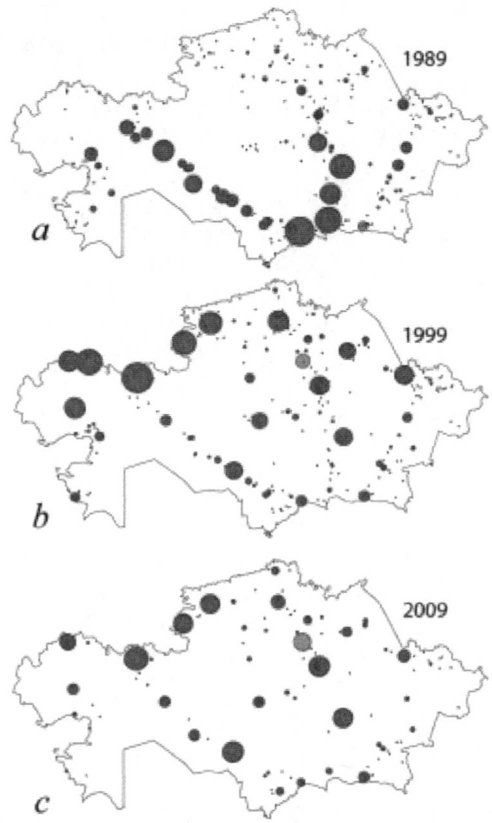

Figure 5. The betweenness centrality of the given networks. The size of the node reflects the value of betweenness. The capital is marked by orange color

4. Conclusion

Summarizing the results, we should note, that the information presented in this paper can be important for optimizing the structure of city networks, traffic flows, demographic and internal migration policies. Further development of mathematical models of such socio-geographic systems will allow to more accurately determine the geographic and demographic characteristics of a network of urban settlements of a particular region or country, and to develop recommendations to government based on rigorous mathematical approaches to improve the efficiency of such systems.

Acknowledgements
This work has been supported by the Ministry of Education and Science of Russian Federation (Grants No. 3.861.2017/4.6 and 3.4593.2017/6.7) and the Council for Grants of the President of the Russian Federation by the supporting program of leading scientific schools (project NSh-2737.2018.2).

References

[1] Strano E, Shai S, Dobson S and Barthelemy M 2015 *J. R. Soc. Interface* **12** 20150651
[2] Molinero C, Murcio R and Arcaute E 2017 *Sci. Rep.* **7** 4312

[3] Nicosia V and Latora V 2015 *Phys. Rev.* E **92** 03280
[4] Pagani G A, and Aiello M 2013 *Physica A: Statistical Mechanics and its Applications* **39211** 2688–2700
[5] Wang J, Mo H, Wang F and Jin F 2011 *Journal of Transport Geography* **194** 712–21
[6] Arcaute E, Hatna E, Ferguson P, Youn H, Johansson A and Batty M 2015 *J R Soc Interface* **12** 20140745
[7] Capello R 2000 *Urban Studies* **37** 1925–45
[8] Leo Y, Fleury E, Alvarez-Hamelin J I, Sarraute C and Karsai M 2016 *J R Soc Interface* **13** 20160598
[9] Zhen F, Wang X, Yin J and Zhang M 2013 *Chinese geographical science* **23** 274–85
[10] Haynes K E and Fotheringham A S 1984 *Gravity and spatial interaction models* Vol 2 (Beverly Hills, CA: Sage)
[11] Makse H A, Havlin S and Stanley H E 1995 *Nature* **608** 3776550
[12] Barabási A L, and Albert R 1999 *Science* **2865439** 509–12
[13] Ebel H, Mielsch L I and Bornholdt S 2002 *Phys. Rev. E* **663** 035103
[14] Barthelemy M 2004 *The European physical journal B* **382** 163–68
[15] Crucitti P, Latora V and Porta S 2006 *Phys. Rev. E* **733** 036125
[16] Puzis R, Altshuler Y, Elovici Y, Bekhor S, Shiftan Y and Pentland A 2013 *Journal of Intelligent Transportation Systems* **171** 91–105
[17] Pastor-Satorras R and Vespignani A 2001 *Physical review letters* **86** 3200

IOP Conf. Series: Earth and Environmental Science **177** (2018) 012018 doi:10.1088/1755-1315/177/1/012018

Modelling Urban Route Transport Network Parameters with Traffic, Demand and Infrastructural Limitations Being Considered

A Kazhaev[1], Z Almetova[2], V Shepelev[2] and K Shubenkova[3]

[1] Department of logistics, Moscow Automobile and Road Construction State Technical University, 64 Leningradsky prospect, Moscow, 125319, Russia
[2] Department of Automobile Transport, South Ural State University, 76 Lenin prospekt, Chelyabinsk, 454080, Russia
[3] Department of Automobile Transport, Kazan Federal University, 10a prospekt Syuyumbike, Naberezhnye Chelny, 423800, Russian Federation

e-mail: shepelevvd@susu.ru

Abstract: In recent years, there has been a significant increase in urban public transport. This has led to overabundant route networks; deteriorated conditions for moving route transport within cities; conflict situations at stop points occurring between vehicles of duplicating routes. Conflict situations occurring between vehicles of different routes at stop points when loading and unloading passengers prove the problem of optimizing urban route networks to be insufficiently investigated, with traffic, demand and infrastructure limitations being considered. Conflict situations at stop points are found to be dependent on route configurations and network parameters, with the number of conflict situations being inversely dependent on the number of combined intervals for route transportation through stop points and their throughout capacity. There has been developed an imitation model for moving vehicles of adjacent routes, with a model for optimizing urban route network parameters being presented. It reduces conflicts in number due redistributed vehicles of various types for routes, with the population's demand for transportation being absolutely satisfied. Models and methods can be used by transportation companies, city administrations to develop and improve urban route public transport networks.

1. Introduction

Today the state and quality of urban transport systems depends to a large extent on the efficiency and safety of urban public transport as the most important system that ensures the economic development of cities and the population's social well-being [1, 2]. For effective and safe operation of urban transport systems, it is required to ensure the coordinated operation of route transport, with the limitations of urban transport infrastructure being considered [3–5]. To create new routes in modern conditions of free competition in the urban public transport sector it is necessary to consider the parameters of route networks to avoid possible conflict situations between vehicles of different routes at stop points during passengers' loading and unloading [6]. Conflicts at stop points adversely affect the transport process safety, lead to significant time losses for carriers and passengers, increase the costs of transportation companies.

Content from this work may be used under the terms of the Creative Commons Attribution 3.0 licence. Any further distribution of this work must maintain attribution to the author(s) and the title of the work, journal citation and DOI.
Published under licence by IOP Publishing Ltd

The analysis of scientific works [7, 8] shows the issues of reducing conflict situations of urban route networks to have been insufficiently studied. There are no well-founded methods for these issues to increase the route transport operation efficiency as there exists a "gap" between the levels of automobilization and urban transport infrastructure development [9–12]. Therefore, the aim of increasing the efficiency and safety of urban public transport operation and route networks is of primary importance from a scientific and practical points of view, with conflict situations at stop points being thoroughly considered.

The aim of the work is to increase the efficiency and safety of functioning urban route networks by reducing conflict situations at stop points.

Thus, the research tasks are:

- to study the causes and mechanisms of conflict situations on route networks of urban public transport;
- to develop an imitation model for identifying conflict situations occurring at stop points of urban route public transport networks;
- to develop a model for optimizing the parameters of route networks to avoid conflict situations at stop points of urban route public transport networks.

2. Research methodology

The system approach, mathematical statistics, probability theory, mathematical modelling approach, mass service theory and optimization approach are the theoretical and methodological basis of the research to draw reliable conclusions.

3. Main part

Due to analysed scientific works concerning the problem of optimization of urban route networks the concept of "conflict situation" is specified [13, 14].

Conflict situations at stop points of urban route transport networks are understood as cases of conflicts of interests occurring between vehicles of competing carriers concerning their priority of used transport infrastructure facilities (stop points) as the throughput capacity of stop points makes their simultaneous and free usage difficult.

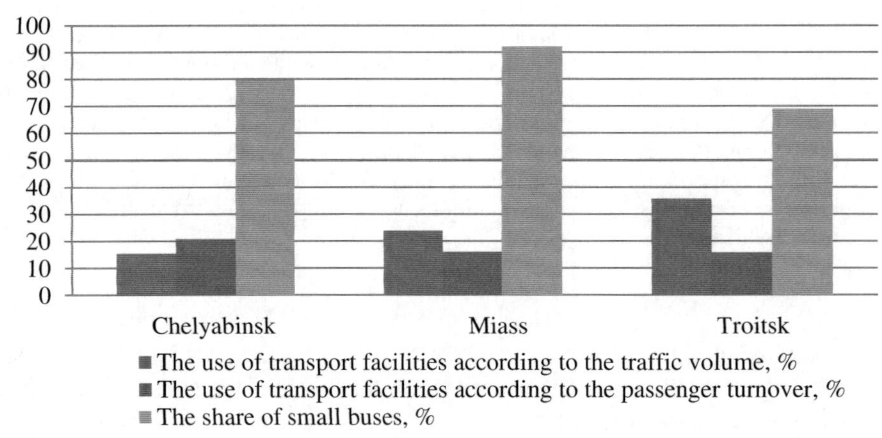

Figure 1. The use of transport facilities and the share of small buses in the cities of Chelyabinsk region

As researches have established, time losses due to conflicts at stop points in the cities of Chelyabinsk region on average per year are the following: the losses of vehicles caused by waiting are 575 (thousand vph/year); the passengers' losses caused by trips increased in duration are 1.43 (million pph/year).

Recently in the large cities of the region (Troitsk, Miass and Chelyabinsk), there has been an increase in the capacity of route networks due to increased public transport routes and vehicles for these routes. During the period from 2005 to 2016 the number of public transport routes doubled and more, with the number of vehicles for these routes increasing more than fivefold.

Fig. 1 provides the information on the use of transport facilities according the traffic volume and passenger turnover, as well as the share of small buses in the structure of traffic fleets in the cities of Chelyabinsk region.

The situation caused by overabundant transport facilities of duplicating routes is worsened by the tendency of decreasing public traffic in the municipalities of Chelyabinsk region.

According to [15], there is a steady tendency of decreasing passenger traffic on all types of public transport in the region.

Table 1. Passenger traffic according to the public transport types in Chelyabinsk region, (million people)

Transport types	2009	2010	2011	2012	2013	2014	2015	2016
Transport (total), including:	1229.4	1115.0	656.0	550.8	591.4	508.2	426.6	424.7
trains	30.1	21.6	16.1	16.4	15.2	14.7	11.4	9.3
trams	396.8	355.6	245.1	193.3	143.6	128.9	115.0	109.9
trolleybuses	280.6	173.1	101.6	85.3	68.0	66.3	53.0	47.9
airplanes	0.5	0.2	0.0	–	–	–	–	–
buses	521.4	564.5	293.2	255.8	364.6	298.3	247.2	257.6

Over the previous eight years, the traffic volume decreased by 2.63 times. The route networks has a significant surplus of transport facilities: for the traffic volume it is over 60%, for the passenger traffic it is over 70%

Considering the results of the analysed dynamics of the transport facilities of urban route networks, the passenger traffic, the amount of public transport, the following is established:

- there is a steady tendency to reduce the passenger traffic volume along urban route networks in whole, due to high rates of automobilization;
- there is an increase in the transport facilities of urban route networks as private carriers prefer to purchase small vehicles;
- there is an increase in the total number of passenger transport routes with insignificant differences in the configurations of their patterns to satisfy the population's needs within existing transport networks.

These tendencies as a whole being the main reason for such a negative phenomenon characteristic for urban route networks, conflicts between passenger vehicles of duplicating routes, routes duplicating each other lead to conflict situations between vehicles at stop points. Transportation delays caused by conflicts lead to significant time losses in the economy: up to 50 hours per year for one passenger vehicle while working on the route (about 12 minutes per day), up to 3 hours per year per passenger (about 20 seconds per one trip). Conflict situations at stop points impair the safe operation of transport and passengers' loading-unloading at stop points, worsen the quality of transport services for urban population.

The problem of conflict situations of urban route networks is complex and is to be solved systemically [16, 17]. In this connection, it is necessary to develop theoretical aspects to identify duplicating routes, models for assessing conflict situations at stop points, and a method for optimizing the parameters of route networks to reduce conflict situations at stop points.

The causes and mechanisms of conflict situations on urban route passenger transport networks are investigated. The main prerequisite for conflict situations at stop points of urban route networks is the competition between different carriers of duplicating routes, this being caused by shortcomings in the management of urban passenger transport: when new routes appear, there are no information of their influence on the transport facilities of route networks, the competition and economic results of carriers' activities, on occurring conflict situations at stop points.

The main competitive models in the market of urban passenger transportation (commercial monopoly, free competition, municipal regulation), the features and stages of their formation, the specifics of carriers' work: the conditions for their access to the market, the principles of route network formation and fare regulations are investigated. In the passenger transport sector there are two main forms of competition occurring on regular urban routes: competition on one route; competition on various (alternative and duplicating) routes.

Due to the results of factor analysis carried out with the help of mass service theory, the influence of the combined frequency of traffic and the time necessary for passenger transport servicing at stop points, as well as the facilities of stop points on the probability of conflict situations was established. To reduce conflict situations at stop points is possible by changing the combined interval for moving route transport provided that the passenger traffic is fully serviced.

The analysis of scientific works [18–21] revealed the absence of effective methodological developments for assessing and identifying duplicate routes, models and methods for optimizing the parameters of route networks, with the formation peculiarities of conflict situations at stop points being taken into account to ensure their reduction.

4. Calculations

The duplicating routes are characterized by the predominance of combined parts and stop points. To estimate the degree of such predominance of combined elements on adjacent routes, the corresponding coefficients are found:

$$\zeta_{ij} = \frac{l_{ij}}{l_{mi}} \tag{1}$$

where ζ_{ij} is the coefficient of combining the segments for the *i*th route ($i= 1, 2, ... n$) with respect to the combined part of the *j*th route ($j = 1, 2, ... m$); l_{ij} is the length of the combined part of the *i*th route relative to the *j*th route, km; l_{mi} is the total length of the considered *i*th route, km.

$$\zeta_i = \frac{\sum_{j=1}^{m} \zeta_{ij}}{m} \tag{2}$$

where ζ_i is the generalized coefficient of combined parts for the *i*th route with respect to all *j*th adjacent routes; *m* is the number of adjacent routes which have combined parts and stop points, units.

$$\xi_{ij} = \frac{S_{ij}}{S_{mi}} \tag{3}$$

where ξ_{ij} is the coefficient of combining the stop points for the *i*th route ($i= 1, 2, ... n$) with respect to the combined stop points of the *j*th route ($j = 1, 2, ... m$); S_{ij} is the number of stop points of the combined part of the *i*th route relative to the *j*th route, units; S_{mi} is the total number of stop points of the *i*th route, units.

$$\xi_i = \frac{\sum_{j=1}^{m} \xi_{ij}}{m} \tag{4}$$

where ξ_i is the generalized coefficient of combining stop points for the *i*th route with respect to all *j*th adjacent routes.

For duplicating routes, the coefficients of combining for parts and stop points approach to unity. The route duplication factor in the complex includes these coefficients and estimates the degree of combining the route elements:

$$d_{mij} = \frac{\zeta_{ij}}{2}\left(\frac{\zeta_{ij}+\xi_{ij}}{2}\right) + \frac{\xi_{ij}}{2}\left(\frac{\zeta_{ij}+\xi_{ij}}{2}\right) = \left(\frac{\zeta_{ij}+\xi_{ij}}{2}\right)^2 \tag{5}$$

where d_{mij} is the route duplication factor.

Due to the route duplication factor and its value, the duplication degree of routes was estimated. The work of duplicating routes will be difficult, with the difficulties caused by passengers' loading and unloading at intermediate combined stop points being also included. The value of the route duplication factor over 0.64 was considered as a criterion for identifying duplicating routes.

Adjacent routes can be considered as duplicating routes as route transport uses them in whole or partially ($\zeta_{ij} \geq 0.8$) as combined with the transport of other (duplicated) routes. These two transport types use the same stop points ($\xi_{ij} \geq 0.8$) along the combined road parts. The duplication factor on them is $d_{mij} \geq 0.64$.

On the basis of theoretical analysis and generalization of actual data on configurations of route networks, methods of forming adjacent routes and combining elements, a new standard model for urban route schemes was developed, i.e. a "route scheme with combined route elements", which includes three basic types of route schemes:

1) route schemes without combined elements that consist of separate routes without combined areas and stop points;
2) route schemes with combined elements on all routes that consist of adjacent routes;
3) route schemes with combined elements on separate routes that include adjacent and isolated routes.

In modern conditions, urban route networks include several duplicating routes. For an integrated assessment of the duplication degree of urban route network in whole, the route duplication factor has been developed:

$$D_m = \frac{M_d}{M_c} \tag{6}$$

where D_m is the duplication factor of a route network; M_d is the number of duplicating routes along a route network, units; M_c is the total number of routes to make up an urban route network, units.

For urban networks with duplicating routes, the value of potential conflict situations at stop points was found.

An imitation model has been developed to characterize quantitatively conflict situations at stop points (frequency, number and duration of conflicts, etc.) on the basis of reproducing real working conditions of route passenger transport in cities, with the speed and intervals of its movement, as well as the capacity of stop points being considered.

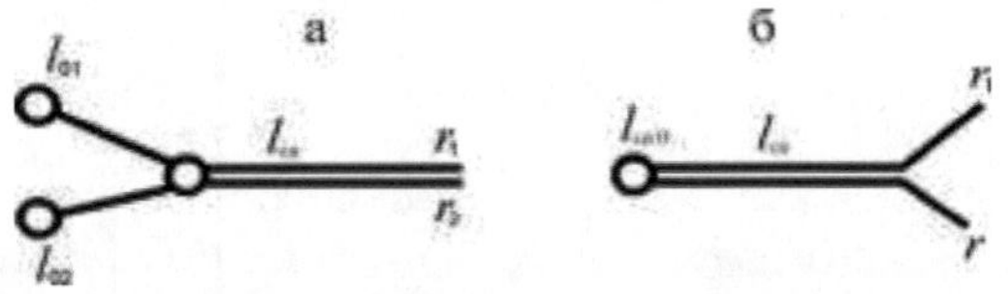

Figure 2. Schemes of combined parts of routes

The starting point of a possible conflict situation on duplicating routes is the beginning of the combined part. On "converging" routes the starting points of vehicle movements from duplicating routes does not coincide, i.e. before the combined part they cover a certain distance separately l_{cdi} from the beginning of the *i*th route (Fig. 2a). On "converging" routes, the combined movement of vehicles starts at the junction (as a rule, the intersection of roads) of their routes $l_{cd0}=l_{cd01}=l_{cd02}$. On "diverging" routes the initial point of a conflict situation is the common starting point for both routes $l_0=l_{01}=l_{02}$ (Fig. 2b). Conflict situations are possible only on the part of combined movement.

The length of the combined part used by vehicles A_{1i} and A_{2j} is determined by the route network configuration and can not exceed the length of any route ($l_{cd} < l_{M1}$, $l_{cd} < l_{M2}$).

At a distance lк from the beginning of the combined part of the route, there is a place of a probable conflict where the vehicles A_{1i} и A_{2j} simultaneously arrive in some time period after moving along the routes t_{k1} and t_{k2}, respectively: $l_k = \upsilon_1 \cdot t_{k1} = \upsilon 2 \cdot t_{\kappa 2}$.

The possible variants of the combined movement of vehicles on duplicating routes are considered (Figures 3–6).

The variant when route transport moves along two duplicating routes with $V_1=V_2$, $t_{02}=t_{01}$ is a special case of the variant shown in Fig. 6, so it is not given separately.

The number of conflicts is determined by:

$$n_k = \begin{cases} \frac{(l_1-l_{cd})}{l_2} \text{ when } (t_k + n_l \cdot l_2) \leq t_{cd1} \\ \frac{(t_{cd1}-t_{k1})}{l_2} \text{ when } (t_k + n_l \cdot l_2) > t_{cd1} \end{cases} \quad (7)$$

where n_k is the number of conflicts, units; l_1, l_2 and l_{cd} are the intervals of movement for vehicles A_{1i} and A_{2j} along the first and second routes and the combined interval of motion, respectively; t_{cd1} is the time of A_{1i} movement along the combined part; n is the number of conflicting A_{2j} vehicles during A_{1i} movement along the combined part:$n = \frac{l_1-l_{cd}}{l_2}$;.

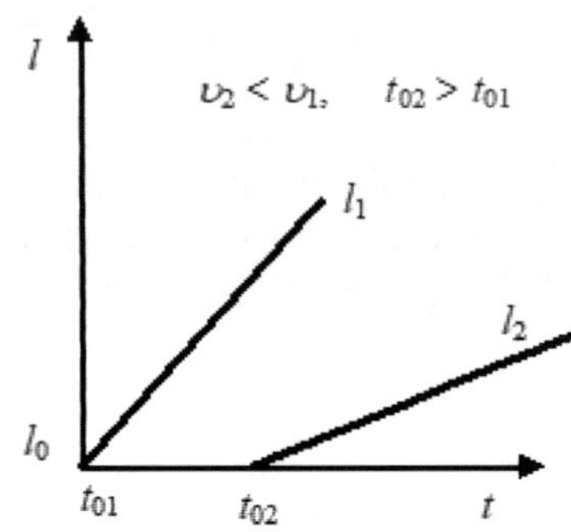

Figure 3. Conflict situation does not occur

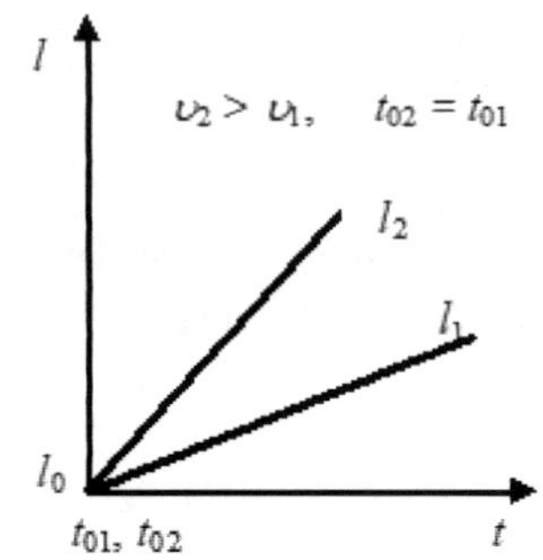

Figure 4. Conflict situation does not occur

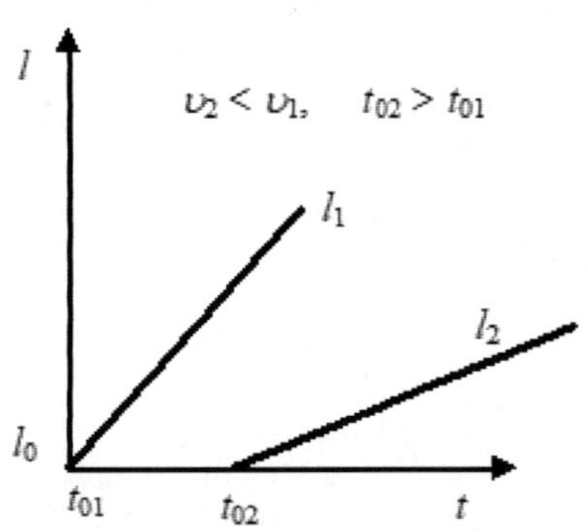

Figure 5. Conflict situation does not occur

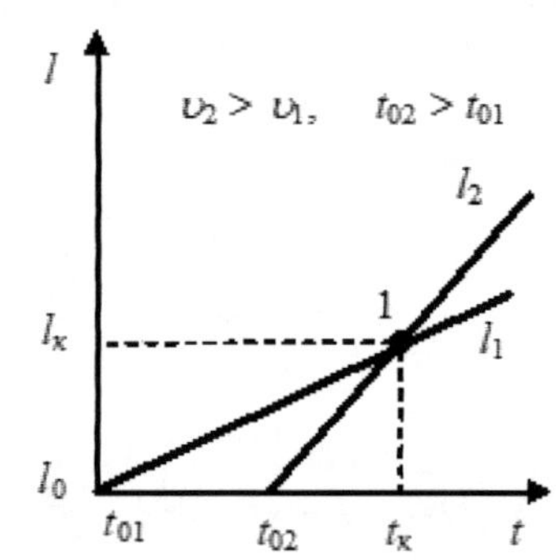

Figure 6. At point 1 a conflict situation occurs

If a conflict situation occurs at a stop point, then its loading is simulated, with the throughout capacity of the stop point and the time of passenger transport servicing being considered. The conflict situation is fixed if the vehicle can not be serviced in proper time. The results of simulation showed that the number of conflict situations at the stop points of the object in question exceeds 10 thousand per day. The deviation between the calculation results and full-scale surveys does not exceed 10% on average. The total loss of route transport caused by conflict situations in Troitsk reached 4.1 million rubles a year.

A model for optimizing urban route networks has been developed. The route network $R = \{r_i\}$ ($i = 1, 2, \ldots, n$) is the totality of the ith number of routes $r_i \in R$. On the route network R there is a set

$S = \{s_j\}(i = 1, 2, \ldots, m)$ of stop points included in the route network. In this case, each stop point s_j is identified separately, i.e. the same stop points located on different sides of streets, intersections, and also used by different routes and transport types, etc. are considered. In general, the *j*th stop points considered as belonging to the *i*th routes can be represented as the following matrix *B*:

$$B = \|b_{ij}\| = \begin{Vmatrix} b_{11} & b_{12} \cdots & b_{1m} \\ b_{21} & b_{22} \cdots & b_{2m} \\ \cdots & \cdots\cdots & \cdots \\ b_{n1} & b_{n2} \cdots & b_{nm} \end{Vmatrix} \tag{8}$$

in which the elements b_{ij} can be equivalent to 0 or 1:

$$b_{ij} = \begin{cases} 1, \text{if } S_{ij} \in r_i; \\ 0, \text{otherwise} \end{cases} \tag{9}$$

The frequency of moving vehicles $v_i(s_j)$ for each the *i*th route having the *j*th stop point to ensure passenger traffic to the full extent is calculated due to the maximum hourly capacity of passenger traffic Q_{hi} on the most used part of the route for a characteristic period of time and the estimated capacity q_p of a vehicle to be used on the route:

$$v_i = \frac{Q_{hi}}{q_p} \tag{10}$$

The value of capacity is determined due to the established level of capacity utilization according to the number of seats for passengers to travel standing and sitting and meets the requirements for the due quality of passengers' trips.

The combined frequency of moving vehicles through the s_j stop point is calculated as the sum of the frequency of $v_i(s_j)$ moving vehicles for each the *i*th route passing through the *j*th stop point:

$$v_{sj} = \sum_{i=1}^{n} (S_j),\ S_j \in S \tag{11}$$

The conflict index ω_{sj} at the *j*th stop point is characterized by the deficit (reserve) amount of the throughout capacity of stop points, which is calculated as the difference between the value of the combined frequency v_{si} of a moving vehicle through the *j*th stop point and the value of its throughout capacity ω_{sj}:

$$\omega_{sj} = v_{sj} - w_{sj} \tag{12}$$

The indicator of the conflicting nature of a route network in whole ω_s is calculated as the sum of the values of the level of conflict situations at all *j*th stop points:

$$\omega_s = \sum_{j-1}^{m} \omega_{sj} \tag{13}$$

Thus, the indicator of the conflict nature of a route network is chosen as an optimality criterion of the parameters of the route network, which ensures a minimum number of conflict situations at stop points:

$$\omega_s = \sum_{j-1}^{m} \omega_{sj} \to \min \tag{14}$$

The limits are set on:

- the number of vehicles available A_m:

$$\sum_{i=1}^{n} A_{mi} \leq A_m \tag{15}$$

- the value of established intervals lyi when vehicles move along the routes:

$$l_{yi} \leq l_{di} \tag{16}$$

where l_{di} is the maximum allowable interval of moving route transport on the route.

A mathematical model for optimizing the parameters of a route network is developed:

$$F(\omega_s) = F(q_p, l_{yi}, A_{mi}) \rightarrow \min$$
$$\text{when } q_p \in G, q_p \geq 0,\ l_{yi} \leq l_{дi}, l_{yi} \geq 0, \tag{17}$$
$$\sum_{i=1}^{n} A_{mi} \leq A_m,\ A_{mi} \geq 0.$$

Thus, to solve the problem of forming optimum parameters for a route network to avoid possible conflict situations at stop points, vehicles of q^z_p capacity are chosen and distributed over the route network so as the combined frequency of traffic at the v_{sj} stop points will not exceed the w^k_{sj} throughout capacity of the due stop points or this excess (due to the existing fleet structure A_m^z of passenger vehicles or permissible intervals of movement l_{di} of route transport) will be minimal for the entire route network in whole w^k_s.

5. Research Results

Using the developed provisions, the route network was optimized using the example of the city of Troitsk, during which the total number of routes was reduced from 28 to 13, the number of rolling stock was reduced from 102 to 58 units, the structure of the rolling stock operating on routes to increase the number of rolling stock the total number of conflict situations at the stopping points for new parameters of the route network will decrease by 33.46 times, from 10163 to 313 conflicts. The total time of delays in queues before stopping points decreased by 32.4 times, from 18.8 to 0.6 hours. The duplication factor of the route network was 0.464. Consequently, the share of duplicate routes in Troitsk is almost half (about 46% of the total) of the total number of urban routes. The effect of reducing the time losses of carriers in the event of downtime due to conflict situations at the stopping points amounted to 3,985.8 thousand rubles.

6. Conclusion

Thus, the developed simulation model of vehicle traffic on adjacent routes takes into account the parameters of routes and stop points, allows to evaluate possible conflict situations at the combined stop points of route networks, to identify potentially conflicting stops and routes, and make informed decisions on the reconstruction of infrastructure facilities and optimization of route parameters network.

The optimized parameters of the route network ensure the full realization of passenger traffic in compliance with the requirements for the quality of transport services for the population on the use of the capacity of the rolling stock and the intervals of its movement.

The obtained results testify to the usefulness of the developed optimization model for practice. With its use, new variants of urban route networks can be developed, in which a minimum number of conflict situations at the stopping points will be ensured, with full satisfaction of the population's requests for transport by urban public transport.

References

[1] Baloian N, Frez J, Pino J A et al 2015 Efficient planning of urban public transportation networks *Lecture Notes in Computer Science* **9454** 439–48

[2] Teichmann D, Dorda M and Olivkova I 2013 Optimal assignment of vehicles to urban transport routes *Komunikacie* **15 (1)** 34–8

[3] Al-Mudhaffar A, Nissan A and Bang K 2016 Bus stop and bus terminal capacity *Transportation Research Procedia* **14** 1762–71

[4] Horváth B 2012 A simple method to forecast travel demand in urban public transport *Acta Polytechnica Hungarica* **9(4)** 165–76

[5] Larin O N and Dosenko V A 2014 Use of a phase transition concept for traffic flow condition estimation *Transport and Telecommunication* **15 (4)** 315–21

[6] Nagy V and Horváth B 2017 Hidden content of passenger data in public transport *Procedia Computer Science* **109** 506–12
[7] Nagy V, Horváth B and Horváth R 2017 Land-use zone estimation in public transport planning with data mining *Transportation Research Procedia* **27** 1050–57
[8] Barabino B, Di Francesco M and Mozzoni S 2015 Rethinking bus punctuality by integrating Automatic Vehicle Location data and passenger patterns *Transportation Research Part A: Policy and Practice* **75** 84–95
[9] Padrón G, García C R, Quesada-Arencibia A et al 2014 Using massive vehicle positioning data to improve control and planning of public road transport *Sensors* (Switzerland) **14 (4)** 7342–58
[10] Xiaobao Y, Mei H and Ziyou G 2013 Car delay model near bus stops with mixed traffic flow *Journal of Applied Mathematics* **437637**
[11] Barabino B, Di Francesco M and Mozzoni S 2013 Regularity analysis on bus networks and route directions by automatic vehicle location raw data *IET Intelligent Transport Systems* **7 (4)** 473–80
[12] Li Y, Yu T and Yu Y 2016 Modeling Public Bus Scheduling Based on IC Cards Data *Green and Multimodal Transportation and Logistics* 876–85
[13] Wang L and Zhang L 2014 Research of road saturation based on queuing-time graph *Journal of Wuhan University of Technology (Transportation Science and Engineering)* **38(6)** 1198–1202
[14] Tollazzi T 2015 First developments of different roundabout types *Springer Tracts on Transportation and Traffic* **6** 11–55
[15] 2016 *Transport in the Chelyabinsk region* (Chelyabinskstat)
[16] Adhvaryu B 2006 Design of bus station: A case study in brighton *Traffic Engineering and Control* **47(5)** 182–6
[17] Horbachov P, Naumov V and Kolii O 2015 Estimation of the bus delay at the stopping point on the base of traffic parameters *Archives of Transport* **35(3)** 15–25
[18] Hernández D, Muñoz JC et al 2015 Analysis of real-time control strategies in a corridor with multiple bus services *Transportation Research Part B: Methodological* **78** 83–105
[19] Ibarra-Rojas O J, Delgado F, Giesen R et al 2015 Planning, operation, and control of bus transport systems *Transportation Research Part B: Methodological* **77** 38–75
[20] Schmidt M and Schöbel A 2015 The complexity of integrating passenger routing decisions in public transportation models *Networks* **65(3)** 228–43
[21] Carro-Pérez I and González-Hernández H G 2016 Topological characterization of the public transportation complex network in Puebla City, México *Proc. Int. Conf. Electronics, Communications and Computers* **7438561** 113–17

Providing thermal protection when replacing window blocks in historical buildings

A Konstantinov and A Mukhin

Moscow State University of Civil engineering, 26 Yaroslavl highway, Moscow, 129337, Russia

E-mail: apkonst@yandex.ru

Abstract. The work presents the prerequisites to replace the window units in the historical buildings (primarily increased energy efficiency standards of buildings and significantly changed performance requirements of window units). The analysis of possible replacement options (full/partial) of window units of the historical buildings is done according to their architectural and historical values. A review is made of the existing technological capabilities of modern window systems and translucent filling (double-glazed windows), as well as the most common solutions used in practice. The issue of providing the temperature-humidity conditions of the nodes of adjunctions of the window blocks to the openings of the external walls in the winter operating conditions is considered. For this purpose, a numerical simulation of the temperature fields of the junction points is made for various structural embodiments of window frames. It is established that in the case of a full replacement of window blocks and maintaining the existing structure of external walls (the calculation is made using the example of the most common external wall for Central Russia – a brick wall 770 mm thick), condensation may form on the inner surface of window slopes. Different options are suggested for window blocks installation, which ensure the fulfillment of sanitary and hygienic requirements. Further directions for the research on the use of modern types of translucent structures in the historical buildings are outlined.

1. Introduction

Currently, many historical buildings built in Russia in the early 20th century and earlier require major repairs / restoration. Often this is due both to the unsatisfactory technical condition of such buildings and their discrepancy with modern standards of comfort and energy efficiency. One of the most cost-effective ways to preserve historical buildings is their adaptation to modern use [1]. This is reflected, among other things, in the implementation of a set of measures aimed at achieving regulatory requirements of thermal protection and internal parameters of indoor climate in these buildings. Issues related to the restoration or replacement of windows require special attention. This is due to the fact that a wide range of requirements are applied to the window units used in historical buildings [2–5]:

- the architectural appearance of buildings;
- thermal protection of buildings and specified parameters of indoor climate (natural lighting, temperature-humidity conditions, acoustic microclimate);
- operational safety and protection against unauthorized access;
- ease of operation and maintenance.

It is necessary to understand that the possibilities of increasing the energy efficiency of historic buildings due to increased thermal protection of external walls are significantly limited. This is due to

Content from this work may be used under the terms of the Creative Commons Attribution 3.0 licence. Any further distribution of this work must maintain attribution to the author(s) and the title of the work, journal citation and DOI.
Published under licence by IOP Publishing Ltd

the fact that changes in the appearance of the facades of such buildings in many cases is not allowed. Therefore, insulation of external walls of historical buildings with energy efficient insulators is not possible in most cases. Alternative options for insulation (for example, the use of thermal insulating plaster [6]) in the climatic conditions of Russia will not obviously have a significant effect. Therefore, the use of modern designs of window units with high thermal resistance is the most effective way to increase the thermal protection of the outer shell of buildings of this type.

Modern technological capabilities of manufacturers of translucent structures allow the mass production of window blocks with the thermal resistance of 1.0 m^2 $^0C/W$ and higher, which exceeds the thermal resistance of external brick walls with a thickness of 770 mm (3 bricks). When they are used, the heat loss of buildings through the outer enclosure in the winter period of operation differs from traditional significantly [7–15]. In this case, from the point of view of providing thermal protection, the windows will not be the most vulnerable part of the outer shell of the building.

The use of energy-efficient window units along with the implementation of measures for the installation of modern heating and ventilation systems with recuperation, as well as, possibly, the use of alternative energy sources (solar collectors, solar panels, etc.) are integral elements of improving the energy efficiency of historical buildings in modern conditions [16–20].

2. The possibility of replacing window blocks in historical buildings

Depending on the historical and architectural value of the historical buildings, there are two categories of them:

- buildings without special historical and architectural value. They are the most common ones and are currently used to accommodate residential, office and retail space, hotels and theaters;
- objects of cultural heritage, which are monuments of architecture. They may have the status of a monument of architecture of local, regional or federal significance, or be a World Heritage site.

In buildings without special historical and architectural value, special requirements to the constructive solution of the replaced window blocks are usually not presented. The most appropriate way is the use of modern energy-efficient window units (see Figure 1a). Thus, when choosing an architectural design and the color of the window frames, the characteristic features of the architectural style of the building should be taken into account.

For architectural monuments, the legitimacy of the use of modern window units in order to replace old ones is allowed based only on the historical and cultural examination and/or the security obligation of the user of the cultural heritage object [21]. The constructive solution of new window blocks (the material of bindings, translucent filling, etc.) is made based on the object of protection of cultural heritage reflected in the security commitment or developed as a part of the restoration project.

Typically, the historical buildings used to have isolated window units installed with the outer and inner window sash made of one glass. The following options are possible for the given design of windows:

A) The subject of protection is both the facade and the interior of the building. In this case, it is necessary to restore the existing window blocks or completely recreate the damaged/lost items. Due to the fact that it is necessary to use materials and technologies applied in the old window blocks, a qualitative improvement of their operational characteristics, as well as energy efficiency of buildings, will not generally occur.

B) The subject of protection is the facade of the building. In this case, it is necessary to preserve (restore) the outer part of the existing window block, or to recreate it from identical materials according to the measurement drawings. Instead of an internal frame, it is possible to install a modern design of window units (see Figure 1b). In this case, the outer part of the window unit will provide architectural requirements of security obligations and the inner part – regulatory requirements for the thermal protection of buildings and indoor climate. In order to increase the natural lighting, reduce the cost of production, as well as maximum preservation of visual perception of the interior of the building when viewed from the side of the street, it is most optimal to use window units without additional decorative elements.

C) The subject of protection is the interior of the building or its separate room. In this case, it is necessary to save/recreate the internal part of the window, as well as possible replacement of its outer part with a modern one (see Figure 1c). At the same time, the architectural pattern, color and texture of the external part of the window block must comply with the architectural style of the building and the color palette of its facades. The outer part of the window unit in this case ensures compliance with the regulatory requirements for thermal protection of buildings and the inside climate, anti-burglary.

D) The facade and interior of the building are not eligible for protection. If the material and appearance of the window blocks are not specifically negotiated in the security obligation, thus, as in the case with buildings without a special historical and architectural value, it is possible to completely replace the old window units with modern ones (see Figure 1a).

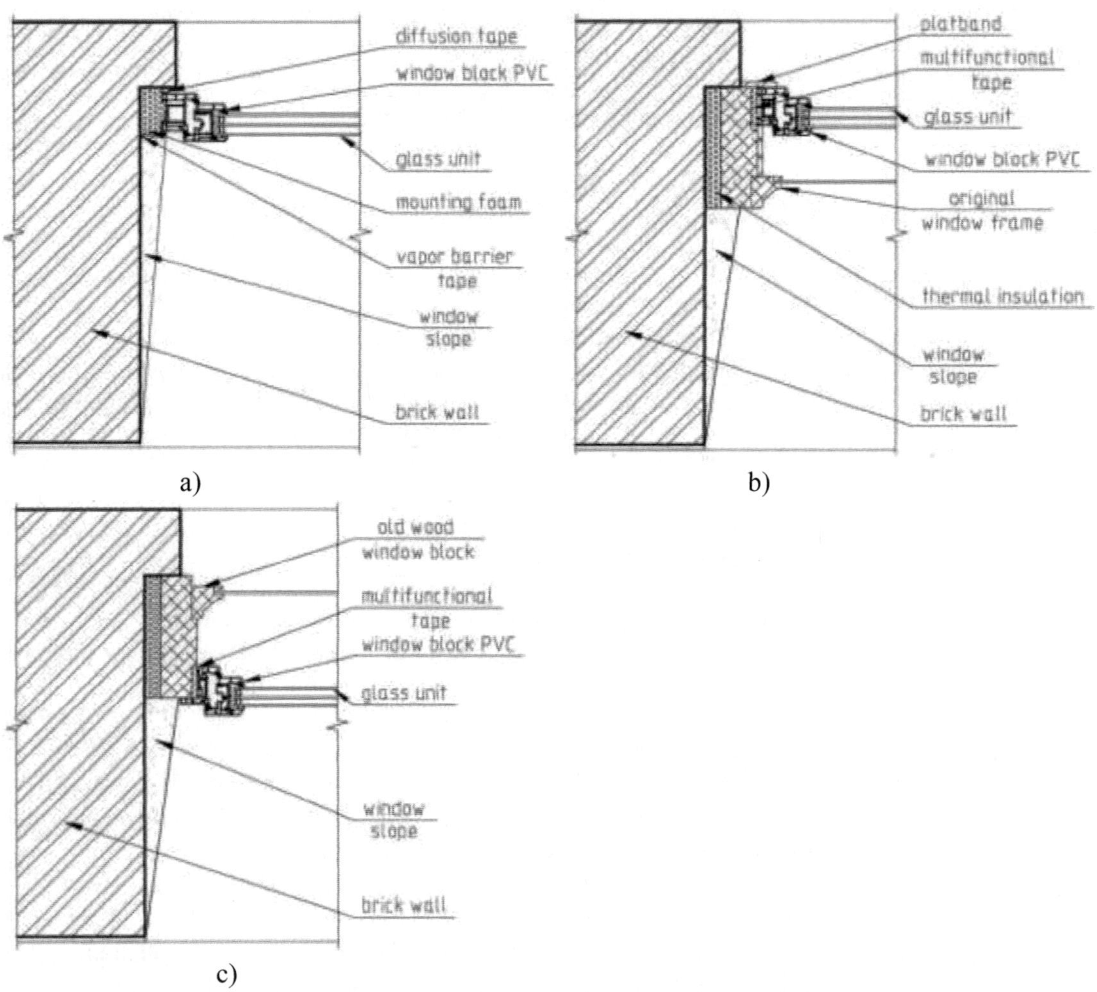

Figure 1. Possible schemes for replacing window blocks in historical buildings
a – for buildings without special historical and architectural value and architectural monuments (the facade and interior of the building is not a subject to protection);
b – for monuments of architecture (the subject of protection – the facade of the building);
c – for monuments of architecture (the subject of protection – the interior of a building)

3. Technological capabilities of modern window systems and translucent filling

Replaceable window units can be made based on modern window systems from wood, PVC, aluminum, combined materials (wood-aluminum, aluminum with wooden plates), etc. Thus, there are technical solutions for restoration in the developed window systems of leading manufacturers (for example, restoration of the frame, which can be installed in the opening on top of the old ones without dismantling them). Modern windows can be painted in any color, and PVC profiles can have lamination of different colors and textures that allows to make window frames to fit any color scheme of the facade. Their use in conjunction with modern designs of double-glazed windows allows to achieve high rates of thermal resistance of window units and significantly improve the thermal protection of buildings.

The most common type of window units used in historical buildings and objects of cultural heritage due to their relatively low cost are PVC window frames (see Figure 2).

Figure 2. Examples of the use of PVC window frames in the objects of cultural heritage

A perspective solution for historical buildings is the use of vacuum insulating glass as a translucent filling. Due to the high thermal resistance [22, 23], low weight and the overall thickness of the insulating glass unit (8 mm), they can also be used in the storable/reconstructed window blocks (after additional milling of the window fillets).

4. Temperature-humidity conditions of the junction points of window blocks during the winter period of operation. Analysis of results

Let's consider the temperature conditions of the junction points of different types of window blocks in the opening of the outer wall. The analysis of temperature-humidity conditions of the junction points will be produced based on the calculations of the temperature fields of the considered nodes performed in the Flixo 6 software package. Calculations will be made for the conditions of Moscow city (outside temperature of the coldest five-day period of -25 °C) and internal air temperature of + 20 °C.

IOP Conf. Series: Earth and Environmental Science **177** (2018) 012019 doi:10.1088/1755-1315/177/1/012019

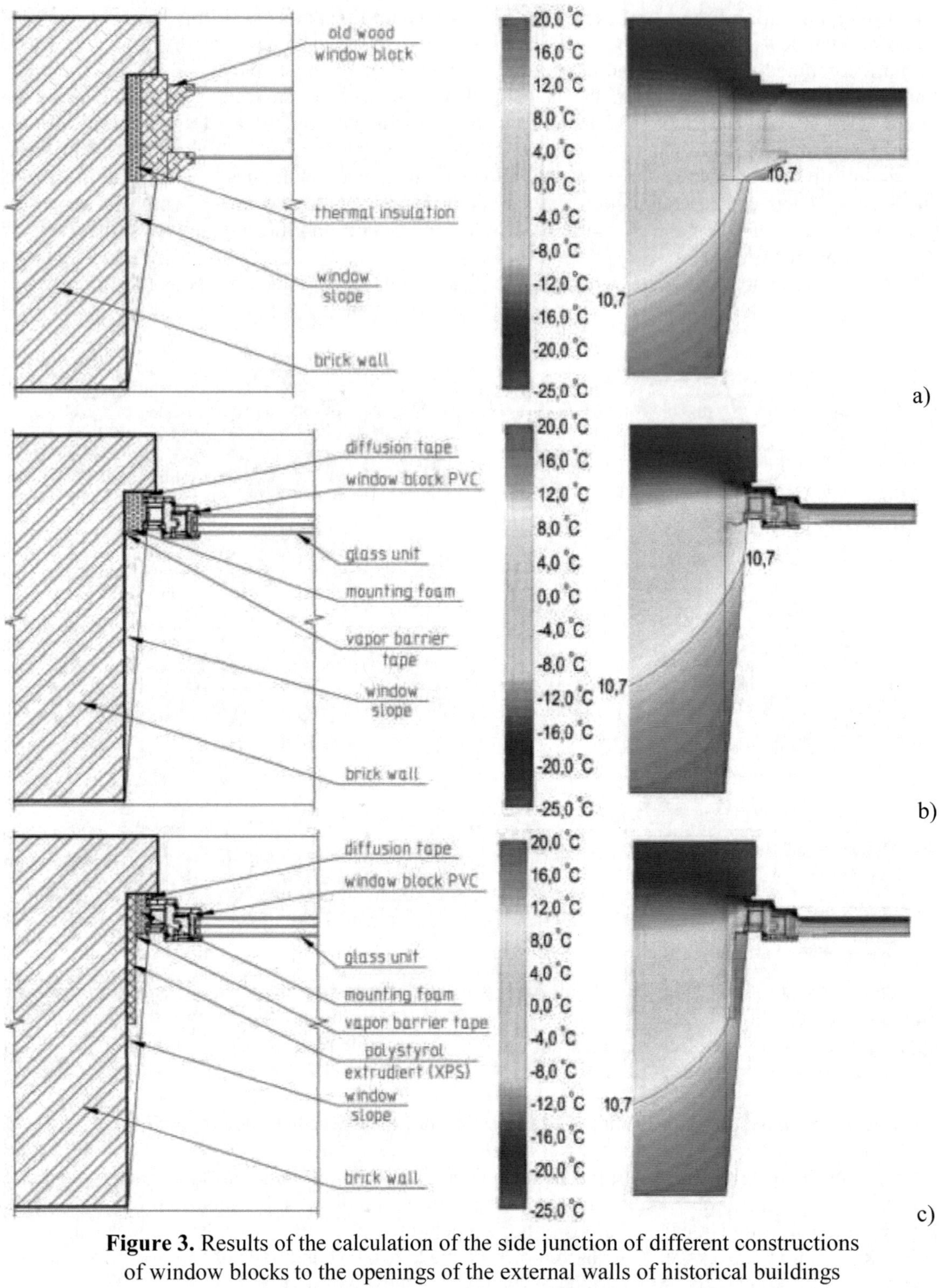

Figure 3. Results of the calculation of the side junction of different constructions of window blocks to the openings of the external walls of historical buildings
a – for the old window unit;
b – for a window unit made of PVC (without additional measures);
c – for window units made of PVC with the insulation of a window slope

The outer walls of most buildings built on the territory of Russia are made of ceramic bricks and, typically, have a thickness of at least 770 mm (excluding plaster). The old window blocks are usually made of a single window frame with the hanging outer and inner window sashes on it. Generally, window units are installed in the outer wall outside of the window quarter with a thickness of one brick (120 mm). Due to the homogeneous construction of the external wall, in terms of heat engineering, such positioning of the window unit in the outer wall is optimal because the internal part of the window unit is in the zone of positive temperatures of the external wall. Thus, the formation of condensation on the window slopes does not occur (see Figure 3a).

When the old window unit is replaced with a new one (for example, PVC), the temperature-humidity mode of this unit significantly changes. Before installing a new window block, the old block is dismantled or cut off. Installation of a new window unit is made directly outside of the window quarter. Thus, the new window unit is located in the zone of negative temperatures of the outer wall, which leads to condensation of the internal surface of the window slopes (see Figure 3b).

In order to eliminate the above-described phenomenon and to ensure the regulatory requirements for the temperature-humidity mode in accordance with [2], it is necessary to perform one of the following measures before installing the window units:
- insulation of the window slope with subsequent plastering;
- the device of a warmed slope from a sandwich panel (not always possible from an architectural point of view);
- forming groove in wall with its subsequent warming (with an unsatisfactory technical condition of the masonry, this measure is not possible).

These measures make it possible to change the temperature-humidity mode of the considered junction nodes significantly. This statement is confirmed by calculations (see Figure 3c).

5. Discussion

Conducted studies in the article certainly do not affect all the factors that must be taken into account when replacing window blocks in historical buildings. It is planned to conduct further research in the future:
- thermal characteristics of combined structures of window blocks (for the case of a partial replacement of the old window unit) based on winter and summer operating conditions (overheating of the space between the old and the new window unit);
- the issue of providing natural illumination of rooms when using combined constructions of window blocks;
- thermal characteristics of window blocks when using vacuum insulating glass.
- economic efficiency of design options discussed above.

Based on the data of the research, it is planned to develop a methodology for assessing the economic and technical feasibility of various designs of windows in historical buildings.

6. Conclusions

The existing possibilities for replacing window blocks in historical buildings of different historical and architectural value were analyzed. The technological capabilities of modern window systems with their application in buildings of this type were considered.

A computer simulation of temperature conditions of junction points of window blocks to openings of external walls of a typical external wall was conducted. It was established that the use of standard window blocks in the external walls of historical buildings without additional measures for the insulation of walls in the junction zone may lead to a violation of the temperature and humidity conditions during the winter period of operation. Proposed technical solutions eliminates this problem.

References

[1] Spigliantini G, Fabi V and Corgnati SP 2017 *Energy Procedia* **134** 376–385

[2] SP 50.13330.2012 *Thermal performance of the buildings. The updated edition of SNIP 23-02-2003*
[3] GOST 23166-99 *Windows. General specifications*
[4] DIN EN 14351-1:2016-12 *Windows and doors – Product standard, performance characteristics – Part 1: Windows and external pedestrian doorsets; German version EN 14351-1:2006+A2:2016*
[5] DIN 18055:2014-11 *Criteria for the use of windows and exterior doors in accordance with DIN EN 14351-1*
[6] Bianco L, Serra V, Fantucci S, Dutto M and Massolino M 2015 *Energy and Buildings* **95** 86–91
[7] Huber A, Korjenic A and Bednar T 2013 *Bauphysik* **35**(2) 107–18
[8] Kain G, Gschwandtner F and Idam F 2013 *Bauphysik* **39** (2) 144–47
[9] De Berardinis P, Rotilio M, Marchionni C and Friedman A 2014 *Energy and Buildings* **80** 415–23
[10] Mozzarella L 2015 *Energy and Buildings* **95** 23–31
[11] Milone D, Peri G, Pitruzzella S and Rizzo G 2015 *Energy and Buildings* **95** 39–46
[12] El Mankibi M, Cantin R and Zoubir A 2015 *Energy Procedia* **78** 2470–75
[13] Ciampi G, Rosato A, Scorpio M and Sibilio S 2015 *Energy Procedia* **78** 2669–74
[14] Garau G and Rosa-Clot M 2017 *Energy Procedia* **134** 244–55
[15] Litti G, Audenaert A and Lavagna M 2018 *Journal of Building Engineering* **17** 135–53
[16] Ascione F, De Rossi F and Vanoli GP 2011 *Energy and Buildings* **43** 1925–36
[17] Lopes CSP and Frontini F 2014 *Energy Procedia* **48** 1493–02
[18] Vietes E, Vassiliva I and Arias JE 2015 *Energy Procedia* **75** 1679–85
[19] Pisello AL, Petrozzi A, Castaldo VL and Cotana F 2016 *Applied Energy* **162** 1313–22
[20] Tomšič M, Mirtič M, Šijanec Zavrl M and Rakuščck A 2017 *Procedia Environmental Sciences* **38** 212–19
[21] GOST R 55528-2013 *Composition and content of scientific and project documentation for the conservation of cultural heritage. Monuments of history and culture. General requirements*
[22] Colins RE and Simko TM 1998 *Solar Energy* **62**(3) 189–213
[23] Cuce E and Cuce PM 2016 *Renewable & Sustainable Energy Reviews* **54** 1345–57

Stoichiometric analysis of air oxygen consumption in modern vehicles using natural and synthetic fuels

S E Shcheklein and A M Dubinin

Ural Federal University named after the first President of Russia B. N. Yeltsin, Ekaterinburg, Russia

e-mail: s.e.shcheklein@urfu.ru

Abstract. The study presents analysis results obtained for stoichiometric reactions of fuel oxidization in air for various motor fuels (such as gasoline, diesel fuel, gaseous fuel) and for biomass-converted and coal-based artificial fuels (such as ethanol, methanol, etc.) and also for hydrogen. Based on the analysis results, comparison is carried out for the specific air consumptions and carbon dioxide emissions of the combustion products.

At a temperature of, which is an ultimate combustion process temperature to be dictated by the achievable properties of structural materials made available for the state-of-the-art engine engineering, the air excess factors are found, the actual air and fuel consumptions are determined, and carbon dioxide emissions are evaluated for the variety of fuels.

The calculated data are presented for the greenhouse effect caused by the use of the various natural and synthetic fuels in transport applications.

In the cases of using hydrocarbon and hydrogen fuels, an abnormally high consumption of the atmospheric air is theoretically evaluated. To reduce the environmental impact, methane-based fuels, bio-ethanol and bio-methanol-based fuels are to be recommended for wider transport vehicle applications.

The analysis method presented in the study offers a universal tool to be instrumental in performing analysis of any fuel compositions both already existing and to be developed for future applications.

Keywords: stoichiometry, motor fuels, gasoline, diesel, ethanol, methanol, hydrogen.

1. Introduction

The use of any fuel occurs as a result of their huge air flow oxidation with formation of polluting carbon dioxide and water vapor. Known environmental impact studies environment emissions engines operating on refined petroleum products [1]. However, in a world increasingly acquire new fuels derived from plant materials, synthesis gas (methanol), fermentation of glucose (ethanol), the decomposition of water into hydrogen and oxygen (including renewable energy) of natural and anthropogenic hydrocarbons by their conversion [2–6]. These types of fuels for harmful emissions from their air-oxidation on the environment not investigated. In this work, performed a quantitative analysis of the air pollution of the atmosphere carbon dioxide and water vapor for common synthetic fuels. The influence of air flow coefficient on output of harmful emissions. The results are compared with the emissions from the oxidation of fuels derived from petroleum.

Content from this work may be used under the terms of the Creative Commons Attribution 3.0 licence. Any further distribution of this work must maintain attribution to the author(s) and the title of the work, journal citation and DOI.
Published under licence by IOP Publishing Ltd

2. Methods of analysis of the influence of oxidation of fuels pollution Environment

2.1. Analysis of the stoichiometric oxidation of fuels when air flow coefficient equal to 1 (α = 1).
It is known that complete oxidation of fuel in the air stehiometricheskimi is described by equations that depend on the chemical composition of the fuel. Composition when α = 1 is the theoretical ratio of fuel and air, which in the process of oxidation of oxygen of the air and fuel to be spent completely without residue.

In a general form [7], the stoichiometric equations for hydrocarbon fuels and alcohols are expressed with the Equations (1), (2) as follows:

$$C_mH^n + (m + n/4)\,(O_2 + 3.76N_2) = mCO_2 + n/2\ H_2O + 3.76\,(m + n/4)\,N_2 + Q \qquad (1)$$

$$C^mH_{2m+1}\ OH + 1{,}5m\,(O_2 + 3.76N_2) = mCO_2 + (m+1)\,H_2O + 3.76\,(1.5\ m)\,N_2 + Q \qquad (2)$$

Table 1 summarizes the air to fuel ratio data for the most common types of motor fuels.

Table 1. Stoichiometric equation oxidation for common fuels when α = 1.

Fuel	Formula	Stoichiometric ratio
Hydrogen	H_2	$H_2+0.5(O_2+3.76\ N_2)=H_2O+1.88N_2$
Ethanol	$C_2\ H_5\ OH$	$C_2\ H_5\ OH+3(O_2+3.76\ N_2)=2CO_2+3H_2O+11.28N_2$
Methanol	$C\ H_3\ OH$	$C\ H_3\ OH+1.5(O_2+3.76\ N_2)=CO_2+2H_2O+5.64N_2$
Gasoline	$C_7\ H_{16}$	$C_7\ H_{16}+11(O_2+3.76\ N_2)=7CO_2+8H_2O+41.36N_2$
Diesel	$C_8\ H_{18}$	$C_8\ H_{18}+12.5(O_2+3.76\ N_2)=8CO_2+9H_2O+47N_2$
Methane	CH_4	$CH_4+2(O_2+3.76N_2)=CO_2+2H_2O+7.52N_2$

2.2. Analysis of the stoichiometric oxidation of fuels with an air flow of larger units of α > 1.
However, modern internal combustion engines are designed to operate at much lower combustion temperatures, which is fed into the working volume of the engine fuel mixtures, fortified with excess air. The ratio of the actual flow rate of oxidant (air) to the required for complete oxidation is characterized by excess air coefficient-α. In this case, the formula (1) and (2) acquire the following form:

$$C_mH_{n+}\alpha\,(m+n/4)\,(O_2 + 3.76\ N_2) = mCO_2 + n/2\ H_2O + 47N_2$$
$$+ (\alpha - 1)(m + n/4)\,(O_2 + 3.76N_2) \qquad (3)$$

$$C_mH_{2m+1}OH + \alpha(1.5m)\,(O_2 + 3.76N_2) = mCO_2 + (m+1)H_2O$$
$$+ 3.76(1.5m)N_2 + (\alpha - 1)\,(1.5m)\,(O_2+3.76N_2) \qquad (4)$$

2.3. Analysis of the influence of air flow coefficient on temperature oxidation of fuels (α = 1).
According to the stoichiometric equation, oxidation of fuel is characterized with a certain temperature of combustion process described with Equation (5) [8]

$$Q^p_H = C^o_d(t^c_d)\cdot t^c_d\,, \qquad (5)$$

where $C^o_d(t^c_d)$) is the true isobaric specific heat of oxidation products at oxidation temperature, measured in MJ/(·kg of fuel), and t^c_d is theoretical oxidation temperature, ºC.

The method for determining theoretical oxidation temperature actually means [8, 9] finding a temperature, in the adiabatic conditions, at which the equality of combustion products enthalpies occurs $C^o_d(t^c_d)\cdot t^c_d$, and finding the lowest heat of fuel combustion Q^p_H.

Table 2 summarizes the calculation results, oxidation temperatures, air consumptions and oxidation product yields at α = 1.

It is seen from Table 2 that at $\alpha = 1$ complete oxidation of fuel occurs at elevated temperatures, with a large amount of free air being consumed. Fuels that consume especially great volumes of air are hydrogen, gasoline, diesel fuel, and methane.

Table 2. Air consumption and oxidation product yields at $\alpha = 1$

Fuel	Heat of combustion, MJ/kg	Theoretical oxidation temperature, °C	Air consumption, kg/kg of fuel	CO_2 emission, kg/kg of fuel	Specific fuel consumption, kg/MJ	Specific air consumption, kg/MJ	Specific CO_2, emission , kg/Mj	Specific H_2O yield, kg/MJ
Hydrogen	121	2020	34.32	0	0.0083	0.284	0.000	0.061
Ethanol	27.785	1880	8.95	1.91	0.0360	0.322	0.069	0.152
Methanol	21.131	1900	6.44	1.38	0.0473	0.305	0.065	0.252
Gasoline	44	1850	15.1	3.08	0.0227	0.343	0.070	0.074
Diesel	41.8	1780	15	3.08	0.0239	0.359	0.074	0.082
Methane	49.09	1860	17.16	2.75	0.0204	0.350	0.056	0.093

2.4. Analysis of the influence of air flow coefficient on temperature oxidation of fuels when $\alpha > 1$

In the process of oxidation of the fuel mixture with excess air combustion of the fuel used for heating of the oxidation products (if $\alpha = 1$) and excess air to a temperature of t_d. Let us consider oxidation at a temperature t_d =1400 °C, which corresponds to the level of temperatures achievable in the state-of-the-art engine designs [10]. In this case, fuel oxidation reaction takes somewhat different form (Table 3).

Table 3. Dependence of excess air ratio of fuel

Fuel	Excess air factor α	Formula
Hydrogen	1.7	$H_2 + 0.85(O_2 + 3.76\ N_2) = H_2O + 3.196N_2 + 0.35O_2$
Ethanol	1.55	$C_2H_5OH + 4.66(O_2 + 3.76\ N_2) = 2CO_2 + 3H_2O + 17.52N_2 + 1.66O_2$
Methanol	1.5	$CH_3OH + 2.25(O_2 + 3.76\ N_2) = CO_2 + 2H_2O + 8.46N_2 + 0.75O_2$
Gasoline	1.6	$C_7H_{16} + 17.6(O_2 + 3.76\ N_2) = 7CO_2 + 8H_2O + 66.176N_2 + 6.6O_2$
Diesel	1.4	$C_8H_{18} + 17.5(O_2 + 3.76\ N_2) = 8CO_2 + 9H_2O + 65{,}8N_2 + 5O_2$
Methane	1.43	$CH_4 + 2.86(O_2 + 3.76N_2) = CO_2 + 2H_2O + 10.7536N_2 + 0.86O_2$

Heat loss for dissipation into environment and for dissociation of CO2 and H2O and initial fuel and air temperatures are assumed negligible. In this case, Equation (3) takes the form:

$$Q_{\text{H}}^{\text{p}} = [C_{\text{d}}^{\text{o}}(t_{\text{d}}) + (\alpha - 1)C_{\text{a}}^{\text{o}}\ (t_{\text{d}})]t_{\text{d}}\ , \tag{6}$$

where $C_d^o(t_d)$ is the true specific heat of oxidation products at a temperature of t_d, kJ/K·kg of fuel, $C_{\text{a}}^o(t_d)$ is the true isobaric heat of air at a temperature t_d, kJ/K·kg of fuel, Q_{H}^{p} is the lowest fuel oxidation heat, kJ/kg of fuel, α is the air excess factor for fuel oxidation; and t_d is the temperature of oxidation products, °C.

From Equation (6) we find the value of α factor at the initial temperature of fuel and air as t_{a}= 0, t_{f}= 0 and the temperature of combustion products t_d in the form

$$\alpha = 1 + [Q_{\text{H}}^{\text{p}} - C_{\text{d}}^{\text{o}}(t_{\text{d}})]/[C_{\text{a}}^{\text{o}}(t_{\text{d}})]. \tag{7}$$

Table 4 summarizes the results of calculating the excess air factor, air consumption and oxidation product yields.

Table 4. Excess air factor and air consumption, and product yield at an oxidation temperature of 1,400 °C.

Fuel	Air excess factor α	Air consumption kg/kg of fuel	Steam, kg/kg of fuel	CO_2 emission, kg/kg of fuel	Specific fuel consumption, kg/MJ	Specific air consumption, kg/MJ	Specific steam yield, kg/MJ	Specific CO_2 emission, kg/MJ
Hydrogen	1.7	58.3	9	0	0.0083	0.482	0.074	0.000
Ethanol	1.55	13.9	1.17	1.91	0.0360	0.500	0.042	0.069
Methanol	1.5	9.65	1.125	1.375	0.0473	0.457	0.053	0.065
Gasoline	1.6	24.16	1.44	3.08	0.0227	0.549	0.033	0.070
Diesel	1.4	21	1.42	3.09	0.0239	0.502	0.034	0.074
Methane	1.43	24.53	2.25	2.75	0.0204	0.500	0.046	0.056

3. The results of the study

Analysis of the calculation results demonstrates that there is a considerable difference in the environmental impact caused by various fuels used for a combustion engine. Figures 1–4 display absolute and specific characteristics of fuels such as air consumption, carbon dioxide emissions and steam generation at $\alpha = 1$ and $\alpha > 1$.

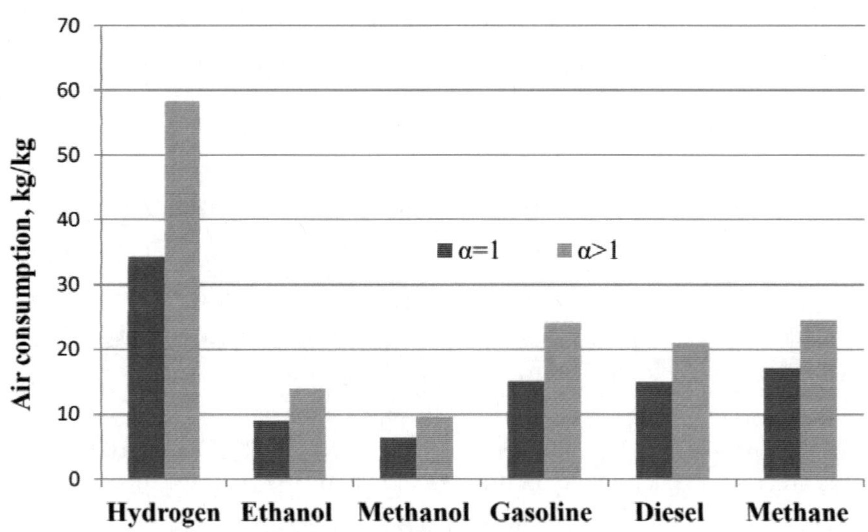

Figure 1. Free air consumption

It can be seen from Fig. 1 that the absolute free air consumption is the highest for hydrogen and hydrocarbon types of fuels. Combustion at $\alpha > 1$ leads to an increase in air combustion for all types of fuel.

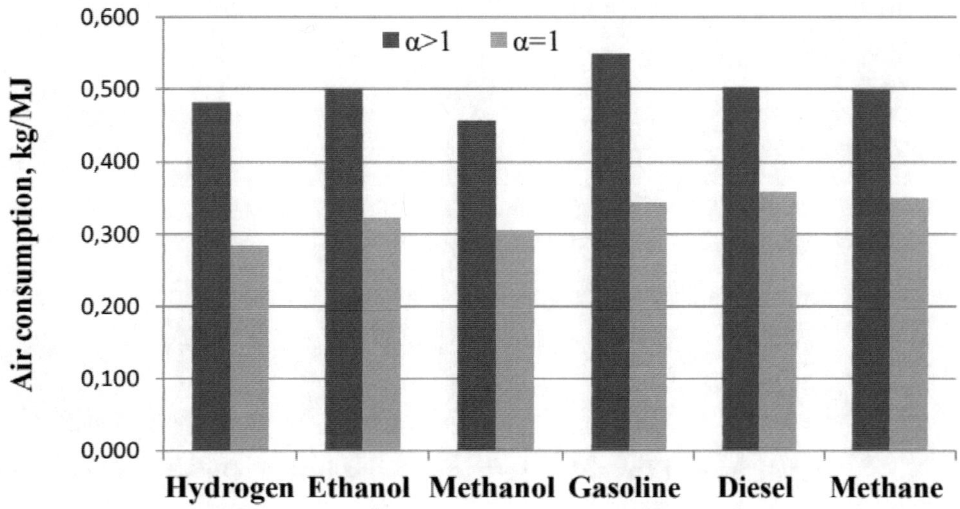

Figure 2. Specific free air consumption

However, the difference in the combustion heat (Table 2) leads to the free air consumptions being similar in order to produce similar amounts of energy. The least air consumption is observed for methanol, hydrogen and ethanol. The greatest air consumption is observed for the cases of using gasoline, diesel fuel and methane.

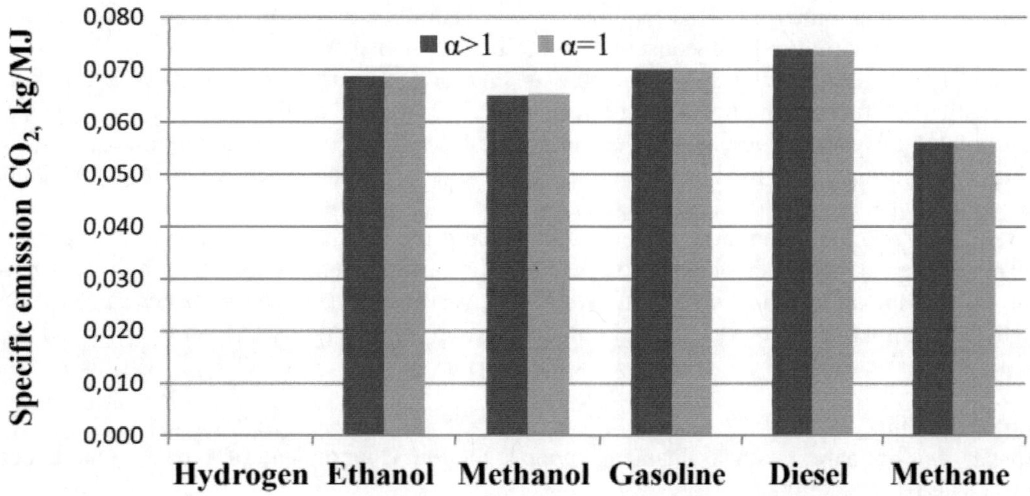

Figure 3. Specific CO2 emission

Carbon dioxide emission is the same for $\alpha = 1$ and $\alpha > 1$, both for hydrocarbon fuels and for alcohols. There is no CO_2 emission for hydrogen oxidation; small values of carbon dioxide emissions are observed for methane and methanol; and the highest emissions are observed for diesel fuel and gasoline.

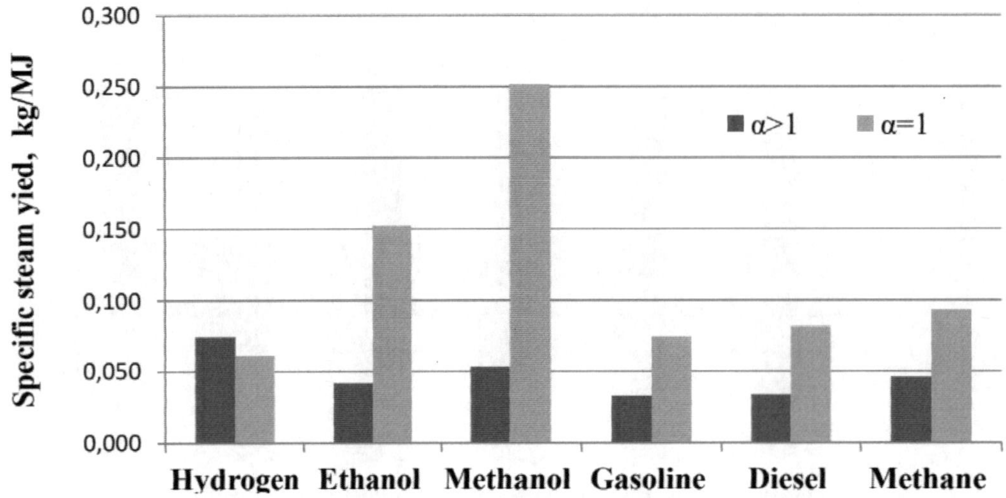

Figure 4. Specific steam yield

Steam is generated most intensely at $\alpha = 1$. At $\alpha > 1$, the least amount of steam is generated by hydrocarbon fuel and alcohols; and the greatest steam yield is observed for hydrogen fuel.

4. Conclusion

1. The analysis showed that, in an effort to raise the temperature of oxidation of all fuels to the limits defined by the stoichiometric ratio $\alpha = 1$, decreased absolute and specific air consumption increases engine efficiency and reduced fuel consumption for 1 kWh of energy.
2. Actual capacities offered by the state-of-the-art engine engineering are limited with the durability characteristics of structural materials and require increased air consumption.
3. The highest carbon dioxide emission occurs in the case of using the most common fuels, such as diesel fuel and gasoline; and the lowest carbon dioxide emission is observed for hydrogen and alcohol-based synthetic fuels.
4. In terms of reducing greenhouse effect and increasing thermodynamic efficiency of engines, most promising appears development of high-temperature combustion engines operating on hydrogen fuel.
5. For state-of-the-art combustion engine designs, a positive environmental effect can be achieved through the use of hydrogen and alcohol-containing fuel compositions and methane-based gaseous fuels.

Acknowledgments
The paper was prepared with a financial support by the Government of the Russian Federation (Contract No. 02.A03.21.0006).

References

[1] Kulchitsky A R 2010 *Toxicity reciprocating internal combustion engine. The formation of harmful substances during combustion of fuels* (Vladimir: Vladimir State University Press)
[2] Nadir Yilmaz 2017 Sustainable alternative fuels in aviation *Energy* **140** (2) 1378–86
[3] Yu Bicer and I Dincer 2018 Life cycle environmental impact assessments and comparisons of alternative fuels for clean vehicles *Resources, Conservation and Recycling* **132** 141–57
[4] Shcheklein S E and Dubinin A M 2018 Production of liquid fuels based on co-current gas generator and nuclear reactor *Atomic energy* **124 (2)** 76–9
[5] Romanova E D, Romanov I D, Vasilyev V A and ChernyshovE A 2011 The development of technology to produce synthetic and life alternatives fuels *Alt. energy and ecology* **6** 81–3

[6] Veziroglu T N and Sahin S 2014 21-century Energy: hydrogen power *Alt. energy and ecology* **2 (142)** 12–28
[7] Baehr H B 1988 *Thermodynamik* (Berlin/Heidelberg: Springer-Verlag)
[8] Lewis B 1968 *Flame Burning. Explosions in gases* (Moscow: Mir)
[9] Stromberg A G and Semchenko D G 2001 *Physical chemistry* (Moscow: Vysshaya Shkola)
[10] 1982 *Environmental chemistry* (Moscow: Chemistry)

IOP Conf. Series: Earth and Environmental Science **177** (2018) 012021 doi:10.1088/1755-1315/177/1/012021

Unsteady-state moisture behavior calculation for multilayer enclosing structure made of capillary-porous materials

V G Gagarin[1], V K Akhmetov[2] and K P Zubarev[1]

[1] Department of Heat and Gas Supply and Ventilation, Moscow State University of Civil Engineering, 26 Yaroslavl highway, Moscow, 129337, Russia
[2] Department of Applied Mathematics, Moscow State University of Civil Engineering, 26 Yaroslavl highway, Moscow, 129337, Russia

e-mail: gagarinvg@yandex.ru, vadim.akhmetov@gmail.com, zubarevkirill93@mail.ru

Abstract. The paper studies unsteady-state moisture behavior of multilayer enclosing structure using discrete-continuous method. Modification of Gagarin's differential moisture transfer equation is proposed. The proposed moisture transfer equation within one layer of a multilayer enclosing structure is a second-order differential parabolic equation with constant coefficients. Third-kind boundary conditions of moisture exchange are set for enclosing structure boundaries. Fourth-kind boundary conditions represented by moisture potential flows equality are set at the joint of two different materials. An analytical expression for moisture potential calculation in any enclosing structure section, at any time, under continuous control for temperature distribution, has been derived using discrete-continuous method. Calculations for multilayer enclosing structure consisting of clay brick base and lime brick lining have been made. The obtained results have been compared to calculation results obtained by well-known Gagarin's unsteady-state method and Kozlov's engineering method. It has been shown that the discrete-continuous method allows calculating unsteady-state moisture behavior of multilayer enclosing structures by analytical expression, and obtaining moisture distribution similar to Gagarin's unsteady-state method, thus it can be used by design engineers in practice.

1. Introduction

Moisture behavior of enclosing structure is extremely important for construction. Error in construction material moisture condition assessment can result in premature deterioration of the enclosing structure [1, 2], high heat loss of the building [3, 4], and negative humidity effect on human life [5].

Moisture behavior calculations for multilayer enclosing structures are the most important and challenging ones [6]. In multilayer enclosing structure moisture transfer speed is different for every material due to its porous structure and heat-transfer properties. Mathematical models for capillary-porous bodies are considered in the given paper [7]. In such construction materials vapor moisture moves according to vapor permeability mechanism in sorption wetting zone, and condensed moisture also moves by means of capillary pressure gradient in over-sorption wetting zone.

Gagarin's mathematical model [8] based on moisture potential is known:

$$F(w,t) = E_t(t) \cdot \varphi(w) + \frac{1}{\mu}\int_0^w \beta(\zeta)\,d\zeta. \tag{1}$$

Content from this work may be used under the terms of the Creative Commons Attribution 3.0 licence. Any further distribution of this work must maintain attribution to the author(s) and the title of the work, journal citation and DOI.
Published under licence by IOP Publishing Ltd

where F – moisture potential, *Pa*; E_t – maximum water vapor tension, *Pa*; φ – relative air humidity; μ – vapor permeability coefficient, *kg/(m· s· Pa)*; β – moisture conductivity coefficient, kg/(m· s· kg/kg), depending on humidity .

Differential thermal conductivity equation governs temperature distribution of a multilayer enclosing structure:

$$c \cdot \gamma_0 \cdot \frac{\partial t}{\partial \tau} = \frac{\partial}{\partial x}(\lambda \cdot \frac{\partial t}{\partial x}). \tag{2}$$

where t – temperature, ℃; τ – time, *s*; x – coordinate, *m*; γ_0 – enclosing structure dry material density, kg/m^3; c – specific material heat capacity, *J/(kg·℃)*.

Differential moisture transfer equation governs moisture potential distribution of a multilayer enclosing structure:

$$\gamma_0 \cdot (\frac{1}{\mu}\beta(w) + \frac{\partial \varphi(w)}{\partial w}E_t(t))^{-1} \cdot \frac{\partial F(w,t)}{\partial \tau} = \frac{\partial}{\partial x}(\mu \frac{\partial F(w,t)}{\partial x}). \tag{3}$$

Total moisture flow in materials of a multilayer enclosing structure is a sum of liquid and vapor moisture flows:

$$g_{tot} = g_v + g_L. \tag{4}$$

where g_{tot} – total moisture flow, $kg/(m^2 \cdot s)$, g_v – vapor moisture flow, $kg/(m^2 \cdot s)$, g_L – liquid moisture flow, $kg/(m^2 \cdot s)$.

Unsteady-state moisture behavior calculation by means of (1) – (3) equations demands computer use, thus its use in construction practice is complicated.

For engineering calculation, it is much more convenient to use Kozlov's quasistationary method, which appears to be a study of temperature and moisture potential steady-state distribution within every month.

Steady-state temperature distribution takes the following form for temperature pattern of a multilayer enclosing structure:

$$\frac{\partial}{\partial x}(\lambda \cdot \frac{\partial t}{\partial x}) = 0. \tag{5}$$

where λ – thermal conductivity coefficient, *W/(m·℃)*.

Steady-state moisture potential distribution takes the following form for moisture potential pattern of a multilayer enclosing structure:

$$\frac{\partial}{\partial x}(\mu \cdot \frac{\partial F}{\partial x}) = 0. \tag{6}$$

Equations (5) and (6) are much more convenient for engineering calculation, however this method has much less potential as compared to unsteady-state calculation.

2. The problem

To develop calculation method for unsteady-state moisture behavior of a multilayer enclosing structure, which, on one hand, is easy to use, and, on the other hand, allows unsteady-state processes calculation to a high precision.

3. Materials and methods

Modified differential moisture transfer equation for multilayer enclosing structure is proposed:

$$\gamma_0 \cdot \frac{\xi_{F0}(w,t)}{E_t(t)} \cdot \frac{\partial F(w,t)}{\partial \tau} = \frac{\partial}{\partial x}(\mu \cdot \frac{\partial F(w,t)}{\partial x}). \tag{7}$$

where ξ_{F0} – "material relative potential capacity coefficient", kg/kg.

"Relative potential capacity coefficient" is calculated by formula:

$$\xi_{F0} = \frac{dw}{d\varphi_F}. \tag{8}$$

where φ_F – "relative elasticity of moisture potential".

The term "heat-to-humidity material property coefficient" has been introduced:

$$\kappa = \frac{\mu}{\gamma_0 \cdot \xi_{F0}}. \tag{9}$$

Product of "heat-to-humidity material property coefficient" by maximum water vapor elasticity is a potential capacity coefficient.

"Relative elasticity of moisture potential" is calculated by formula:

$$\varphi_F = \varphi + \frac{1}{\mu}\int_0^w \beta(\sigma)d\sigma / E_t. \tag{10}$$

The equation (7) takes the following form for a separate layer:

$$\frac{\partial F(w,t)}{\partial \tau} = \kappa(w,t) \cdot E_t(t) \frac{\partial^2 F(w,t)}{\partial x^2}. \tag{11}$$

Temperature pattern is a steady-state equation (5) within a month. Temperature pattern effects on equation (7) by means of maximum elasticity of water vapor determined by Clausius-Clapeyron correlation:

$$E_t = 1.84 \cdot 10^{11} \cdot \exp(-5330/(273+t)). \tag{12}$$

As temperature distribution appears to be steady within a month (5), distribution of maximum elasticity of water vapor is also constant within a month.

Average values of relative potential capacity coefficients are calculated by equation (9) for every layer of a multilayer enclosing structure for every month. These coefficients are taken to be constant within an assessment month.

Thus, moisture transfer equation (11) is a differential parabolic equation with constant coefficients.

Third-kind conditions for moisture exchange of surface and ambient air are set for the enclosing structure boundaries.

Moisture exchange boundary condition for outside air and structure surface takes the form:

$$-\mu_1 \left.\frac{\partial F}{\partial x}\right|_{i=1} = \beta_{ext}\left(F_{ext} - F_1\right). \tag{13}$$

where F_{ext} – outside air moisture potential, Pa; F_1 – moisture potential of the enclosing structure section which contacts with outside air, Pa; β_{ext} – moisture exchange coefficient of outside air and enclosing structure surface, $kg/(m^2 \cdot s \cdot Pa)$, μ_1 – vapor permeability coefficient of the enclosing structure layer, which is the closest to the outer surface of the building, $kg/(m \cdot s \cdot Pa)$.

Boundary condition of moisture exchange between inside air and structure inner surface takes the form:

$$\mu_2 \frac{\partial F}{\partial x}\bigg|_{i=N} = \beta_{in}\left(F_{in} - F_N\right). \tag{14}$$

where F_{in} – inside air moisture potential, Pa; F_N – moisture potential of the enclosing structure surface which contacts with inside air, Pa; β_{in} – moisture exchange coefficient of inside air and enclosing structure surface, $kg/(m^2 \cdot s \cdot Pa)$, μ_2 – vapor permeability coefficient of the enclosing structure layer, which is the closest to the inner surface of the enclosing structure, $kg/(m \cdot s \cdot Pa)$.

Boundary condition of flow continuity is set for the joint between enclosing structure layers:

$$-\mu_1 \frac{\partial F}{\partial x}\bigg|_{i=v-0} = -\mu_2 \frac{\partial F}{\partial x}\bigg|_{i=v+0}. \tag{15}$$

where v – multilayer enclosing structure section with material joint.

Let's consider solution of the equation (7) with boundary conditions (13), (14), (15) using discrete-continuous method [9, 10]. Let's carry out standard procedure of spatial-time domain discretization for double-layer enclosing structure [Fig. 1].

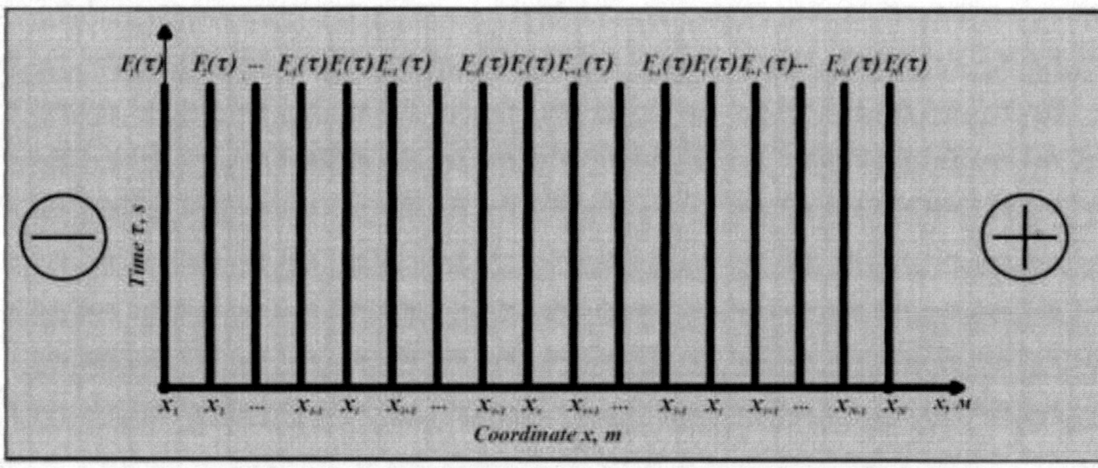

Figure 1. Spatial-time domain discretization for double-layer enclosing structure

In this case initial boundary value problem (7), (13), (14), (15) can be represented by the system of simultaneous equations:

$$\begin{cases} \dfrac{\partial F_1}{\partial \tau} = \dfrac{\kappa_1}{h_1^2} \cdot E_{t1} \cdot (-(1+\dfrac{\beta_{ext}}{\mu_1} \cdot h_1) \cdot F_1 + F_2) + \dfrac{\kappa_1}{h_1^2} \cdot E_{t1} \cdot \dfrac{\beta_{ext}}{\mu_1} \cdot h_1 \cdot F_{ext},\ i=1 \\ \dfrac{\partial F_i}{\partial \tau} = \dfrac{\kappa_1}{h_1^2} \cdot E_{ti} \cdot \left(F_{i-1} - 2 \cdot F_i + F_{i+1}\right), i = 2,3,4,\ldots,v-1 \\ \dfrac{\partial F_v}{\partial \tau} = d_{v-1} \cdot F_{v-1} - d_v \cdot F_v + d_{v+1} \cdot F_{v+1},\ i = v \\ \dfrac{\partial F_i}{\partial \tau} = \dfrac{\kappa_2}{h_2^2} \cdot E_{ti} \cdot \left(F_{i-1} - 2 \cdot F_i + F_{i+1}\right), i = v+1, v+2,\ldots,N-1 \\ \dfrac{\partial F_N}{\partial \tau} = \dfrac{\kappa_2}{h_2^2} \cdot E_{tN} \cdot (F_{N-1} - (1+\dfrac{\beta_{in}}{\mu_2} \cdot h_2) F_N) + \dfrac{\kappa_2}{h_2^2} \cdot E_{tN} \cdot \dfrac{\beta_{in}}{\mu_2} \cdot h_2 \cdot F_{in},\ i=N \\ F(x,0) = u_f(x), 0 \le x \le l \end{cases} \tag{16}$$

$$d_{v-1} = 0{,}024 \cdot 2 \cdot E_{tv} \cdot \mu_1 / (h_1(h_1 \cdot \gamma_{01} \cdot \xi_{F01} + h_2 \cdot \gamma_{02} \cdot \xi_{F02})). \tag{17}$$

$$d_v = 0{,}024 \cdot 2 \cdot (\mu_1 / h_1 + \mu_2 / h_2) / (h_1 \cdot \gamma_{01} \cdot \xi_{F01} + h_2 \cdot \gamma_{02} \cdot \xi_{F02}). \tag{18}$$

$$d_{v+1} = 0{,}024 \cdot 2 \cdot E_{tv} \cdot \mu_2 / (h_2(h_1 \cdot \gamma_{01} \cdot \xi_{F01} + h_2 \cdot \gamma_{02} \cdot \xi_{F02})). \tag{19}$$

where i – enclosing structure section number; h_1 – first layer partition step (closer to outer surface) of the enclosing structure in coordinate *x, m*; h_2 – second layer partition step (closer to inner surface) of the enclosing structure in coordinate *x, m*; γ_{01} – dry material density in the first layer of an enclosing structure, kg/m^3; γ_{02} – dry material density in the second layer of an enclosing structure, kg/m^3; ξ_{F01} – relative potential capacity of the first layer material of an enclosing structure, kg/kg; ξ_{F02} – relative potential capacity of the second layer material of an enclosing structure, kg/kg; κ_1 – "averaged heat-to-humidity material property coefficient" of the first layer of an enclosing structure, $m^2/(s \cdot Pa)$; κ_2 – "averaged heat-to-humidity material property coefficient" of the second layer of an enclosing structure, $m^2/(s \cdot Pa)$; u_f – initial moisture potential distribution, Pa.

Inside air moisture potential appears to be constant during the whole calculation period, and is equal to partial pressure of inside air water vapor:

$$F_{in} = e_{in}. \tag{20}$$

where e_{in} – partial pressure of inside air water vapor, Pa;

Outside air moisture potential is taken to be equal to partial pressure of outside air water vapor:

$$F_{ext} = e_{ext}. \tag{21}$$

where e_{ext} – partial pressure of outside air water vapor, Pa;

Assume that outside air moisture potential changes continuously within every month in linear fashion:

$$F_{ext} = m \cdot \tau + n. \tag{22}$$

where m – boundary conditions slope ratio within a month; n – boundary conditions graph rise within a month, Pa;

Taking into account (20), (21), and (22), simultaneous equations (16) can be represented in matrix form:

$$\begin{cases} \overline{F}_{\tau}^{'} = (G + K \cdot E_t \cdot A) \cdot \overline{F} + p \cdot \tau \cdot \overline{L} + \overline{B} \\ \overline{F}_0 = \overline{\chi}, 0 \le x \le l \end{cases} . \tag{23}$$

where $p = \kappa_1 \cdot E_{t1} \cdot \beta_{ext} \cdot h_1 \cdot m / (h_1^2 \cdot \mu_1)$.

$$\overline{F}_{\tau}^{'} = \begin{pmatrix} F_1^{'}(\tau) \\ F_2^{'}(\tau) \\ \dots \\ F_{N-1}^{'}(\tau) \\ F_N^{'}(\tau) \end{pmatrix}.\ G = \begin{pmatrix} 0\ 0\ 0\ 0\ 0 \\ 0\ 0\ 0\ 0\ 0 \\ 0 d_{v-1}\ d_v\ d_{v+1} 0 \\ 0 \dots \dots \dots 0 \\ 0\ 0\ 0\ 0\ 0 \end{pmatrix}.\ K = \begin{pmatrix} \kappa_1/h_1^2 & 0 & 0 & 0 & 0 \\ 0 & \kappa_1/h_1^2 & 0 & 0 & 0 \\ 0 & \dots & 0 & \dots & 0 \\ 0 & 0 & 0 & \kappa_2/h_2^2 & 0 \\ 0 & 0 & 0 & 0 & \kappa_2/h_2^2 \end{pmatrix}.$$

$$E_t = \begin{pmatrix} E_{t1} & 0 & 0 & 0 & 0 \\ 0 & E_{t2} & 0 & 0 & 0 \\ \dots & \dots\dots & \dots & \dots \\ 0 & 0 & 0 & E_{t(N-1)} & 0 \\ 0 & 0 & 0 & 0 & E_{tN} \end{pmatrix}.\ A = \begin{pmatrix} -(1+\beta_{ext} \cdot h_1/\mu_1) & 1 & 0 & 0 & 0 \\ 1 & -2 & 1 & 0 & 0 \\ 0 & \ddots & \ddots & \ddots & 0 \\ 0 & 0 & 1 & -2 & 1 \\ 0 & 0 & 0 & 1 & -(1+\beta_{in} \cdot h_2/\mu_2) \end{pmatrix}.$$

$$\overline{F} = \begin{pmatrix} F_1 \\ F_2 \\ \dots \\ F_{N-1} \\ F_N \end{pmatrix}.\ \overline{L} = \begin{pmatrix} 1 \\ 0 \\ \dots \\ 0 \\ 0 \end{pmatrix}.\ \overline{B} = \begin{pmatrix} \kappa_1 \cdot E_{t1} \cdot \beta_{ext} \cdot h_1 \cdot n/(h_1^2 \cdot \mu_1) \\ 0 \\ \dots \\ 0 \\ \kappa_2 \cdot E_{tN} \cdot \beta_{in} \cdot h_2 \cdot F_{in}/(h_2^2 \cdot \mu_2) \end{pmatrix}.\ \overline{F}_0 = \begin{pmatrix} F_1(0) \\ F_2(0) \\ \dots \\ F_{N-1}(0) \\ F_N(0) \end{pmatrix}.\ \overline{\chi} = \begin{pmatrix} F_{ext} \\ F_{ext} \\ \dots \\ F_{ext} \\ F_{ext} \end{pmatrix}.$$

The equation (23) is a Cauchy problem, which can be solved in matrix form:

$$\overline{F} = p \cdot \left((G + K \cdot E_t \cdot A)^{-2} \cdot e^{(G+K \cdot E_t \cdot A) \cdot \tau} - \tau \cdot (G + K \cdot E_t \cdot A)^{-1} - (G + K \cdot E_t \cdot A)^{-2} \right) \cdot \overline{L} + .$$
$$+ (G + K \cdot E_t \cdot A)^{-1} \left(e^{(G+K \cdot E_t \cdot A) \cdot \tau} - E \right) \cdot \overline{B} + e^{(G+K \cdot E_t \cdot A) \cdot \tau} \cdot \overline{F}_0 . \tag{24}$$

where E – unit matrix.

4. Results and discussion

Multilayer enclosing structure consisting of subbase made of clay brick 0.38 m thick and lining made of lime brick 0.12 m has been examined. Construction area is Moscow. Inside air temperature 22 °C and relative inside air humidity 60 % have been maintained constant in the building. Calculation results of the given multilayer enclosing structure moisture behavior obtained using different moisture potential theory methods have been compared. The following moisture potential theory methods have been used: the proposed discrete-continuous method, Gagarin's unsteady-state method, and Kozlov's engineering method.

An analytical formula (6) has been used for calculation of moisture potential distribution along multilayer enclosing structure thickness according to Kozlov's engineering method. Numerical finite difference method using explicit difference scheme has been used for calculation of moisture potential distribution along multilayer enclosing structure thickness according to Gagarin's unsteady-state method.

Kozlov's experimental data [8] on vapor permeability coefficient, static moisture conductivity and sorption coefficient has been used for the proposed calculation method by equation (24). This data is the basis for moisture potential scales [Fig. 2, Fig. 3] for all materials of a multilayer enclosing structure by expression (1).

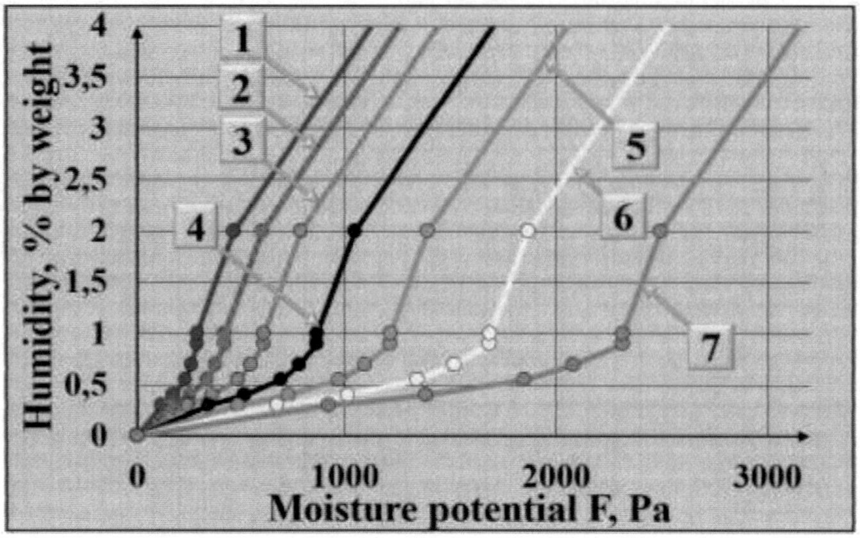

Figure 2. Moisture potential scale of lime brick
(1 – -10 °C temperature; 2 – -5 °C temperature; 3–0 °C temperature; 4–5 °C temperature; 5–10 °C temperature; 6–15 °C temperature; 7–20 °C temperature)

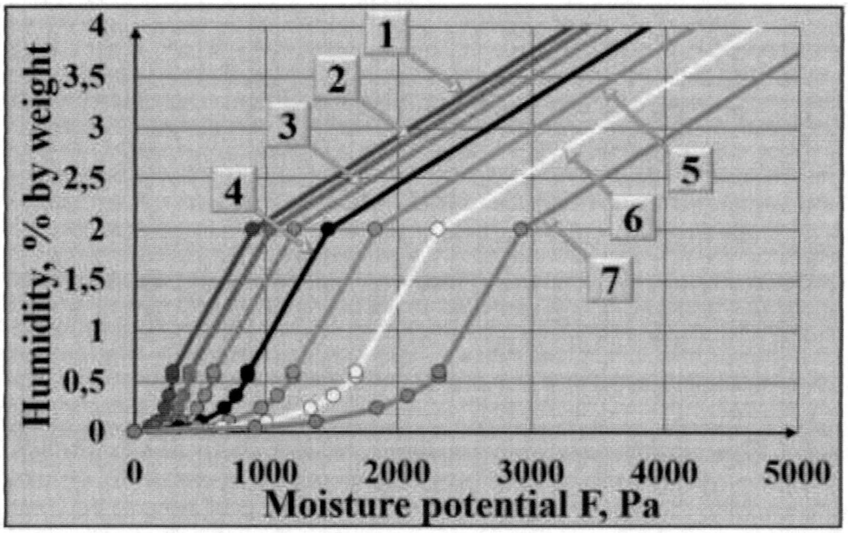

Figure 3. Moisture potential scale of clay brick
(1 – -10 °C temperature; 2 – -5 °C temperature; 3–0 °C temperature; 4–5 °C temperature; 5–10 °C temperature; 6–15 °C temperature; 7–20 °C temperature)

The proposed material moisture dependence on "relative elasticity of moisture potential" has been shown on the basis of equation (10) for all layers of a multilayer enclosing structure [Fig. 4, Fig. 5].

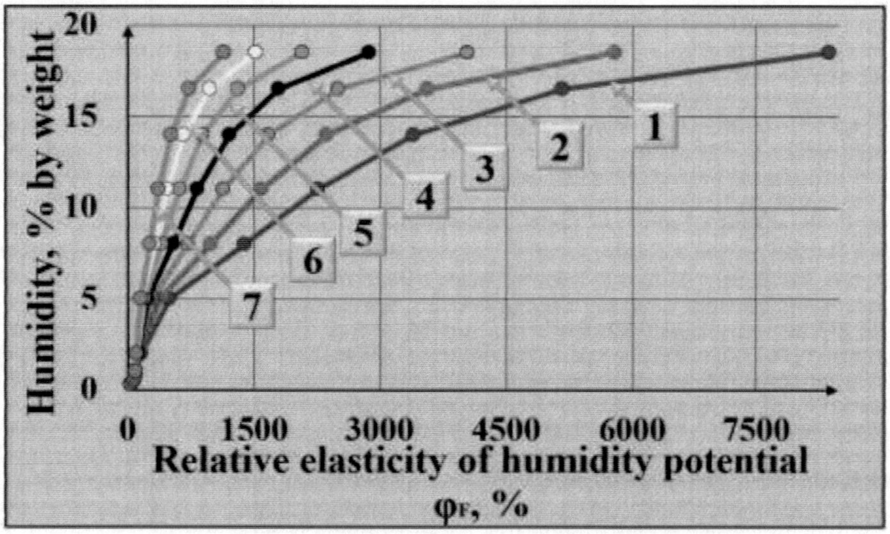

Figure 4. Lime brick material humidity dependence on relative elasticity of moisture potential (1 – -10 °C temperature; 2 – -5 °C temperature; 3–0 °C temperature; 4–5 °C temperature; 5–10 °C temperature; 6–15 °C temperature; 7–20 °C temperature)

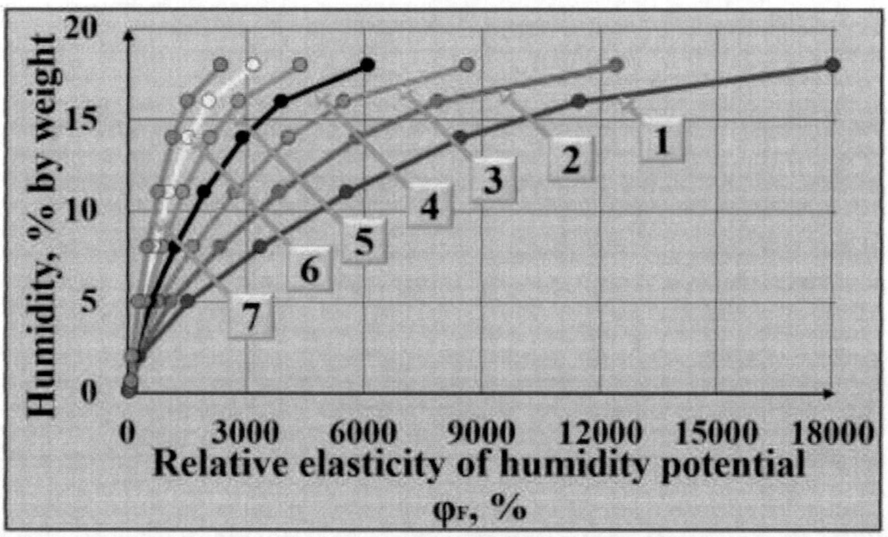

Figure 5. Clay brick material humidity dependence on relative elasticity of moisture potential (1 – -10 °C temperature; 2 – -5 °C temperature; 3–0 °C temperature; 4–5 °C temperature; 5–10 °C temperature; 6–15 °C temperature; 7–20 °C temperature)

Moisture potential distribution at month end has been calculated by formula (24). Material moisture has been determined by obtained moisture potential distribution according to moisture potential scales [Fig. 2, Fig. 3]. Moisture distribution at month start has been taken as equal to moisture distribution at the end of the previous month. Comparison of calculation results of moisture potential distribution along multilayer enclosing structure thickness obtained by means of different moisture potential theory methods for the coldest month – January – is given in [Fig. 6]. Comparison of calculation results of moisture distribution along multilayer enclosing structure thickness obtained by means of different moisture potential theory methods for the coldest month – January – is given in [Fig. 7].

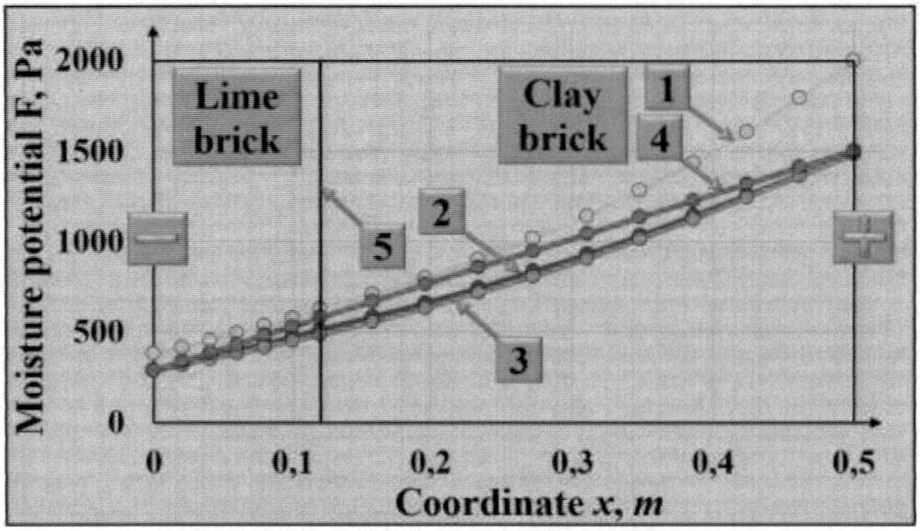

Figure 6. Comparison of moisture potential theory methods for moisture potential distribution along multi-layer enclosing structure thickness in January
(1 – maximum water vapor pressure, 2 – moisture potential distribution in enclosing structure according to Gagarin's unsteady-state method, 3 – moisture potential distribution in enclosing structure according to discrete-continuous method, 4 – moisture potential distribution in enclosing structure according to Kozlov's engineering method, 5 – border between layers)

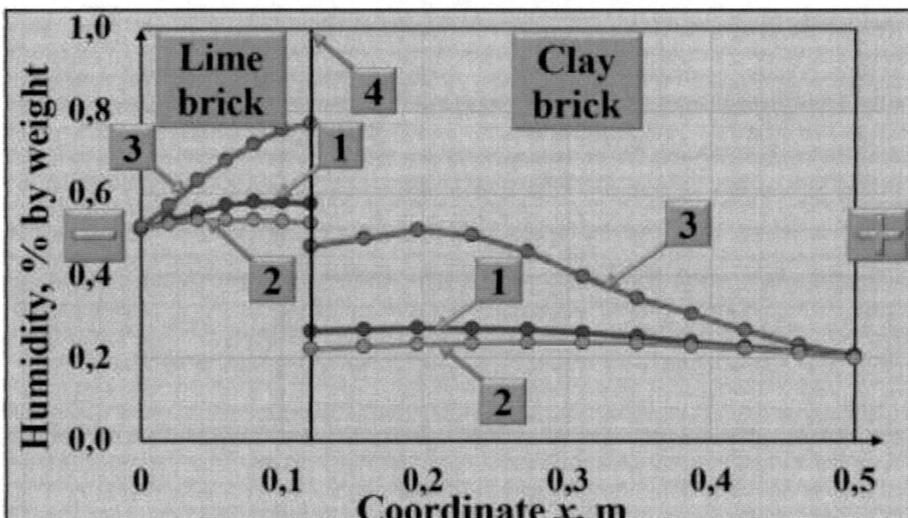

Figure 7. Comparison of moisture potential theory methods for moisture regime calculation of single-layer enclosing structure in January
(1 – moisture distribution in enclosing structure according to Gagarin's unsteady-state method, 2 – moisture distribution in enclosing structure according to discrete-continuous method, 3 – moisture distribution in enclosing structure according to Kozlov's engineering method, 4 – border between layers)

The graphs show that moisture distribution in multilayer enclosing structure obtained by means of discrete-continuous method gives quantitative and qualitative results similar to moisture distribution results, obtained by means of Gagarin's unsteady-state method. However, this distribution is calculated from analytical expression, thus, the calculation is easier.

5. Conclusion

Unsteady-state moisture behavior determination method using discrete-continuous approach has been developed for multilayer enclosing structures. It has been shown that moisture behavior calculation results obtained by the proposed method are close to results obtained by existing Gagarin's unsteady-state method. Moisture behavior is finally calculated from equation (24) without numerical method application. Due to this fact, the proposed method is convenient for design engineers in practice.

The proposed method can be used for calculation of multilayer enclosing structures consisting of two layers and more.

References

[1] Wu Z, Wong H S and Buenfeld N R 2017 Transport properties of concrete after drying-wetting regimes to elucidate the effects of moisture content, hysteresis and microcracking *Cement and concrete research* **98** 136–54

[2] Liu Z C, Hansen W and Wang FZ 2018 Pumping effect to accelerate liquid uptake in concrete and its implications on salt frost durability *Construction and building materials* **158** 181–8

[3] Jin H Q, Yao X L, Fan L W, Xu X and Yu Z T 2016 Experimental determination and fractal modeling of the effective thermal conductivity of autoclaved aerated concrete: Effects of moisture content *International journal of heat and mass transfer* **92** 589–602

[4] Hoseini A and Bahrami A 2017 Effects of humidity on thermal performance of aerogel insulation blankets *Journal of building engineering* **13** 107–15

[5] Girault F and Perrier F 2012 Estimating the importance of factors influencing the radon-222 flux from building walls *Science of the total environment* **433** 247–63

[6] Skerget L, Tadeu A and Ravnik J 2017 BEM numerical simulation of coupled heat, air and moisture flow through a multilayered porous solid *Engineering analysis with boundary elements* **74** 24–33

[7] Arfvidsson J and Claesson J 2000 Isothermal moisture flow in building materials: modelling, measurements and calculations based on Kirchhoff's potential *Building and environment* **35** L6 519–36

[8] Gagarin V G, Kozlov V V and Zubarev K P 2016 Analysis of the Zone Location of Maximum Moistering in the Wall System with Different Thickness of Insulation Layer *Housing Construction* **6** 8–12

[9] Zolotov A B, Mozgaleva M L, Akimov P A and Sidorov V N 2010 About one discrete-continual method of solution of one-dimensional heat conductivity problem *Academia. Architecture and Construction* **3** 287–91

[10] Sidorov V N and Matskevich S M 2016 Discrete-analytical solution of the unsteady-state heat conduction transfer problem based on the finite element method *IDT 2016 – Proceedings of the International Conference on Information and Digital Technologies* 241–4

IOP Conf. Series: Earth and Environmental Science **177** (2018) 012022 doi:10.1088/1755-1315/177/1/012022

Study of feasibility of local renewable resources for substitution of fossil fuels in the Far North of Russia

Z B Namsaraev[1], A M Konovalov[2] and G V Baturova[2]

[1] NRC Kurchatov Institute, Moscow, 123182, Russia
[2] Moscow Technological University (MIREA), Moscow, 119454, Russia

e-mail: zorigto@gmail.com

Abstract. The problem of replacing fossil fuels is extremely important for the sustainable development of the Far North of Russia. At present, the power supply of this region is solved by delivering fossil fuel by water and road transport ("Northern delivery" or "Severniy zavoz" in Russian), which significantly increases its cost. The goal of this paper is to estimate the volume of fossil fuel delivery to the Far North and compare it with potential technical bioenergy resources for a transition of the local power systems to the local renewable resources. The analysis of the Northern delivery volume in 2015 showed that fuel constitutes 95.1 % of the total volume of the Northern Delivery. The estimated energy content of the different types of fuels is equal to 112.7 PJ (diesel 81.5 PJ, gasoline 9.6 PJ, coal 21.6 PJ). The total technical potential of forest waste in the northern Federal Districts is 355.9 PJ. Thus, the bioenergy potential of forest residues of these Federal Districts is more than 3 times higher than the energy content of fuels delivered to Far North. This makes the partial transition of the Far North to renewable energy sources theoretically possible, which will also help to solve a number of environmental problems related to the accumulation of waste from a forest industry. The use of the existing infrastructure of the Northern Delivery will allow creating a new local market for the forest residues and increasing the efficiency of the bioenergy industry. The possibility of transition to renewable energy resources should be taken into account during the modernization of the energy infrastructure in the Far North, including the possibility of using renewable resources for heating. The process of replacement of fossil fuels by renewable biofuels has already begun in a number of regions of the Far North, such as the Arkhangelsk region, which is one of the largest producers of wood pellets in Russia.

1. Introduction

The problem of replacing fossil fuels is extremely important for the "areas of the Far North and equated localities with limited delivery periods" of Russia (further in the text "Far North"). Such areas include completely or in part the territories of 25 regions of the Russian Federation. The area of regions completely or partially belonging to the territory of the Far North is 11.25 million square km that is equal to 65.7 % of the total territory of the Russian Federation (figure 1). The number of permanent residents living in the territories of the Far North is 9.9 mln people or 6.8 % of the total population of the Russian Federation [1]. The Russian economy essentially depends on the resources obtained in the territories of the Far North. 76 % of Russian oil, 93 % of natural gas, 95 % of coal, 95 % of gold, 100 % of diamonds, 100 % of salmon, and many other useful resources are produced on these territories [2]. A presence of large cities in the regions of the Far North is characteristic for Russia. For example, in the territory to the north of the Arctic Circle, there are 5 large cities with a

Content from this work may be used under the terms of the Creative Commons Attribution 3.0 licence. Any further distribution of this work must maintain attribution to the author(s) and the title of the work, journal citation and DOI.
Published under licence by IOP Publishing Ltd

population of more than 50 thousand people: Murmansk (298 thousand people), Norilsk (178 thousand), Vorkuta (58 thousand), Apatity (56 thousand), Severomorsk (51 thousand). Murmansk is the largest city in the world beyond the Arctic Circle, and Norilsk is the most northern city with a population of more than 100 thousand people [3]. To the south of the Arctic Circle, it is necessary to note Yakutsk with a population of 311 thousand people, which is the largest city in the Far North of Russia [1].

The energy supply of the Far North is a very difficult task. 33 % of territory of Russia, mainly on the Far East and Siberia, are not connected to the Unified Energy System of Russia and located on the Far North Territories [4]. At present, this problem is solved by delivering fuel by water and road transport ("Northern delivery" further in the text ND, "Severniy zavoz" in Russian), which significantly increases its cost. In remote settlements with extremely expensive power supply from local power plants operating on fossil fuels, the energy costs for end-users are up to 50 % of local GRP. The energy supply is possible only because of the huge state subsidies (more than 50 billion rubles per year) [2]. Currently, a correct calculation of the total volume of the ND is complicated due to change in the accounting system that occurred in 2014 and the methodology of calculation requires development. This complicates the assessment of the volume of renewable resources that could replace fossil fuels in the ND.

Given the high cost of electricity and heat in the Far North, the use of local renewable energy resources is an important way to increase the energy efficiency and diversify energy sources. This is reflected in the major documents in the field of energy in Russia: "The Energy Strategy of Russia for the period up to 2030" and "The State Program on Energy Efficiency and Energy Development" [5, 6]. Russia leads the world in the forest area that is located mainly in the Far North territories (809 million hectares as of 2010, 20 % of the world's forest area) [7]. Given a huge amount of biomass in Russia, the bioenergy industry could play an important role in the power supply of the Far North territories. A number of regions of Russia located in the Far North have large bioenergy resources. Nevertheless, the role of bioenergy in Russia is now surprisingly low [8]. The goal of this paper is to estimate the volume of fossil fuel delivery to the Far North and compare it with potential technical bioenergy resources for a transition of the local power systems to the local renewable resources.

2. Materials and methods

The estimation of the ND composition as well as volume of the main categories of industrial goods and products represents a methodological problem. Until 2014, the Federal plan for statistical works included a separate section for statistical survey in areas of the Far North and equated localities with limited delivery periods. It was formed in accordance with the list of areas adopted by the Government of the Russian Federation [9] and published in the official statistical yearbook of the Federal State Statistics Service (Rosstat) [10]. According to mentioned list the data were aggregated by the form No. 1-SB sever (urgent) "Data on incoming delivery of consumer goods and industrial products to areas of the FN and equated localities with limited delivery periods for supply of the population, the enterprises and the organizations of the social sphere, housing and utilities infrastructure".

Currently these data collected only by the Federal Agency for Maritime and River Transport (Rosmorrechflot) and published in the aggregated form in the statistical yearbook "Transport and communication in Russia". Data on the volumes for some industrial goods and products are published in statistical yearbook "Regions of Russia. Socio-economic indicators" as well. In this regard the authors developed a method for ND volume estimation in 2015. The method is based on the following principles.

Firstly, for the regions of the Russian Federation that are completely or partially located on the territory of the Far North, we used the information from a statistical yearbook of Rosstat (data on volumes for the main categories of industrial goods and products in the context of subjects of the Russian Federation [11]). Secondly, for each region of the Russian Federation that is located in the FN, we calculated the weighted-average coefficients for 2005–2014. These coefficients were defined as the ratio of volumes of the major categories of industrial goods and products delivered to the Far

North to the total volume of industrial goods and products delivered to the region. These coefficients for each region were extrapolated for 2015. According to these coefficients, we determined ND volume in 2015 based on data of statistical yearbooks of Rosstat [11, 12]. For verification of the method we compared the official statistical data for the period between 2010 and 2014 [10] with ND volumes calculated by our method for the same period. It showed the minimum deviation and reliability of elaborated method. For analysis of the composition of water transport to the Far North we used the official data of Rosstat for the relevant time periods [11, 12]. For the assessment of budget expenditures for purchase and delivery of fuel and food products to the areas of the Far North we used the information from the published budgets of Russian regions for the corresponding years taking into account the regional funds allocated for these purposes and subsidies from the budget of the Russian Federation. Federal budget expenditures were determined using the data on the "equalisation transfer" from federal to regional budget in accordance with article 131 of the Budget code of the Russian Federation. The data on the Mezen district were obtained from the report prepared by the Government of the Arkhangelsk region on the results of the Northern delivery in 2015 and the data obtained by Alsufiev & Shnayder [13].

The primary energy content of different fuels was calculated using the following lower heating values: gasoline – 43.6 MJ/kg, winter and arctic diesel fuel – 43.6 MJ/kg, summer diesel fuel – 43.4 MJ/kg, coal – 27 MJ/kg [14]. The technical bioenergy potential of Federal Districts of Russia data was obtained from the paper published by Namsaraev and coauthors [8]. The technical biomass potential considers all biomass theoretically available in the Russian Federation that can be technically supplied during one year.

Figure 1. The map of the areas of the Far North.
Bold lines show the borders of the Federal Districts of Russia: 1 – Central, 2 – North Western, 3 – Southern, 4 – Volga, 5 – Ural, 6 – Siberian, 7 – Far Eastern. The boundaries of the Federal Districts are shown as of 19 January 2010, before the North Caucasian Federal District was separated from the territory of the Southern Federal District. Blue area – regions of the Russian Federation referred in the list of the areas of the FN. A – Arkhangelsk region. Dark blue area on the territory of Arkhangelsk region – municipal districts belonging to Far North. Shaded area – the area technologically isolated from the Unified Energy System of Russia [8, 9].

3. The analysis of the ND volume

The analysis of the ND volume in 2015 showed that the major categories of goods delivered to the Far North are goods for the technical and industrial purposes (3048.3 thousand tons), including fuel, and food (69.3 thousand tons). The total volume of various types of fuel is 2897.2 thousand tons or 95.1 % of the total volume of the ND. The major type of fuel is a diesel fuel, which makes up to 61.6 % of the ND. It includes summer (34.8 %), winter (9.8 %) and Arctic diesel fuel (17 %). Coal is the second most important type of energy source and accounts for 26.3 % of the ND. Motor gasoline is in the third place (7.2 %) (figure 2). The estimated energy content of the different types of fuels is equal to 112.7 PJ (diesel 81.5 PJ, gasoline 9.6 PJ, coal 21.6 PJ).

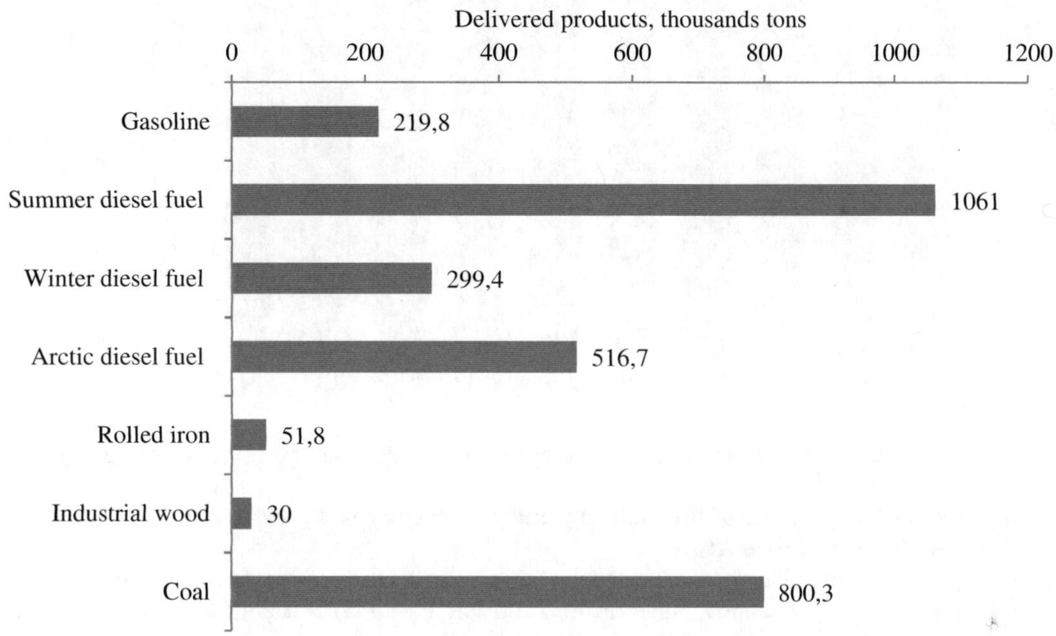

Figure 2. The composition of the ND of industrial goods by major categories in 2015 in thousands tons

The major type of transport used for ND is a water transport. In 2015 by the water transport 20383.6 thousand tons were delivered to the Far North that is 5 % less than in 2014 and 27 % less than in 2008. The volume of deliveries by water transport considerably exceeds the volume of ND because it also includes goods that are not considered as ND (building materials, consumer goods, etc.). Most of the cargo (84 %) was delivered using the inland waterway transport. The dry cargo dominated in the ND and accounted for a little less than 80 %, the share of the liquid bulk cargo was 12 % and the rest was delivered in rafts (figure 3). In recent years, the dynamics of the volume of delivered products has a downward trends.

Also, it is necessary to note the heterogeneity of the regions of the Far North in terms of existing transport infrastructure and type of cargo delivery. In most regions of the Far North, the transport system does not provide year-round accessibility of settlements by a land transport, and the delivery of goods is carried out mainly during the navigation period by a water transport.

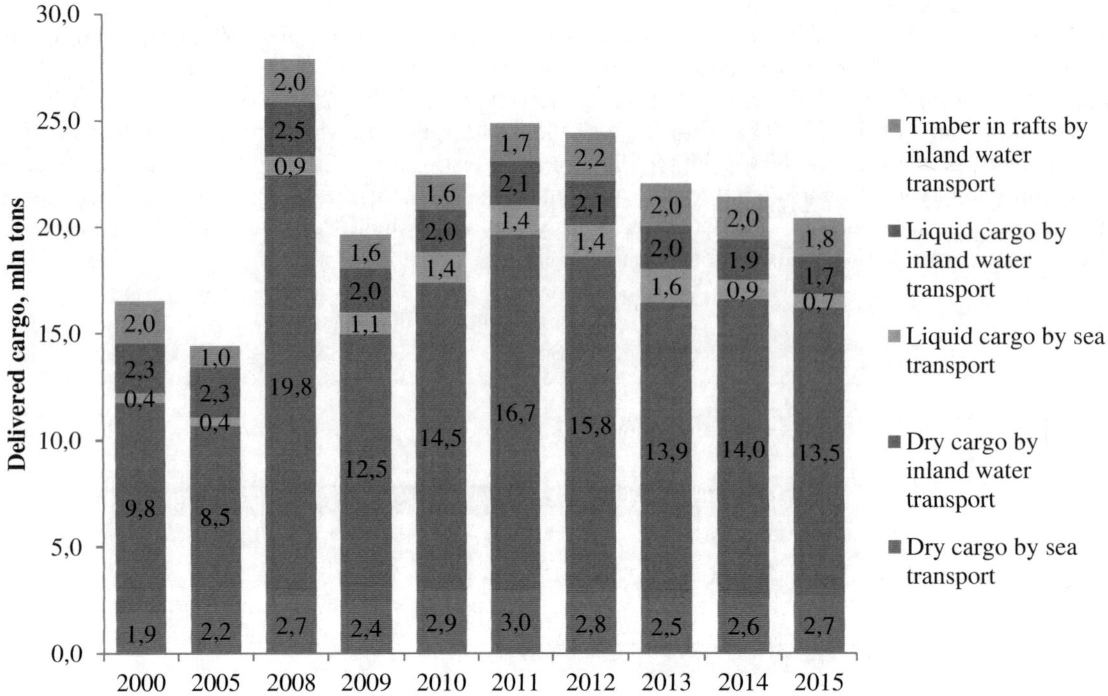

Figure 3. Delivery of cargo to the Far North by a water transport (millions of tons)

4. Comparison of the volume of fuel delivery and bioenergy resources of the Far North Federal districts

The territories of the Far North are located in the northern parts of the four Federal districts (FD) including Northwest FD, Ural FD, Siberian FD, and Far East FD. The dominant types of bioenergy resources in these districts are forest residues, which are also the most convenient for transportation among other types of biomass resources [8]. Therefore, only forest residues data will be used for analysis in this article. In the Northwest FD, the technical potential of forest waste is 133.5 PJ, Ural FD 50.2 PJ, Siberian FD 130.9 PJ, and Far East FD 41.3 PJ. The total technical potential of forest waste in these districts is 355.9 PJ. Thus, the bioenergy potential of forest residues of these FD is more than 3 times higher than the energy content of fuels delivered to Far North (112.7 PJ).

Substitution of fossil fuel supplies by the renewable resources has a number of positive aspects. Firstly, it will solve a number of environmental problems associated with the accumulation of waste in areas with a developed forest industry. Secondly, the use of the existing infrastructure of the ND will create a new domestic market for forest residues, which can stimulate a further development of the bioenergy industry. The possibility of transition to renewable energy resources should be taken into account during the modernization of the energy infrastructure in the Far North, including the possibility of using renewable resources for heating. This process has already begun in several regions of Russia, in which the number of boiler stations using forest residues (wood pellets) is gradually increasing.

5. Arkhangelsk region as an example of the transition to renewable energy sources

The most active process of the transition to the renewable energy resources takes place in regions with a well-developed wood pellet industry, such as the Arkhangelsk region [15]. The northeastern part of the Arkhangelsk region which includes Mezensky, Pinezhsky, and Leshukonsky municipal districts is classified as the territory that belongs to the Far North and Mezensky region also belongs to Arctic zone of the Russian Federation [9]. The transition to the local renewable energy sources in these areas will increase the local market for pellets and support the local bioenergy industry since the opportunities to increase export of pellets from Russia are currently limited. Proskurina and coauthors analyzed the pellet-export potential of the Northwestern FD [16]. This FD has the highest level of the pellet production; it is close to seaports and major export markets (Denmark and Sweden). However, when considering an increase, the pellet exports from this region face lots of problems throughout the supply chain.

Local energy resources in the Arkhangelsk region are mainly represented by low-grade wood, wood processing waste (chips, sawdust), pulp and paper industry waste, wood pellets. In 2013, they accounted for 36 % of energy resources used in municipal energy generation. In 2013, 320 boiler stations (out of 750) used them as the main fuel, and 72 more as additional. However, in the northernmost Mezensky district, coal-fired boiler stations generate 96 % of the thermal energy and wood-burning boiler stations – only 4 %. In 11 municipal districts and cities of the region, the share of firewood, pellets, chips and wood waste exceeds 40 % of the energy balance of the boiler stations, and by 2030 the Arkhangelsk region plans to completely abandon the use of fossil fuel in the municipal boiler stations by maximizing the use of forest residues [17, 2]

6. Conclusion

Optimization of the economic mechanisms of the ND through the use of renewable local energy resources, modernization of outdated power plants, and the increase of energy efficiency are one of the major priorities of the "Strategy for the development of the Arctic zone of the Russian Federation and ensuring national security for the period until 2020" [18]. However, on the regional level the major measures to support the ND are focused on creating favorable conditions for the delivery of fuel. Only in rare cases it is planned to develop the local production of biofuels and promote the use of renewable energy resources. Our analysis showed that the bioenergy potential of the FD that includes the territories of the Far North is 3 times higher than the volume of the ND of fuel. This makes the partial transition of the Far North to renewable energy sources theoretically possible, which will also help to solve a number of environmental problems related to the accumulation of waste from a forest industry. The use of the existing infrastructure of the ND will allow creating a new local market for the forest residues and increasing the efficiency of the bioenergy industry. The possibility of transition to renewable energy resources should be taken into account during the modernization of the energy infrastructure in the Far North, including the possibility of using renewable resources for heating. The process of replacement of fossil fuels by renewable biofuels has already begun in a number of regions of the Far North, such as the Arkhangelsk region, which is one of the largest producers of wood pellets in Russia.

Acknowledgements
This work was supported by Ministry of Education and Science of the Russian Federation (project No. 14.574.21.0148, identifier RFMEFI57417X0148).

References

[1] 2016 *The population on municipal districts of the Russian Federation for January 1, 2006*: statistical yearbook (Moscow: Rosstat)
[2] *Analysis of the current situation of isolated energy systems with high energy costs* (Center for Energy Efficiency (CENEf))
[3] Laruelle M and Hohmann S 2017 *Polar Geography* **40** 306–323
[4] 2014 Russian Government Regulation No. 854 *On the approval of the Rules of operational dispatching management in the electric-power industry*
[5] 2009 Russian Government. Decree No. 1715-r *On Approval of the Energy Strategy of Russia for the period up to 2030*
[6] 2014 Russian Government. Decision No. 321 *On Approval of the State Programme on Energy Efficiency and Energy Development*

IOP Conf. Series: Earth and Environmental Science **177** (2018) 012022 doi:10.1088/1755-1315/177/1/012022

[7] 2010 *Food and Agriculture Organization of the United Nations*. The global forest resources assessment Main report

[8] Namsaraev Z B, Gotovtsev P M, Komova A V and Vasilov R G 2018 *Renewable and Sustainable Energy Reviews* **81** 625–34

[9] 2000 Russian Government. Decree No. 402 *On Approval of the list of areas of the Far North and equated localities with limited delivery periods* (*consolidated*)

[10] 2014 *Regions of Russia. Socio-economic indicators* (section 2.4.4. Delivery of products (goods) to areas of the Far North and equated localities) (Rosstat, Moscow)

[11] 2016 *Regions of Russia. Socio-economic indicators 2016*: statistical yearbook (Moscow: Rosstat)

[12] 2016 *Transport and communications in Russia 2016*: statistical yearbook (Moscow: Rosstat)

[13] Alsufiev A V and Schneider A G 2016 *Arctic: ecology and economy* **3** 58–66

[14] Abyutin A A 1998 *Thermal calculation of boiler units. Normative method* (Saint Petersburg: NPO CKTI)

[15] 2015 *Complex wood utilization in Arkhangelsk region*

[16] Proskurina S, Rimppi H, Heinimö J, Hansson J, Orlov A, Raghu K C and Vakkilainen E 2016 *Renew Sustain Energy Rev* **56** 38–50

[17] 2015 Analytical report. *Assessment of the prospects and expediency of the transition of the subjects of the Russian Federation using petroleum products for the purpose of heat supply to local and renewable fuels* (Analytical Center under the Government of the Russian Federation)

[18] 2013 Russian President. Decree No. Pr-232 *On Approval of the Strategy for the development of the Arctic zone of the Russian Federation and ensuring national security for the period until 2020*

Improvements to motor fuel taxation in Russia as impetus for sustainable development of cities

E Iadrennikova

Department of Financial and Tax Management, Ural Federal University named after the first President of Russia B. N. Yeltsin, 19 Mira str., Ekaterinburg, 620002, Russia

e-mail: elena.yadrennokova@mail.ru

Abstract. The article considers the role of the transportation system in the sustainable development of cities. The article pays special attention to the development dynamics of road transport in Russia and of the emissions of air pollutants by motor vehicles. The role and significance of fuel taxes in the system of government revenue are investigated. The author analyzes the existing system of fuel taxes in the Russian Federation and concludes that the mechanism of taxing motor fuel is based on consumption and does not take into account the adverse impact on the environment and the energy content of fuels. The author provides arguments for changing the mechanism of calculating the rate of the excise tax on motor fuel so that it factors in its energy efficiency and carbon dioxide emissions. A conclusion is drawn that the introduction of the proposed adjustments would produce a substantial fiscal impact on government revenues and would make it possible to eliminate the existing distortions for two competing sources of energy (petrol versus diesel fuel); would encourage the use and development of public transport and help reduce the emissions of air pollutants. The objective of the study is to work out proposals for improving fuel taxation that would also contribute to the sustainable development of cities.
Keywords: excise tax, petrol, diesel fuel, tax rates, road transport, CO_2 emissions, fuel efficiency.

1. Introduction

The sustainable development of cities has become a widely accepted strategy over the past decade as the larger part of the world's population. In Russia, for example, 74 % of people are city dwellers. Sustainable development is the sum total of multiple factors including
- economic well being that acts as the development driver for modern cities;
- urban environment and infrastructure (including the transport system) that form the living conditions and standards of the population;
- environmental conditions.

The role of the transport system in the harmonious development of cities cannot be overestimated. The transport system determines the look and viability of a modern city [1]. According to Vuchic [2], transport is the "the life force" of cities that connects all other subsystems and functions, including economic, social and other ones. Consequently, the efficiency of the urban transport system determines to a large extent the effective and reliable operation of other systems in a modern megalopolis.

Content from this work may be used under the terms of the Creative Commons Attribution 3.0 licence. Any further distribution of this work must maintain attribution to the author(s) and the title of the work, journal citation and DOI.
Published under licence by IOP Publishing Ltd

Road transport plays the most important role in the transport systems of modern cities. From 2000 to 2016, the car fleet in Russia grew 110 %, primarily because of an increase in the number of passenger cars (Table 1). In 2016, the share of cars in the total vehicle stock was 87.5 %. As the passenger car fleet grows, the use of public transport declines considerably. From 2000 to 2016, the number of public transport passenger journeys declined by 59.6 %, while passenger journeys by tram and trolleybus – the most environmentally friendly modes of public transport – decreased 5.6-fold. As a result, privately owned passenger cars account for 90 % of road traffic today. The share of public transport and commercial vehicles is below 10 % [3].

Table 1. Motor vehicles and passenger journeys made by public transport in Russia

Indicator	2000	2005	2010	2015	2016	% of growth 2016 on 2000.
1. Motor vehicles, in thousand units.	24863	30497	39926	50605	51635	207.7
incl.:						
- passenger cars	20353	25570	34354	44200	45163	221.9
- trucks	4401	4848	5414	6230	6300	143.1
- buses	109	79	158	175	172	157.8
2. Passenger journeys made by all modes of public transport, in million.	43367	28724	21013	17953	17488	40.3
incl.:						
- by bus	23001	16374	13434	11523	11296	49.1
- tram, trolley bus	16180	8776	4285	3094	2880	17.8
- metro	4186	3574	3294	3336	3312	79.1

The growth in the number of privately owned cars brings about negative external effects that are most pronounced in big cities. The quality of atmospheric air has deteriorated considerably in some urbanized areas of the country, worsening public health problems and driving up mortality [4, 5]. Motor vehicles are one of the main sources of toxic pollutants and greenhouse gases (including CO_2) that harmful to health. Cities are more prone to higher concentrations of pollutants because of motor vehicles [6]. Carbon dioxide (CO_2) is the main constituent of the exhaust gases from internal combustion engines. Growing atmospheric concentrations of carbon dioxide causes climate change [7]. Pallavidino, Prandi et al [8] consider passenger cars to be a major source of CO_2 emissions.

Our analysis of the air pollutant emissions in Russia in 2000-16 shows that the volume of emitted pollutant did not increase, but the structure of emissions sources changed. Stationary sources have been able to decrease emissions, while motor vehicles accounted for 45.1 % of air pollutant emissions in the country in 2016, up from 41.8 % in 2000 (Table 2).

Table 2. Emissions of air pollutants in Russia

Indicator	2000	2005	2010	2015	2016	% of growth 2016 on 2000.
Emissions of pollutants into the atmosphere, million tons	32.3	35.8	32.7	31.3	31.6	97.8
Emissions of air pollutants from stationary sources, million tons	18.8	20.4	19.1	17.3	17.3	92.0
Emissions of air pollutants from motor vehicles, million tons	13.5	15.4	13.6	14.0	14.3	105.9
Share of vehicle emissions in the total emissions of pollutants into the atmosphere	41.8	43.0	41.6	44.7	45.1	107.9

Privately owned passenger cars are the biggest contributors to air pollutants emissions. Moreover, Hollantz and Tamms [9] showed that a high concentration of cars in a city not only leads to lower efficiency of the entire transport system of the city, but generally decreases the quality of life and safety for the entire urban community.

Mayburov and Leontyeva [1] posit that all cities in Russia are in serious need of effective taxation and administration measures for regulating the development of various modes of city transport. The main goal of such regulation is to reduce the use of privately owned vehicles in urban areas and to drive up the use of public transport. Some economic studies [10, 11] have proved the effectiveness of indirect taxes as a regulatory tool for decreasing the emissions of CO_2 and other pollutants. Indirect taxes create pricing stimuli that force consumers to change their "polluting" behavior and act in a more eco-friendly and energy efficient way.

The excise tax on energy products (petrol, diesel fuel, motor oils) is included in the price of motor fuel and affect the cost of the journey. Today, the fuel tax rates in Russia do not factor in fuel efficiency and CO_2 emissions. Changing the mechanism of computing fuel tax rates by taking into account the above factors would increase the tax rates and the price of motor fuels and, consequently, provide an incentive for motorists to use public transport. As a result, one should expect a substantial increase in government revenues, lower air pollutants emissions, improvements in public health and better efficiency of the urban transport system. In other words, positive changes would occur in all key components of the sustainable development of cities.

2. Research methods

The author used various theoretical and empirical research methods. The theoretical methods employed included analysis, synthesis, generalization and classification. The analysis of the system of energy taxes in Russia showed that the mechanism of taxing motor fuel is based on the volume consumed but does not take into account negative environmental impacts and the energy content of the fuel.

The empirical methods that included observation and comparison were used for identifying main economic trends and substantiating and selecting a mechanism of computing the fuel tax rate in Russia that would factor in fuel efficiency and the level of CO_2 emissions.

Works by Russian and foreign scholars served as the methodological and theoretical foundation of the study. The list of data sources for the study included statues and regulations, data of the Federal State Statistics Service (gks.ru) and the European Commission (ec.europa.eu), the press, online resources and the author's own research findings.

3. Results and Discussions

3.1. Fiscal importance of fuel taxes in the Russian Federation

Excise taxes on motor fuel fall into the category of indirect energy taxes. Table 3 indicates the role and importance of excise taxes on fuel in the system of government revenues in Russia.

Table 3. Environmental tax revenue by type of tax

Indicator	2012	2013	2014	2015	2016	% of growth 2016 on 2012.
Environmental taxes (excl. VAT and customs duties), % of GDP	4.9	5.1	4.7	4.8	4.5	-0.4
Excise taxes on fuel, % of GDP	0.6	0.6	0.5	0.4	0.6	–
Excise taxes on fuel, billion RUB. incl.:	358.8	412.9	395.3	314.5	485.6	135.3
- petrol	219.9	224.6	239.0	201.1	319.2	145.2
- diesel fuel	126.2	173.0	139.5	102.1	154.8	122.7

Over the assessed period, government revenues from environmental taxes (excluding VAT and customs duties) as a share of GDP varied from 4.5 % to 5.1 %. Over the five-year period, the figure went down 0.4 %, but it still considerably higher than in the EU member states (2.3 % to 2.5 % of GDP). The mineral extraction tax contributes the biggest share to government revenues from environmental taxes, accounting for an average of 3.8 % of GDP. Indirect environmental taxes generate much smaller revenue. For example, fuel tax revenues as a share of GDP vary from 0.4 % to 0.6 %, which is much lower than in the EU (1.9 % of GDP on average across the European Union). It possible to conclude that the potential of environmental excise taxes is not fully utilised in the Russian Federation.

The primary taxable energy products that account for over 95 % of fuel tax revenues in Russia are petrol and diesel fuel. In 2016, revenues from the excise tax on petrol made up 65.7 %, or two thirds of total government revenues from fuel taxes. Meanwhile, the consumption of diesel fuel by all types of motor vehicles is four times higher than the consumption of petrol (18.5 m tonnes versus 4.6 m tonnes). This indicates a competitive distortion between the two main types of motor fuel.

Revenues from the excise tax on petrol grew the fastest between 2012 and 2016 (145.2 %). Revenues from an excise tax are usually an outcome of the consumption of the taxable product and changes in the tax rates. From 2012 to 2016, the actual volume of petrol consumption did not change considerably. In 2013, for example, petrol consumption increased 3.6 %, but then decreased annually by 0.2 to 0.8 %. The tax rate for petrol increased by 48.5 % to 80 % (Table 4) depending on the type of petrol. The slower growth in revenues from petrol tax, by contrast with the increase in the tax rate, is due to a transition to production of better-quality petrol that is taxed at a lower rate. Consequently, there has been an environmental effect of the tax because the consumption of greener petrol reduces the harmful impact on the environment.

In 2015, there was a drop in tax revenues from both petrol and diesel fuel for road use. That was due to a " tax maneuver" that is essentially a cut in export customs duties and excises imposed on oil and oil products along with a hike in mineral resources extraction tax and a decrease in the rates on petrol (by 15 to 35 % depending on its environmental class) and diesel fuel (by 45 %). The maneuver was aimed at preventing an increase in the prices of oil products in the domestic market. The cut in the tax and customs duty rates was executed in line with an agreement of the Eurasian Economic Union on the establishment of a common market for oil products and crude oil and unification of export duties.

3.2. Excise fuel taxes in Russia today

Petrol has been on the list of excisable products in Russia since 1996. Originally, the tax was levied as a percentage of the sales price. Starting from 1998, the tax is levied as a fixed rouble-per-tonne rate and varies based on octane levels. In 2001, diesel fuel was placed on the list of excisable products, and the tax rates grew considerably.

The most significant reform of fuel taxation took place in 2011 when a transition was made from tax rates pinned to octane ratings to tax rates varying by emissions standard type. Additionally, motor oils, straight-run gasoline, and other fuels were included in the list of excisable energy products in 2011. Between 2015 and 2018, there has not been any considerable changes to the list of excisable products (Table 4).

In Russia, like in European countries, fuel taxes are charged on a per unit volume of the consumed fuel that is used as the tax base. The tax base does not reflect the amount of pollutants that the fuel contains, yet the tax rates vary by fuel environmental class. The rates are lower for greener fuels (Euro 5 petrol) and higher for the types of petrol that are below the Euro 5 standard.

Since the existing tax rates are based on fuel consumption, but do not take into account the carbon footprint of the fuel and its energy content, the approach results in distortions for competing sources of energy, for example, petrol versus diesel fuel. Diesel fuel is far more efficient than petrol and causes a lot more damage to the environment. Some expert assessments find that petrol for road use generates negative environmental impacts of 5.9 euro cents per km, while diesel fuel generates 7.7 euro cents worth of environmental damage per km. However, both in Russia and the EU member states diesel fuel is taxed at a lower rate than petrol despite the former's higher energy efficiency and environmental hazard. In Russia, however, the tax rate for petrol and that for diesel fuel vary at a wider margin. In 2018, the tax rate charged on petrol that is below the Euro 5 emissions standard was 70 % higher than the tax rate for diesel fuel, whereas in the EU the minimum tax rate for petrol is 27.6 % higher than that for diesel fuel. The actual tax rate difference in most EU countries is even narrower. In Russia, however, diesel fuel tax rates have been growing faster (up 120 % between 2015 and 2018) than those for petrol and much faster than the rate of inflation.

Table 4. Fuel tax rates in Russia, RUB per tonne

Excisable product	2015	2016	2017	2018	% of growth 2018 on 2015
Petrol:					
✓ below Euro 5 standard	7300	10500	13100	13100	179.5
✓ Euro 5	5530	7530	10130	11213	202.8
Diesel	3450	4150	6800	7665	222.2
Motor oils	6500	6000	5400	5400	83.1
Straight-run gasoline	11300	10500	13100	13100	115.9
Benzene, paraxylene, ortho-xylene, jet fuel	2300	3000	2800	2800	121.7
Middle distillates	x	4150	7800	8662	x

The disproportion is due to the fact that petrol is primarily consumed by individuals using cars, while diesel fuel is largely consumed by transportation businesses (most of freight vehicles and buses are powered by diesel fuel). A lower tax rate for diesel fuel and, consequently, a smaller share of the tax in the selling price work as an incentive for road freight transport and alleviates the tax burden for the cargo industry. Non-commercial consumption of diesel fuel has, however, been growing. Higher demand for diesel fuel has been pushing prices up in comparison with petrol prices.

The cost of producing diesel fuel is much lower than the cost of refining oil into petrol. The reason lies in the technology that apart from straight distillation incorporates such costly processes as isomerization, reforming, catalytic cracking with hydrotreatment, and alkylation. The diesel fuel production process only includes fractional distillation and hydrotreatment [7]. The selling prices of diesel fuel and petrol and the narrow gap between them do not match the relevant cost of production and are largely due to growing demand for diesel fuel in the market. At present, petrol stations sell petrol and diesel fuel at the same prices despite the fact that the cost of diesel fuel production and the excise tax on diesel fuel are much lower than that of petrol. Consequently, it is oil refineries, middlemen and petrol stations who get to keep the additional revenues generated by the higher demand for diesel fuel, whereas the government is losing on the insufficiently low tax rate.

The excise tax accounts for a fairly high share of the selling price of petrol and diesel fuel in European countries – from 30.5 % in Hungary to 46.4 % in the UK for diesel fuel, and 33.6 % in Hungary to 50 % in the Netherlands for petrol. In Russia, the excise tax accounts for an average of 24 % of the selling price of petrol and 16 % of the selling price of diesel fuel. Given the considerably lower tax rates on main fuels in Russia than in Europe, there is enough room for an increase.

In general, the motor fuel taxation mechanism does not appear to be optimal and needs improving.

3.3. Ways of improving excise taxes on motor fuel in Russia

The mechanism of charging environmental taxes in Russia and the EU countries is based on fuel consumption and does not take into account negative environmental impacts and the energy content of the products being consumed. In the Russian Federation, the environmental damage caused by the combustion of fuels is reflected in tax rates that vary by fuel eco-class and fuel use. Consequently, both in Russia and the EU, motor fuels and other energy products are taxed on the basis of their environmental impacts that are, however, not entirely computed in the taxes. As a result, there are distortions in the taxation of competing fuels (petrol versus diesel fuel) that show in unreasonably low tax rates on diesel fuel compared with petrol and the non-receipt of substantial tax revenues by the government.

A design of fuel excise tax that factors in the environmental impacts of fuel consumption should help strike a match between the tax burden and the environmental pollution caused by the taxed product, and therefore make taxation fairer and equitable. The implementation of such an approach calls for a change to the mechanism of motor fuel taxation. It is necessary to eliminate distortions between competing fuels and create incentives for better energy efficiency and emissions reduction.

The author believes that taxes on motor fuels must take into account their energy content and environmental impacts. Some authors [4, 7] propose introducing a carbon tax. Today, a tax on CO_2 emissions is used in Denmark, Ireland, Finland and Sweden, but it is not harmonized at EU level. It is worth taking a look at the European Commission's proposal to revise the Energy Tax Directive and adopt energy taxes that would be split into two components:

- one component would be based on the emissions of CO_2 from the energy product and would be charged on a per-tonne basis;
- the other component would be based on the energy content of the product, rather than its use. The minimum rate was proposed to be set at a euro-per- GJ basis.

In practice, though, it might be difficult for the taxpayer to determine the tax base (to measure CO_2 emissions and the amount of generated energy) in compliance with the proposed mechanism.

That being said, the system of fuel taxation in the Russian Federation should be designed with the following considerations in mind:

1. The tax base remains the same: the volume of consumed fuel expressed in tonnes.
2. When computing the tax rate it is necessary to factor in CO_2 emissions and energy content.
3. CO_2 emissions from the combustion of fuel depends on its density. The density of petrol is around 0.75 kg/L; the density of diesel fuel is around 0.84 kg/L. Carbon dioxide emissions of petrol-powered cars amount to 3.134 tonnes-per-tonne, and 3.174 tonnes-per-tonne in diesel-powered cars. The calorific value of petrol is 30.8 MJ/L, or 41.06 GJ/t. The calorific value of diesel fuel is 36.3 MJ/L, or 43.22 GJ/t. The per-tonne charge for CO_2 emissions is set at 575.6 RUB; the per-GJ charge for emitted energy is set at 11,305 RUB.
4. The tax rate for petrol remains unchanged.
5. The tax rate for diesel fuel will be 13,730 RUB per tonne.

The fiscal impact of the revised tax will be significant provided that the production and consumption of diesel fuel in Russia remain unchanged and the tax rate is set at 13,730 RUB per tonne of diesel fuel (an equivalent of 221.5 dollars per tonne). In that case, annual revenue from the excise tax is projected to grow by 122.5 bln RUB (an equivalent of 1,975.5 m dollars).

If implemented, the proposal would encourage the consumption of energy from sources which produce less CO_2. The energy component of the tax rate would make it possible to eliminate the existing distortion of competition between petrol and diesel fuel. Taxation of fuel and energy products tied to energy content and CO_2 emissions would encourage a more effective use of energy resources and reduce carbon dioxide emissions because the proposed mechanism sends a clear price signal to the consumer about the real energy value of the product he consumes. It would also be unnecessary to introduce a separate tax on carbon dioxide emissions.

The downside of a higher excise tax on diesel fuel is that it would drive up selling prices and, consequently, the expenditures of individuals and transportation companies and transportation costs in other industries, including agriculture. The price growth would be less significant compared to the hike in the excise tax if oil refineries and petrol stations reduced their profit margin. Higher transportation costs incurred by agricultural producers could be smoothed out with subsidies for agriculture businesses in order to keep agricultural commodity prices down.

The higher cost of motor fuel will spur a transition to more efficient and environmentally friendly vehicles and alternative fuels (natural gas, propane, electricity). Motorists will be encouraged to use public transport. Haulage businesses will have to streamline logistics. All of this will reduce car ownership rates and motor traffic in big cities, improve the environmental situation and public health and make the urban transport system more effective.

4. Conclusion

The analysis of the system of excise taxes on motor fuel for road use in Russia shows that the taxation mechanism is based on consumption and does not take into account negative environmental impacts and energy content. Moreover, there is distortion of competition between the two main types of motor fuel – petrol and diesel fuel – that shows in unreasonably low tax rates charged on diesel fuel compared with petrol and the non-receipt of substantial tax revenues by the government. The existing motor fuel tax system fulfills the fiscal and regulatory functions of taxation quite effectively, is not optimal and needs improvement.

Excise tax tied to the energy efficiency and carbon dioxide emissions of fuel could generate additional public revenue to an amount of 122.5 bln RUB (an equivalent of 1,975.5 m dollars) annually provided that the production and consumption of diesel fuel in Russia remain unchanged. Taxes are a powerful instrument for influencing taxpayers' behavior with economic means and encouraging them to use "green" fuel and environmentally friendly vehicles. The implementation of the proposed novelties will drive up tax rates and the selling prices of diesel fuel, thus reducing its consumption and spurring a transition to more efficient and environmentally friendly vehicles and alternative fuels. One should, therefore, expect an increase in public revenue, a decrease in pollutant emissions, a reduction in car traffic, improvements in the environmental situation and public health in cities and better efficiency of urban transport systems, which will eventually contribute to the sustainable development of cities.

Acknowledgements

The work was supported by the Russian Foundation for Basic Research, contract No. 17-22-21001.

References

[1] Mayburov I and Leontyeva Y 2017 Fiscal instruments for regulating the sustainable development of urban transport systems in Russia *International Conference on Sustainable Cities (ICSC 2017) IOP Conference Series: Earth and Environmental Science* **72(1)** 012016

[2] Vuchic V R 1981 *Urban Public Transportation: Systems and Technology* (Englewood Cliffs, NJ: Prentice-Hall)

[3] Mayburov I and Leontyeva Y 2015 Transport tax in Russia as a promising tool for the reduction of airborne emissions and the development of the road network *WIT Transactions on Ecology and the Environment* **198** 391–401

[4] Golubeva A and Magaril E 2013 Improved economic stimulation mechanism to reduce vehicle CO2 emissions *Urban Transport XIX* (C A Brebbia WITPress) 485–95

[5] Magaril E, Abrzhina L and Belyaeva M 2014 Environmental damage from the combustion of fuels: Challenges and methods of economic assessment *WIT Transactions on Ecology and the Environment* **190** 1105–15

[6] Magaril E, Magaril R and Abrzhina L 2017 Environmental assessment of the measures increasing the sustainability of motor transport *International Conference on Sustainable Cities (ICSC 2017) IOP Conference Series: Earth and Environmental Science* **72(1)** 012003

[7] Magaril E and Golubeva A 2016 Economic stimulation to reduce vehicle CO2 emissions *Bulletin of Ural Federal University Series Economics and Management* **15(3)** 359–81

[8] Pallavidino L, Prandi R, Bertello A, Bracco E and Pavone F 2014 Compilation of a road transport emission inventory for the Province of Turin: Advantages and key factors of a bottom-up approach *Atmospheric Pollution Research* **5(4)** 648–55

[9] Hollantz and Tamms 1965 *Die Kommunalen Verhehrsprobleme in der Bundesrepublik Deutschland* (Essen, Germany: Vulkan-Verlag)

[10] Baumol W J and Oates W E 1971 The use of standards and pricing protection of the environment *Swedish Journal of Economics* **73** 42–54
[11] Baumol W J 1972 On taxation and the control of externalities *The American Economic Review* **62(3)** 307–22

IOP Conf. Series: Earth and Environmental Science **177** (2018) 012024 doi:10.1088/1755-1315/177/1/012024

Analysis of the power quality impact in power supply system of Urban railway passenger transportation – the city of Ulaanbaatar

V Z Manusov, U Bumtsend and Yu V Demin

Department of Power Supply System, Novosibirsk State Technical University, Novosibirsk, 630073, Russia

E-mail: uuya@mail.ru

Abstract. The urgency and expediency of additional passenger transportation by urban railway as an extension of the transport infrastructure of megacities has been proved. The fact that the urban railway uses AC 2x25 kV, which is uses to out of the three phases of city's power supply system negatively impacts. Thus, it creates a significant asymmetry. In this paper, a comparative analysis on the negative sequence component of the voltage unbalance, which occurs in the nodes of connection of traction substations through both a three-phase transformer or a Scott transformer are been carried out. A mathematical method was created for determining the negative sequence voltage unbalance factor is based on the phase coordinates and symmetrical components methods. The results show that the Scott transformer provides a more symmetrical load distribution to the three-phase electric grid, when compared to the three-phase transformer and was verified using a simulated model in the Matlab program. Consequently, the use of a Scott transformer has better symmetrising abilities than a three-phase transformer, which gives it a significant advantage in the traction power systems.

1. Introduction

The growth of the world population leads to a significant increase in the urban population in most countries, including new the birth and growth of megacities. One of the problems arising in the development of megacities is the complication of passenger transportation, which requires new types of transport and significant additional material and financial costs. The most approachable solution regarding this task is the development of urban railways, including the connection with other passenger transport systems: subway, bus and air transportations. However, up until now, negative technical problems are not fully examined, such as quality of electric power that arise from the fact that modern electric trains operate on an AC 2×25 kV and use only two out of three phases. This entails an unbalance of voltages and currents in the feeding electric grids of megacities, as well as the higher harmonic components of the alternating current. In this paper, the results of a scientific study are presented specifically the asymmetric load flow caused by feeding electrified urban railway, in connection with the fact that the traction power supply system has a number of phases not multiples of three.

It is proved that among the traction transformer connections the Scott and Le Blanc transformers have the best symmetrising abilities, which ensure the voltage balance on the side of primary voltage, under condition equal currents in the feeding sections on the side of the secondary voltage.

Content from this work may be used under the terms of the Creative Commons Attribution 3.0 licence. Any further distribution of this work must maintain attribution to the author(s) and the title of the work, journal citation and DOI.
Published under licence by IOP Publishing Ltd

In this case, the most simple and practical way to balance the voltage is to tight train schedule.

2. Urgency and brief historical overview

Ulaanbaatar – the capital of Mongolia, is located in the broad valley of the Tuul River, at the foot of the Bogd-Uul mountain, at an altitude of 1300–1350 m. The city's area is 4704.4 km², at the moment the city's population is 1.405.000 people, and it is separated into the independent, administrative, political, commercial, cultural and scientific center of Mongolia and is also an important financial center of Asia. It is a megacity and a major transport hub. The city is served by the international airport of Genghis Khan and two airfields, and this year it is promised to put into operation a new international airport. The central railway station of the Ulaanbaatar is the center of regional and international railway service [1].

Officially, the capital of Mongolia is almost 400 years old, but Ulaanbaatar – looks like a very young city since the master plan for Ulaanbaatar development was only developed in 1975. Then the specialists predicted that the capital's population will not exceed 400 thousand people, and no more than 10 thousand cars will participate in the traffic.

Now Ulaanbaatar is on the 451st place in the list of the largest cities in the world. Forecasts were not justified: according to official data of the Traffic Police Department of Mongolia, in 2017 more than 458 thousand cars were registered in the capital. Every day more than 200 thousand cars leave on the roads of Ulaanbaatar. Since almost all governmental and state organizations are located in the center of Ulaanbaatar, there is morning and evening hours of traffic peaks, the morning peak continues from 7:30 to 9:00 and the evening peak from 17:00 to 21:00 hours. Therefore, one of the headaches of city residents is traffic jams.

To date, one of the most urgency problems for the world's largest cities is traffic jams. Measures are offered in different ways. Some cities increase the number of paid parking lots and automobile interchanges, somewhere they offer drivers to travel by public transport for free. To reduce traffic jam, the city administration has taken several measures. One of them is restriction of movement of vehicles. Since 2014 in Ulaanbaatar cars can move around the city on weekdays taking turn, depending on the data indicated on the state plates. And try to solve this problem there are several ways.

The development of modern megacities also requires the development of the transport system, along with buses, trolleybuses and the subway, in recent years the delivery of passengers has become much simpler in some cities and is getting cheaper by using the urban railway or as we call by municipal train. The essence of railway transport is that it passes located above ground, does not require large expenditures. In many cities, these lines already exist, they just need to be merged and run in circle.

Table 1. Prospects for a new urban railway in the city of Ulaanbaatar

№	Station	Additional benefits
1	Tolgoit	1. Train time en route: from Tolgoit to Amgalan – 45–47 minutes, travel time by public transport is not less than 90 minutes
2	Tavan shar	
3	Bars-2	
4	Ulaanbaatar railway station	2. Number of passengers carried: 350–400 people / one train
5	Dund gol	3. Periodicity of cruising: on the route to each side during the day – 3 trains
6	Narantuul	
7	Janjin club	4. Speed of the passenger train: up to 100 km/h
8	Amgalan	

In 2014, according to the joint project of the Ulaanbaatar City Hall and the Russian-Mongolian Joint Stock Company "Ulaanbaatar Railway", on the 65th anniversary of its founding, it was decided

IOP Conf. Series: Earth and Environmental Science **177** (2018) 012024 doi:10.1088/1755-1315/177/1/012024

to create a new public transport network based on the existing railway infrastructure in the city [2]. At the moment there is only one line, the first stage of construction has been built four new intermediate stations with low platforms, two terminal stations have been reconstructed – the Tolgoit station and Amgalan station, and the central station – Ulaanbaatar.

Unfortunately, this method of transporting passengers was not fully realized in because of the costly system of transportation of people, the rare frequency of trains and the uncomfortable waiting conditions. One of the main reasons for the collapse of the project is that the movement of the trains was carried out on diesel locomotives, operating on expensive diesel fuel.

3. Choice of traction power system for electrified urban railways

In the near future, the Ulaanbaatar Railway JSC plans to electrify the central railway line, to create a transport corridor between Europe and Asia, through Mongolia. This will open the possibility for Ulaanbaatar megacity to electrify the internal urban railway, which will reduce the traffic load in the city by 30 %.

Based on the development of railway electrification in the world and in neighboring countries, Russia and China, are taking into account the increase in the transport of cargo, both domestic and transit for the electrification of the Mongolian railway, it is appropriate to choose the traction system of 2x25 kV, 50 Hz of alternating current [3].

Small changes in phase voltages can lead to more asymmetries in the phase currents. This causes heat and loss and makes the grid less stable. In order to remain within the permissible limits specified by the standard and the electric utility, the consumer is obliged to search for optimal solutions for ensuring the quality of electric energy.

In real conditions, traction substations tend to be located at large railway substations. At the same time, according to profile conditions, inequality of distances between substations, and changes in cargo traffic, even the average loads of the substations are different. When considering the effect of the asymmetrical load of a three-phase system on the operation of electrical power system and consumers, it will be noted that the voltage unbalance is determined to a large extent by the voltage drop in the wires of the transmission line [4–6]. The latter depends on the load and location of the traction substations. It is not difficult to come to the conclusion that, as if the substations were not connected to a three-phase line, it is not possible to obtain identical voltage losses in all phases, since asymmetrical loads are located at different distances from power supplies. The load of traction substations continuously varies over a wide range and almost never the loads of individual substations are equal to each other. Thus, in practice, the problems of current and voltage asymmetry are solved only on the basis of concrete conditions [7].

The most common types of transformers used in Russia are three-phase and single-phase transformers. Lately in Europe transformer Scott and Le Blanc is being commonly used, which transforms the three-phase system into a two-phase 25 kV. There are also two systems of not widely spread so-called V/V system and also a Y/Δ system. As will be shown below, they don't fulfill the symmetry requirement on the primary side.

The main advantage of the Scott transformer is the fact that when the currents in two feeding sections of the secondary winding of the transformer (25 kV) are equal, the voltage on the upper side becomes practically balanced. Figure 1 shows the phasor diagrams of the voltage and current of the Scott transformer [8, 9].

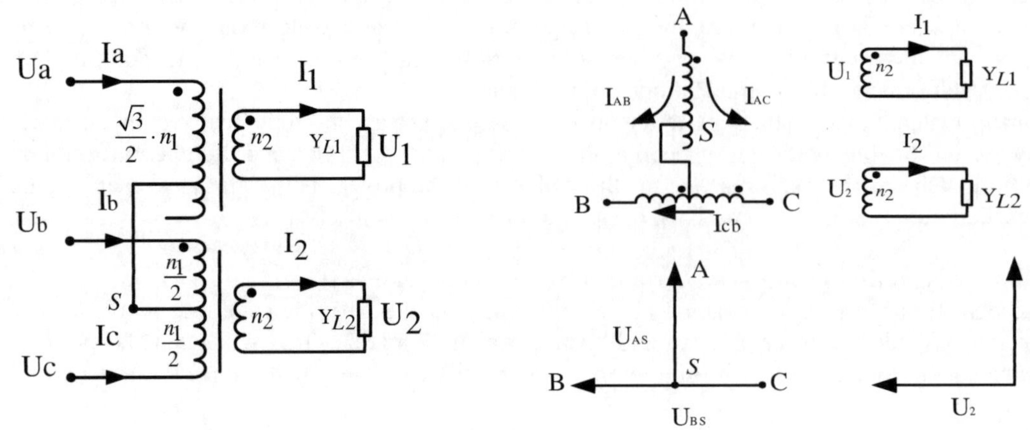

Figure 1. Scott connection transformer and its phasor diagram

4. Calculation of the voltage unbalance factor in the power grid

In order to compare the ability to ensure symmetry of the transformer connections, we use the symmetrical component method. The C.L Fortesque recommended this method in 1918. He states in his paper, the most striking thing about the results of his mathematical investigation of induction motors under unbalanced conditions was their symmetry. The result of his inquiry is the method of symmetrical components, the essential means for analyzing unbalanced conditions in power systems.

The relation between primary and secondary voltages and currents of the transformer are shown in the following equations

$$I_1 = Y_{L1}U_1 = \frac{2}{\sqrt{3}}\beta\cdot(U_A - \frac{1}{2}U_B - \frac{1}{2}U_C)\cdot Y_{L1}, \quad (1)$$

$$I_2 = Y_{L2}U_2 = \beta\cdot(U_B - U_C)\cdot Y_{L2}, \quad (2)$$

where:

Y_{L1} and Y_{L2} are traction loads in the two feeding section

β – factor of transformation, defined by the formula $\beta = \frac{n_2}{n_1}$.

After some transformations, we get the currents in the phases of the primary side, depending on the unbalance of currents on the secondary side of the Scott transformer

$$I_A = \frac{4}{3}\beta^2\cdot Y_{L1}U_A - \frac{2}{3}\beta^2\cdot Y_{L1}U_B - \frac{2}{3}\beta^2\cdot Y_{L1}U_C$$

$$I_B = -\frac{2}{3}\beta^2\cdot Y_{L1}U_A + (\frac{1}{3}\beta^2\cdot Y_{L1} + \beta^2\cdot Y_{L2})U_B + (\frac{1}{3}\beta^2\cdot Y_{L1} + \beta^2\cdot Y_{L2})U_C$$

$$I_C = -\beta^2\cdot Y_{L1}U_A + (\frac{1}{3}\beta^2\cdot Y_{L1} + \beta^2\cdot Y_{L2})U_B + (\frac{1}{3}\beta^2\cdot Y_{L1} + \beta^2\cdot Y_{L2})U_C$$

Given $\beta^2\cdot Y_{L1} = Y_1$ and $\beta^2\cdot Y_{L2} = Y_2$ the relation between the U_{ABC} and the I_{ABC} is defined as

$$\begin{bmatrix} I_A \\ I_B \\ I_C \end{bmatrix} = \begin{bmatrix} \frac{4}{3}Y_1 & -\frac{2}{3}Y_1 & -\frac{2}{3}Y_1 \\ -\frac{2}{3}Y_1 & \frac{1}{3}Y_1 + Y_2 & \frac{1}{3}Y_1 - Y_2 \\ -\frac{2}{3}Y_1 & \frac{1}{3}Y_1 - Y_2 & \frac{1}{3}Y_1 + Y_2 \end{bmatrix} \cdot \begin{bmatrix} U_A \\ U_B \\ U_C \end{bmatrix} \tag{3}$$

Transforming (3) into symmetrical components, the relation between U_{120} and I_{120} can be obtained as follows

$$\begin{bmatrix} I_1 \\ I_2 \\ I_0 \end{bmatrix} = \begin{bmatrix} Y_1 + Y_2 & Y_1 - Y_2 & 0 \\ Y_1 - Y_2 & Y_1 + Y_2 & 0 \\ 0 & 0 & 0 \end{bmatrix} \cdot \begin{bmatrix} U_1 \\ U_2 \\ U_0 \end{bmatrix} \tag{4}$$

The three-phase equivalent circuit represents the traction load on the secondary side of the transformer and can be referred to the primary side. The circuit is directly connected to the power grid.

It can be used, drawing up the circuit of the positive, negative and zero sequences and an equivalent circuit, to obtain a simplified formula for the quickly evaluation of voltage unbalance. If we assume that the power grid is balanced then it can be replaced by the Thevenin's equivalent circuit.

$$\begin{bmatrix} E_A \\ E_B \\ E_C \end{bmatrix} + \begin{bmatrix} E_n \\ E_n \\ E_n \end{bmatrix} = \begin{bmatrix} Z_P & Z_M & Z_M \\ Z_M & Z_P & Z_M \\ Z_M & Z_M & Z_P \end{bmatrix} \cdot \begin{bmatrix} I_A \\ I_B \\ I_C \end{bmatrix} + \begin{bmatrix} U_A \\ U_B \\ U_C \end{bmatrix} \tag{5}$$

where Z_P – phase impedance;

Z_M – mutual impedance;

U_A, U_B, U_C – phase voltages at the point of common coupling.

Thanks to the symmetric transformation

$$\begin{bmatrix} E_S \\ 0 \\ 0 \end{bmatrix} = \begin{bmatrix} Z_S & 0 & 0 \\ 0 & Z_S & 0 \\ 0 & 0 & Z_0 + 3Z_n \end{bmatrix} \cdot \begin{bmatrix} I_1 \\ I_2 \\ I_0 \end{bmatrix} + \begin{bmatrix} U_1 \\ U_2 \\ U_0 \end{bmatrix}, \tag{6}$$

where $Z_1 = Z_2 = Z_S = Z_P - Z_m$ the impedance of the forward and reverse sequences, $Z_0 = Z_P + 2Z_m$ is the zero sequence impedance.

An equivalent circuit of symmetrical components will be compiled by combining equation (6) with equation (3). The results will be used to obtain a formula for evaluation the voltage unbalance at the point of common coupling.

Table 1 shows the evaluation formulas used to calculate the maximum voltage unbalance factor for some types of transformer connections. The advantage of the evaluation formulas is that it provides a quick and easy way to calculate the maximum voltage unbalance factor at the point of common coupling.

In this case, the ratio of the negative and the positive sequence voltages can be approximately defined as the ratio of the load on the secondary side in the two-phase system, the three-phase short-circuit capacity $S_{SC(3p)}$ in the three-phase system [10]. This three-phase short-circuit capacity is an important parameter when designing a power system. For better voltage regulation, the short-circuit capacity must be high enough.

The formulas obtained for the comparison of transformer connections used in traction substations are based on two assumptions. At first, the feeding system is considered to be a three-phase symmetrical system. Second, the equivalent impedance of the traction load (contact network, electric locomotive, rail) is much larger than the impedance of the network (source, transmission line).

Table 2. The formulas of voltage unbalance factor for various traction power supply

Transformer connection	Voltage unbalance factor, %
Single phase	$k_{2U} = \frac{S_{1p}}{S_{SC(3p)}} \cdot 100$
Scott Le Blanc Woodbridge	$k_{2U} = \lvert 1-2k \rvert \left\lvert \frac{S_{1p}}{S_{SC(3p)}} \right\rvert \cdot 100$
V/V Y/Δ Δ/Δ	$k_{2U} = \sqrt{3k^2 - 3k + 1} \left\lvert \frac{S_{1p}}{S_{SC(3p)}} \right\rvert \cdot 100$

In the formulas there is a multiplier besides the power ratio, which determines the difference of the transformers connection and divides all the transformer connections into three categories. The coefficient k is within the limits $0 \le k \le 1$. A comparison of the voltage unbalance factor for various types of transformer is shown in Figure 2.

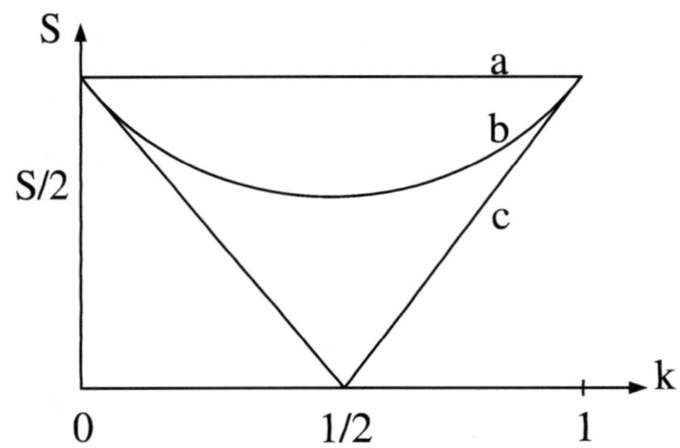

Figure 2. The comparison diagram of voltage unbalance factor of the transformer connections, a – single phase transformer, b – three phase transformer, c – Scott transformer

There is an urgent problem of installing and improving the device for load currents balance in the feeding sections of the Scott transformer. Thus, the usual problem of voltage balancing in a three-phase system is transferred from the grid 110–220 kV to the grid 25 kV. In this case, the balancing problem itself is also simplified, since it is necessary to balance not three phases but two. Other transformer connections of traction power supply have greater asymmetry on the primary side compared to the Scott and Le Blanc connection. This method is simpler and requires less financial costs, since it can partly be realized by the correct train schedule. For example, on the railways in Japan and South Korea, they strictly adhere to the schedule and thus solve the problem of unbalance in high-voltage grids.

5. Symmetrisation of the load flows based on the equalization of currents in the feeding sections

As an example, was calculated the voltage unbalance factor in the node of the city's electric grid, where connected the Tolgoit traction substation, with capacity 40 + j15 MVA, from which feed on 8 urban railway stations of the Ulaanbaatar.

The results of calculation current and voltage unbalance factor, when the current inequality in the traction feeding sections, are given in the table 3. With an inequality of the currents up to 10 %, the current unbalance on the primary side remains at the maximum permissible values. The divergence of the currents in phase instead of 90 degree, creates unbalance in the supply grid also at the level within limit values.

Table 3. Results of the calculation of voltage unbalance factor of three phase and Scott transformer

Parameters of the grid	**Three phase transformer**				**Scott transformer**			
	0	**10 %**	**20 %**	**30 %**	**0**	**10 %**	**20 %**	**30 %**
S_{L1}, MVA	21.3	23.4	25.6	27.7	21.3	23.4	25.6	27.7
S_{L2}, MVA	21.3	19.2	17.1	14.9	21.3	19.2	17.1	14.9
I_A, A	85.6	89.4	93.4	97.5	112.1	100.9	89.6	78.4
I_B, A	32.3	32.8	34.3	36.5	112.1	118.1	124.8	132.1
I_C, A	85.6	82.1	78.7	75.7	112.1	118.1	124.8	132.1
φ_A	170.3	173.0	175.5	177.8	69.4	69.4	69.4	69.4
φ_B	69.4	59.0	50.3	41.9	-50.5	-45.8	-41.6	-37.8
φ_C	-31.4	-28.0	-25.3	-21.8	-170.5	-175.2	-179.5	176.7
I_1	64.7	64.7	64.7	64.7	112.1	112.1	112.1	112.1
I_2	32.3	32.8	34.3	36.5	0	11.2	22.4	33.6
k_{2I}, %	50.00	50.74	52.91	56.34	0	10.0	20.0	30.0
k_{2U}, %	2.34	2.40	2.61	2.96	0	0.49	0.99	1.46

6. Conclusion

It is shown that with the development of the transport infrastructure of megacities, the technical and economic problems of optimizing passenger flows due to above ground urban railway transport. However, due to the fact that the 2×25 kV traction supply system has a different physical basis, the use of two phases instead of three, it introduces unbalance into three-phase city electric grid of the megacity. This reduces the quality of electrical energy and increases losses in electric consumers for household and industrial purposes.

The researches carried out by the authors on the example of the Ulaanbaatar megacity and results show that the Scott and Le Blanc transformers have an advantage in reducing the asymmetry compared to the three-phase transformers. It is very important to ensure the equality of currents in the feeding sections of the balancing transformers, since there is no asymmetry in the electric grids. When an inequality of currents from 10 % to 30 %, the voltage unbalance factor in the 220 kV grid is for the three-phase transformer from 2.34 % to 2.96 %, which exceeds the permissible limit 2 %, while at the same time for the Scott transformer from 0.49 % to 1.46 %, which satisfies the allowable level of 2 %.

The resulting coupling equations for input and output variables for transformer connections utilized in traction substations make it possible to determine k2U, when given the traction load on the secondary side of a transformer (25 kV). Also, using the method determining contributes of voltage unbalance from each elements of electric grid [11], we determined the negative sequence voltage unbalance factor at the point of evaluation. The results of the calculation were adequate with the simulation results, which confirm the correctness of the proposed method [12].

References

[1] National Statistical Office of Mongolia
[2] City Railbus in service from today
[3] Manusov V Z, Palagushkin B V and Bumtsend U 2016 The electrification of international transport railway corridor from Asia to Europe as a way of power system sustainable development of Mongolia *Journal of Transsib Railway Studies* **4** (28) 94–101
[4] Astahov Y N et al 1983 *Electric power systems in examples and illustrations: Manual for higher education institutions* Under V A Venikov's edition (Moscow: Energoatomizdat)

IOP Conf. Series: Earth and Environmental Science **177** (2018) 012024 doi:10.1088/1755-1315/177/1/012024

[5] Manusov V Z, Bumtsend U and Erdenebat E 2017 The analysis and optimization of the states of electric power system of Mongolia taking into account electrification of the Ulanbator main railway *New in the Russian electric power industry* **10** 55–66
[6] Manusov V Z, Bumtsend U and Tretyakova E S 2016 Optimization Compensating Devices in the Power Supply Systems Using Population Algorithms *11th International Forum on Strategic Technology (IFOST)* 276–80
[7] Bei Yu M and Mamoshin R R 1986 *Traction substations* (Moscow: Transport)
[8] Miao Yao-shan and Cao Dongbai 1998 *Electrical Railway Traction Power Supply System Design Manual* (Beijing: China Railway Publishing company)
[9] Burchi G, Lazaroiu C, Golovanov N and Roscia M 2005 Estimation of Voltage Unbalance in Power Systems Supplying High Speed Railway Electrical power Quality and Utilisation **11** (2) 113–9
[10] 2013 IEC TR 61000-3-13 Electromagnetic compatibility (EMC) – Part 3-13: *Limits – Assessment of emission limits for the connection of unbalanced installations to MV, HV and EHV power systems*
[11] Jayatunga U, Perera S, Ciufo P and Agalgaonkar A P 2014 A review of recent investigations on voltage unbalance management: Further contributions to improvement of IEC/TR 61000-3-13:2008 *16th IEEE International Conference on Harmonics and Quality of Power* (*ICHQP*) 268–72
[12] Chernyh I V 2007 *Modeling of electric power devices in MATLAB. Sim Power Systems u Simulink* (Moscow, DMK Press)

IOP Conf. Series: Earth and Environmental Science **177** (2018) 012025 doi:10.1088/1755-1315/177/1/012025

The thermal mode of the covered yard

A Rymarov and M Razakov

Moscow State University of Civil Engineering, 26 Yaroslavl highway, Moscow, 129337, Russia

e-mail: rymarov@list.ru

Abstract. The covered yards in the cities with big durations of the cold period of year are useful for people to stay in rather favorable conditions of a microclimate that increases ability of each person to carry out more useful work, study or other creative activity. The covered yards relieve the person of extra wind loading and children to spend time on the street more comfortably. Formation of the covered yards allows to have not freezing sandboxes in which children can play all the year. The research in this article is dedicated the thermal mode of the covered yards.

1. Introduction

The cities in Russian Federation, during the winter period, are covered with snow, roads in ice, playgrounds are difficult for use, streets round houses are filled with cars, snow on roadsides of roads has a black shade because of receipts of dangerous compounds from motor transport, so how to resolve this problem? Ten thousands of municipal services cars works to clean snow, to bring and melt it at special snow melting stations in cities in winter time [1].

There are many buildings in the center of St. Petersburg with courtyards on which perimeter are usually located low buildings where there are arches for pass of people As it is known the city of St. Petersburg is located on the Neva river and on the shore of the Finnish Gulf which the breeze as the movement of air from a surface of the water towards to the land. The strong wind during the cold period of year with a rain or wet snow causes unpleasant temperature feelings of the people moving on city streets. Existence of double-exit courtyards allows people to take cover from a wind that increases comfort of residence in the city. The wind flow of buildings with courtyards brings to reassignment of a wind stream on streets round the building and above the building with formation of vortex zones of an aerodynamic trace [2] that improves a wind situation in courtyards.

Existence of the courtyard covered with a translucent roof allows to receive, during the cold period of year, comfortable space to rest for adults and children where is not wind, the air temperature is above the external temperature, a minimum of snow, ice and also not freezing sandboxes for children. At desire, visitors from such yards can pass into parks which are actively formed on all districts of the cities such as Moscow and St. Petersburg.

In recent years there were many buildings with atriums and big spaces for people [3, 4] where it is possible to spend with convenience the time throughout the day. But it isn't enough of it for such megalopolises as Moscow and St. Petersburg.

Content from this work may be used under the terms of the Creative Commons Attribution 3.0 licence. Any further distribution of this work must maintain attribution to the author(s) and the title of the work, journal citation and DOI.
Published under licence by IOP Publishing Ltd

2. Methods of calculating the thermal conditions

Let's consider the building with the covered yard with a close to square form in the plan – its width is rather great from all directions and the wind flow for this kind of building has character as for the wide building when the windy vortex zone of an aerodynamic trace is formed partially on a roof of the building and on a back windy facade. A wind flow of various form buildings by air is connected with aerodynamics of city building [2, 5].

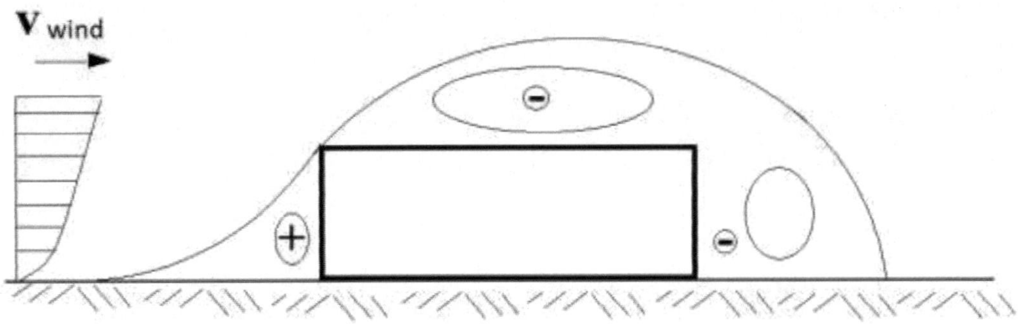

Figure 1. Picture of formation of vortex zones of an aerodynamic trace, where (+) – excessive pressure of air, (-) – small discharge of air

Vortex zones of an aerodynamic trace which are formed at a wind flow of building by air create adverse conditions for finding of the person in these zones that is shown in fig. 1, and the adverse wind situation in the Moscow and St. Petersburg is shown in fig. 2 and 3. It is possible to get rid of these zones at construction of buildings with conveniently streamline shape, or close areas with people by a translucent roof.

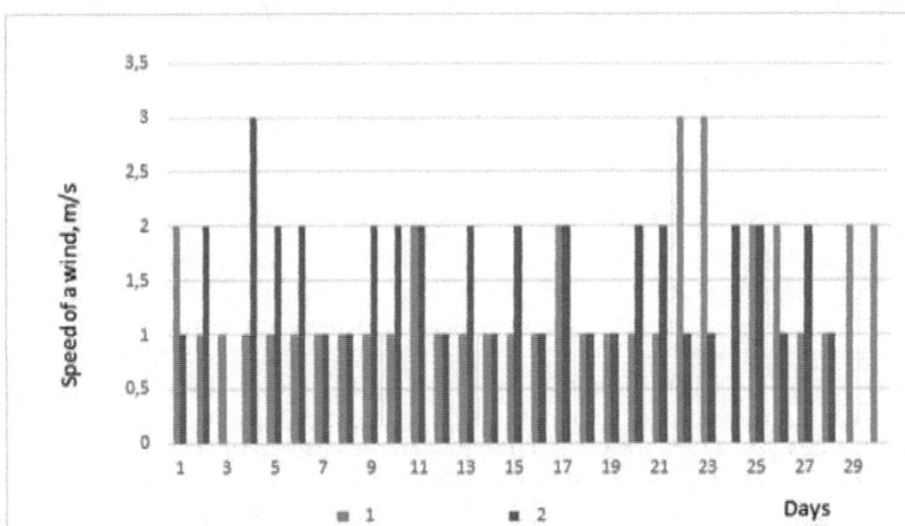

Figure 2. Change of wind's speed in Moscow in November, 2017 (1) and February, 2018 (2)

During formation of the covered yard it is possible to carry out calculations of the thermal mode of internal air to understand the comfort level when the person is inside of this kind of space.

During the cold period of year the loss of warmth from the walls and windows to a courtyard and losses of warmth through translucent roof and ground allow to get the calculations of temperature condition in the covered yard. It is desirable to warm passes through arches for people moving to a courtyard from abutting streets for reduction of warmth losses.

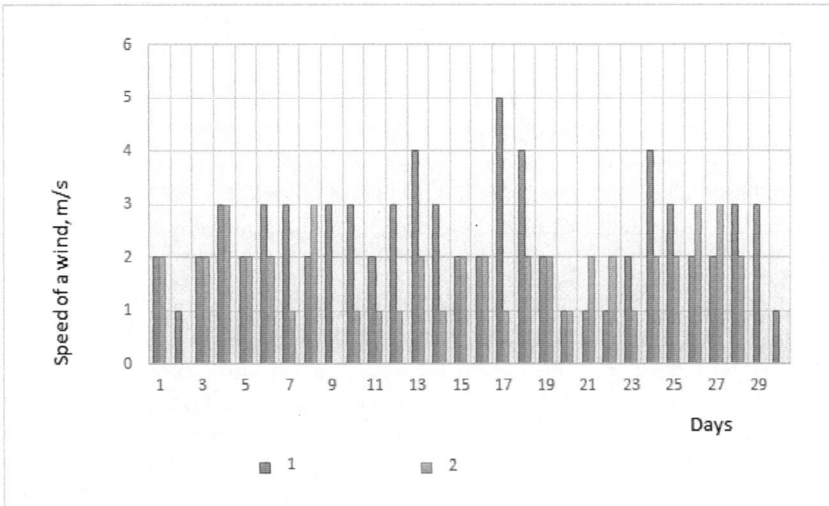

Figure 3. Change of wind's speed in St. Petersburg in November, 2017 (1) and February, 2018 (2)

Management of a microclimate in the covered yard is defined by consideration of the air, thermal and gas modes [6, 7].

3. Results of calculations

Let's consider the building with a square form courtyard, fig. 4, the section of this kind of building is presented in fig. 5.

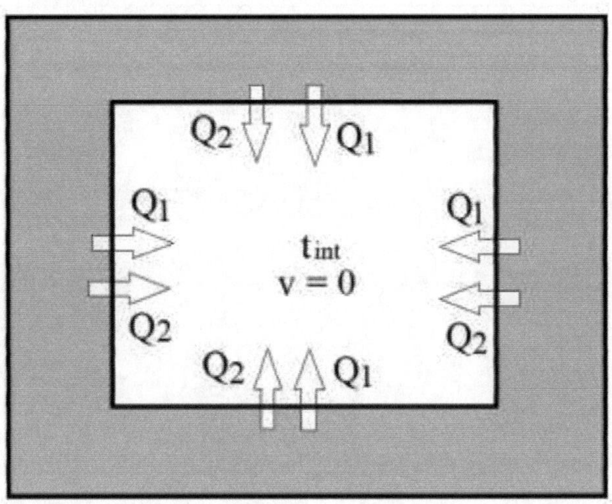

Figure 4. The building plan with the covered yard, where tint – air temperature in the covered yard °C, Q1 и Q2 – thermal streams into the space of the covered yard, W, v – speed of wind, m/s

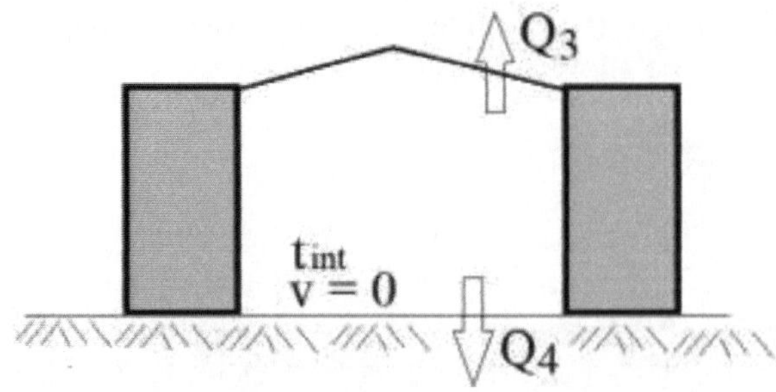

Figure 5. A building section with the covered yard,
where Q3 и Q4 – the thermal streams leaving space of the covered yard, W

The covered yard size is 63 × 63. Its surrounded with the building with external sizes of 87 m on each external facade. The building has 5 floors (with not heated attic). The area of the windows which are leaving in a courtyard is 840 m^2, the area of the walls turned into a courtyard is 3393.6 m^2, the area of a roof and the area of ground which are in the covered yard is 3960 m^2.

The thermal stream from walls and windows to the covered yard (Q1 and Q2), and the warmth losses of the covered yard through a translucent roof and ground (Q3 and Q4) that is shown in fig. 4 and fig. 5.

According of laws in the Russian Federation SP 50.13330.2012 "Thermal protection of buildings" and SP 131.13330.2012 "Construction climatology" resistance to a heat transfer of windows and walls of the building respectively are 0/49 and 2/99 (м^2 × °C)/W and a translucent roof of the covered yard is 0.36 (м^2 × °C)/W.

There are the balance equations (1, 2) of losses and receipts of warmth in the covered yard which allowing to determine temperature in the covered yard.

$$\sum Q_1 + \sum Q_2 = Q_3 + Q_4 \tag{1}$$

$$f(t) = f(Q_1, Q_2, Q_3, Q_4) \tag{2}$$

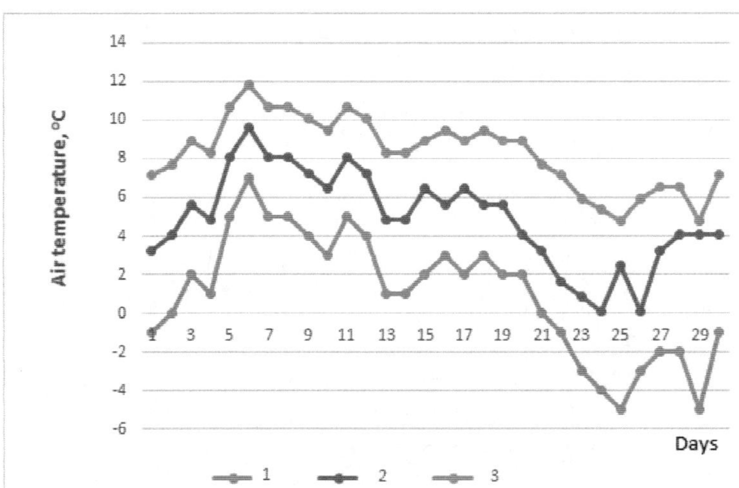

Figure 6. Change of temperature of external air and air in the covered yard
in Moscow in November, 2017,
where 1 – external air, 2 – air in the covered yard and 3 – air in the covered yard 2

Calculations are carried out for temperature conditions of Moscow and St. Petersburg for the real values of temperature of external air in November, 2017 and February, 2018 has taken from the existing meteorological data. The air temperature in the rooms of the building is accepted +20 °C. As the yard is blocked by a translucent covering and this additional resistance to a heat transfer, the protecting designs turned into a courtyard, can have the lowered resistance to a heat transfer that will allow to increase still temperature of internal air in a courtyard during the cold period of year.

Results of calculations are presented on schedules figures 6–10 where "The covered yard" and "The covered yard 2" is the temperature of internal air in the covered yard at the specified values of resistance to a heat transfer and the lowered resistance to a heat transfer respectively.

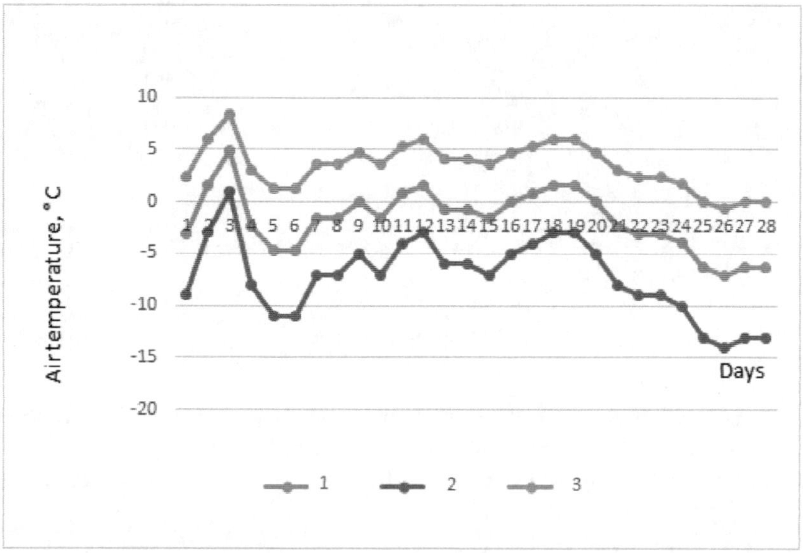

Figure 7. Change of temperature of external air and air in the covered yard in Moscow in February, 2018, where 1 – air in the covered yard, 2 – external air and 3 – air in the covered yard 2

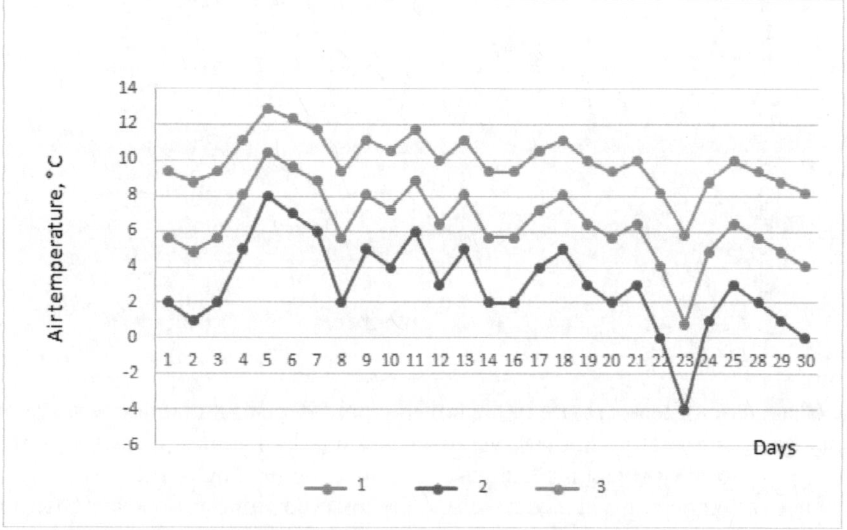

Figure 8. Change of temperature of external air and air in the covered yard in St. Petersburg in November, 2017, where 1 – air in the covered yard, 2 – external air and 3 – air in the covered yard 2

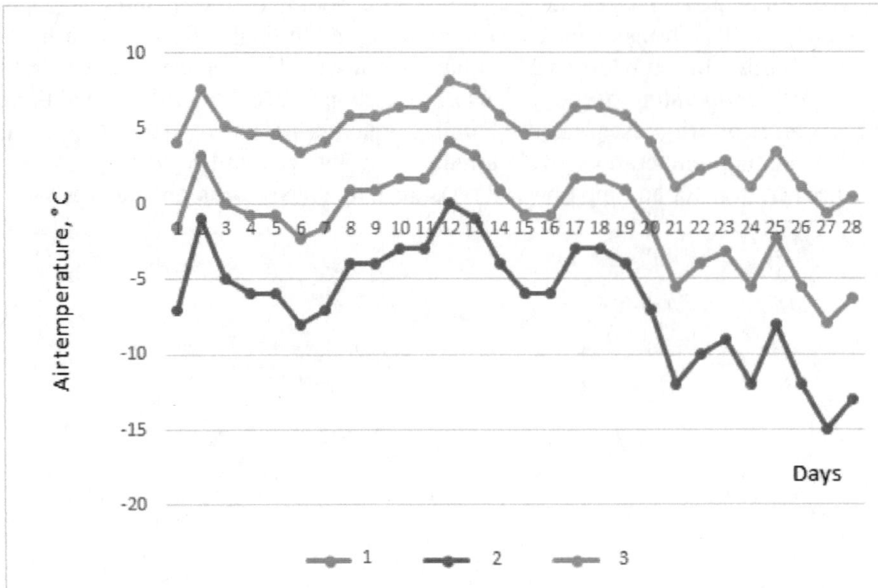

Figure 9. Change of temperature of external air and air in the covered yard in St. Petersburg in February, 2018, where 1 – air in the covered yard, 2 – external air and 3 – air in the covered yard 2

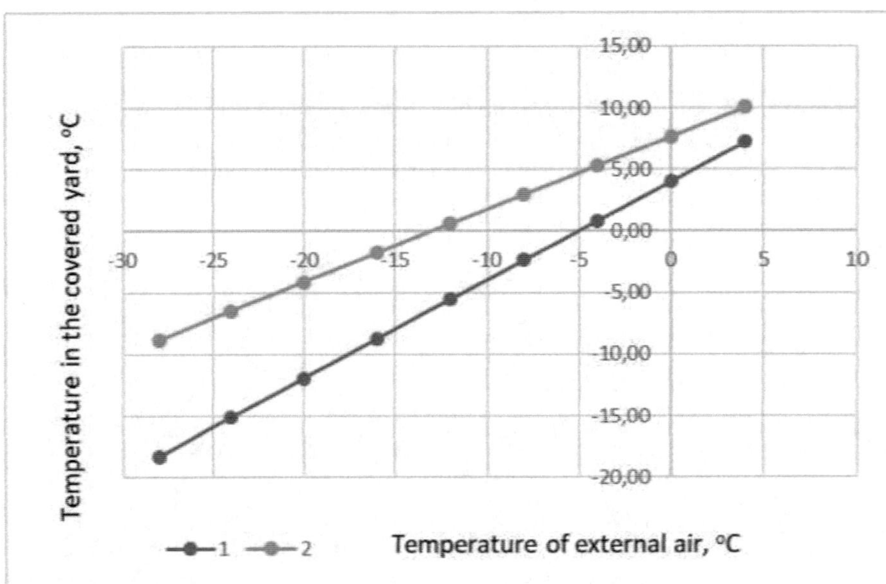

Figure 10. Change of air temperature in the covered yard depending on temperature of external air, where 1 – change of temperature in the covered yard at standard values of resistance to a heat transfer of the protecting designs of the building leaving to the yard and 2 – change of temperature in a courtyard at the lowered values of resistance to a heat transfer of the external protections of the building leaving to a courtyard

The differences in heat physical characteristics of the external protecting designs for Moscow and St. Petersburg have insignificant character therefore fig. 9 is suitable for the considered two cities.

4. Conclusion

According figure 10 (the line 2) it is visible that already at -12 °C, the external air in the covered yard has positive value of temperature of internal air. The increased air temperature in the covered yard allows to receive space, more convenient for rest. The translucent covering shades yard space that can demand additional lighting [8]. Heat losses from the walls and windows to courtyard will be reduced during the cold period of year which will allow to reduce thermal data security of the protecting building designs. The snow which dropped out on a roof should be cleaned up by mechanics or melted away with removal of thawed snow in the sewerage. During the warm period of year in space of the covered yard can be hot that can demands possibility to open the shutters of translucent overlapping for natural airing or it is necessary to provide additional engineering systems for cooling [9]. The big square of a glazing in a translucent covering of the covered yard will lead to formation of the descending streams of cold air which are compensated by the ascending streams of the heated air rising along walls with windows in the form of convective streams. In the covered space of the yard air circulation by means of the ascending and descending convective streams is formed. The microclimate of the covered yard in winter time has cold character, and when cooling air its moisture content aspires to a minimum, and it results in dryness of sand in sandboxes for children as moisture from sand will evaporate and will disappear together with air exchange of the covered yard. To understand all concept of the thermal mode in the covered yard we should carrying out more detailed modelling of the microclimate parameters changing in time is necessary. And the roads which are filled up with snow how to be with them? They can be covered with translucent arches with heating for transformation of snow into water and branch it in the sewerage too. Of course, cars with the electric engine are not fast translation process of huge park of cars on such type of fuel as electricity that improves ecology and allows to solve many infrastructure problems of set.

References

[1] Gogina E S and Derusheva N L 2016 Bases of a method of design of snow of melting constructions taking into account criteria of ecological safety *Water and ecology: problems and decisions* **2** (66) 58–64
[2] Retter A I 1984 *Architectural and construction aerodynamics* (Moskow)
[3] Kim A N 2015 Passages and atriums as elements of formation of public spaces in a modern urban environment *Innovations in sociocultural space Materials VIII of the international scientific and practical conference* (Amur state university) 38–42
[4] Kupriyanov V N and Smetanin D V 2010 History of development and classification of atriums *News of the Kazan state architectural and construction university* **2** (14) 32–39
[5] Egorychev O O and Dunichkin I V 2013 Questions of forecasting of a microclimate of an urban environment for an assessment of wind power potential of building *Vestnik MGSU* **6** 123–31
[6] Rymarov A G 2012 Gas mode of the building *Natural and technical science* **6** (62) 595–9
[7] Rymarov A G 2013 Characteristics of heat-mass exchange modes of mutual influence buildings *Natural and technical science* **1** (63) 380–2
[8] Gorgots S E and Solovyov A K 2006 Calculation of natural lighting in the rooms fronting in an atrium *Lighting engineering* **2** 51–2
[9] Brodach M M and Efremov M N 2014 Systems Ventilation, heating, air conditioning of atriums *AVOK: Ventilation, heating, air conditioning, heat supply and construction thermophysics* **3** 16–24

Published under licence in Journal Title by IOP Publishing Ltd

Comprehensive assessment of options for renovation of buildings and city blocks

K Yu Losev[1], G O Chulkov[2] and V O Chulkov[3]

[1] Department "Information Systems, Technology and Automation in Construction" Research University "Moscow State University of Civil Engineering", 26 Yaroslavl highway, Moscow, 129337, Russia
[2] Scientific and Project Center "City Development", 19 Building 3 prospekt Mira, Moscow, 129090, Russia
[3] Department "Technology and organization of construction production" Research University "Moscow State University of Civil Engineering", 26 Yaroslavl highway, Moscow, 129337, Russia

E-mail: LossevKY@mgsu.ru

Abstract. The study contributes to infographic approaches when providing comprehensive assessment of city areas during reorganization. It presents infografic models and prospective tool for infographic modelling, visualization, analysis and quality evaluation. The object of study is infographic model of "renovation" options of city blocks. The term "renovation" has been actively used in the directive documents and Decrees of the Government of Moscow from 2005 till nowadays. The task of the study was the development of methods for assessing the diversity of parameters of demolished, reconstructed and newly erected buildings and structures of housing stock and social infrastructure. Method that has been used in the study was structural modification of the basis infographic model presented in the article. Conclusions of the study show that a "star" multilayered infographical model produces good results when evaluating the quality of urban area "renovative" variants and it proposes R-language as a development tool for such future models. Also, the study has indicated to the fact that city environment began to be evaluated not only as a socio-technical, but also as a cyber-physical system.
Key words: assessment, renovation, quality, building, cyber-physics, R-language, visualization

1. Introduction

The problem of reconstruction (urbanization) of territories of different countries (including urban areas) seriously worries humanity since the beginning of the twentieth century. It was then that the need to replace (renovate) the physically or morally worn out housing and relocate the residents of demolished buildings and structures rose in full growth and continues to accompany the city territories all over the world, without reducing its heat. Approximately from the middle of the twentieth century, statistical studies of the processes and results of such a reorganization are constantly carried out.

The need to maintain the old housing fund becomes so expensive that it is advisable to demolish it and build it up again. But city authorities cannot afford such kind expenses in full amount.

Content from this work may be used under the terms of the Creative Commons Attribution 3.0 licence. Any further distribution of this work must maintain attribution to the author(s) and the title of the work, journal citation and DOI.
Published under licence by IOP Publishing Ltd

The relevance of this article is that the major cities of Russia have been allocated funds and it is necessary to take an optimal decision based on the parameters of what is going to be demolished and what will be built, and also on the basis of the social criteria of rationality. This should be a comprehensive integrated assessment. This means the need to evaluate the quality of options for reorganization for a very large number of parameters.

The idea of renovation, as a reorganization of city areas, and the term "renovation" were proposed at the Moscow State University of Civil Engineering at the peak of the period of "dotted construction" of the existing town-planning composition of the city of Moscow at the turn of the 20th and 21st centuries. This idea was borrowed from the world practice, which considers the consolidation of urban areas as one of the ways to reduce their spatial growth and the associated environmental impacts. The idea was actively supported by the authorities and Mayor of Moscow Luzhkov Yu. M., and the term "renovation" was included in the directive documents and the Moscow Government Decrees of that period.

Researchers note significant differences between the diagnostic values of indicators in different regions of each country. After the dismissal of the Mayor of Moscow, Luzhkov Yu M in 2010, the idea of total "dotted construction" and "spot development" was criticized, its popularity and intensity of use had been decreased significantly. Recently the term "renovation", the content of which is not directly associated with the idea of "dotted construction" of city blocks, once again has acquired relevance in the activities of the Moscow Government, which is explained by the need to significantly improve the level and quality of life of the population. "Renovation" is now understood as a complex problem, combining new construction, reconstruction and renovation, social and transport aspects of life and work, demolition of buildings and structures at the end of its lifecycle, construction waste recycling and organizational and logistical issues of resettlement of Muscovites as well.

Therefore, the development of scientifically substantiated and coordinated with the population plan of renovating the Moscow urban areas for the coming decades is an urgent point.

2. Methods

One aspect of the scientific justification for such a renovation program is the development of methods for assessing the diversity of parameters of demolished, reconstructed and newly erected buildings and structures of housing stock and social infrastructure. The results of the assessment and forecasting of certain characteristics at the level of global governance should be reduced to integral assessments of the quality of the construction sites themselves, as well as the construction, operation, reconstruction and demolition of the organizational and technological processes of the construction industry.

In other words, we are talking about the advisability of using infographic models of the convolution of a variety of parameters for the integrated evaluation of variants of Moscow city blocks renovation.

Publications of foreign authors, as well as the results of international forums are devoted to the most common a method of reducing the spatial growth of urban areas - the problem of their "compaction" [1–14].

Russia's building science is actively studying the problem of "reorganization" (rearrangement) of urban areas, as evidenced by numerous publications in domestic and foreign professional publications, as well as dissertational research [15–27].

These researchers systematically have been analyzing, competing with each other, different types of reconstruction in the construction industry. This article is devoted to continuing research on this topic.

The abovementioned transformation of semantics of "renovation" term indicates that socio-technical status of a city has changed. Processes of information technologies implementation, trends of common data environment (CDE) and digital economics development have led to the fact that city environment began to be evaluated not only as a socio-technical, but also as a cyber-physical system [28].

It is obvious to authors that in order to improve the quality of the integrated evaluation of options of Moscow city blocks, specific methods of data representation and visualization should be used.

Approaches are just starting to be developed to understand cyber physics of urban buildings life cycles and of a city as a whole [29, 30].

Domestic terminology for determining the cybernetics of building systems is also in the process of design [31].

The basic infographic model is the basis of the systemic scientific study of traditional (repair, reconstruction, restoration) and innovative (renovation, reversion, recombination, retrieval, etc.) types of construction, methods and processes of their use in construction as shown in figure 1.

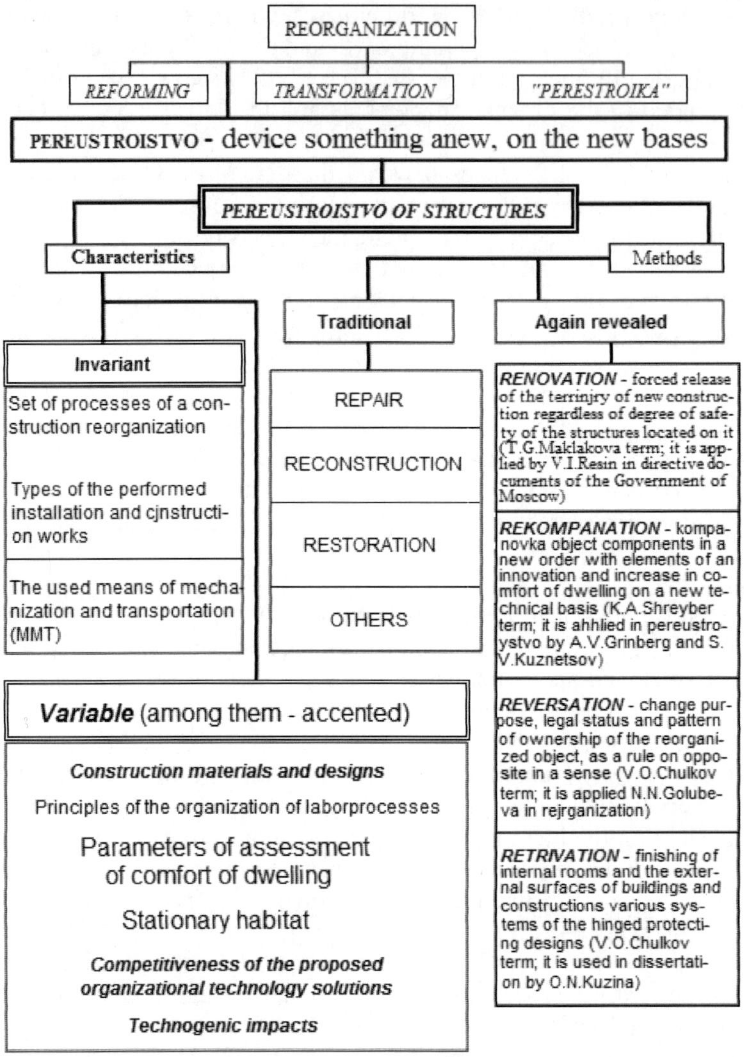

Figure 1. Types of reconstruction in the construction industry [26]

Over the past decade, the world's surface has been reorganized to 1000 km^2 every year, including construction reorganization. Up to four percent of the total number of demolished, erected and reorganized construction sites falls on previously used or already mastered territories for secondary use. This is one of the ways to further reduce the newly developed territories, which characterize the cramped spaces of construction sites, the need to take into account the number of risks and their insurance that is much greater than when developing new territories. The consequent and constantly increasing intensity of air and soil pollution, also associated with a steady increase in the number of vehicles and the complexity of restricting their use, adversely affect people's health and quality of life, increase the level of excessively generated heat. The parameters and characteristics of each of these processes in the society are constantly changing dynamically in the process of their interaction, which requires the development and use of adequate methods, technologies and tools for their diagnosis and monitoring.

3. Results

The variety of values of such parameters of different characteristics of independent and interconnected processes occurring in the construction industry are monitored and analyzed within the framework of operational management by construction. For a global (strategic) management of construction production of design and implementation of renovation of residential urban areas (in particular, in the city of Moscow), it is necessary to perform a reliable integrated assessment of the quality of renovation projects and the processes of its implementation.

When implementing the program of renovation of Moscow's neighborhoods, it is important to choose such a variant of the renovation project for a particular quarter that will ensure its best overall performance. In the preparation of renovation projects, developers consider a number of indicators, some of which have directive fixed values, and the values of other parameters can be varied by the designer in order to achieve the best results of renovation in the context of specific external conditions, the circumstances of the renovation and the limitations that have been adopted.

Thus, already at the stage of developing a renovation project, one can make a prediction about its results based on the application of a flat infographic (visual) model of "folding" the variety of parameter values characterizing the quality of renovation results and the processes of achieving them as shown in figure 2.

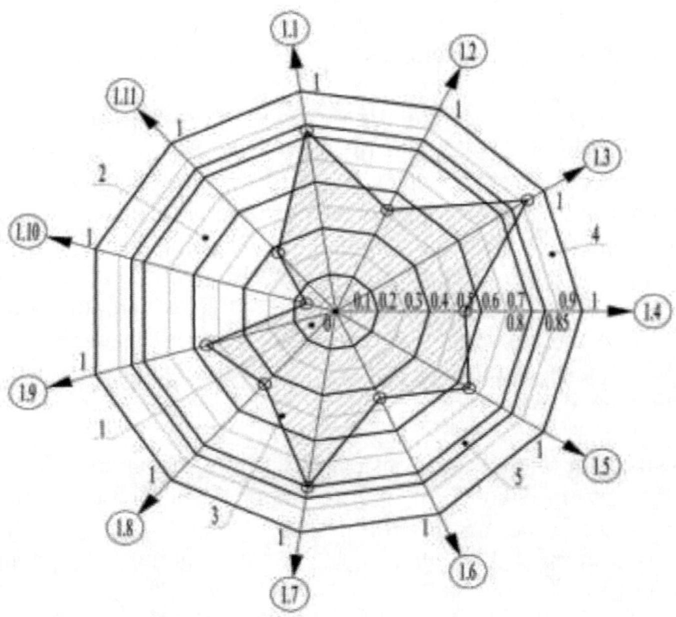

Figure 2. Flat infographic model of "convolution" of the parameters of the object of investigation [16]

The model is executed in the right polar coordinate system, that is, the axes (the number of which corresponds to the number of the investigated parameters of the research object) are arranged clockwise, and the sequence of axes corresponds to the "weight" (significance) of the parameter, which is determined by the expert method. The number of axes of the model is unlimited, each axis must be graduated from zero to one (for example, axis 1.4 in figure 2). If necessary, special zones (for example, zones 1, 2, 3, 4 and 5 in figure 2, which are shown on the callouts).

By connecting successively points on neighboring axes, we obtain a polygon whose area is taken as the integral estimate of the quality of the renovation projects.

Each renovation project has a number of options for combining the values of the original parameters. We set the task of determining the combination that will ensure the best values of the final indicators of renovation and will allow making an informed choice in favor of a specific version of the renovation project.

To do this, the numerical values of all model parameters must be divided into three groups:
- directional fixed values;
- variable values for each variant of the project, which are set by the designer in the framework of existing or accepted by this designer constraints;
- derived values for each version of the renovation project, which are calculated according to the previously developed mathematical formulas.

Parameters within each of the last two groups should be divided into two subgroups:
- striving to minimize their values;
- aspiring to the maximum of their values.

In each subgroup of parameters and their numerical values, we select the minimum or maximum numerical values for the parameters of each denomination, which we take as unity. For the remaining values of the parameters of each denomination, we find their dimensionless quantities (less than or equal to 1) relative to the minimum or maximum value, respectively. All parameters (variables or derivatives) receive dimensionless values of Pi (i = 1, ..., n) on the interval [0,1], where n is the number of parameters considered for the renovation project variant. The received priority sequence of parameters, as was indicated above, determines the ordinal numbers of the axes of the infographic model of "convolution", on which the relative values of the corresponding parameters will be plotted. Thus, the number of axes of an infographic model of convolution will be equal to the number of n names of variables and derived parameters, and the sequence of their location in the model relative to each other will not be arbitrary, but will be chosen by the customer (or developer) consciously.

The infographic model of such a "convolution" (figure 2) is a system of $n > 2$ coordinate axes with a common origin and angular pitch of 360/n (degrees). On each of the axes separate segments are plotted, graphically displaying the reference values of the corresponding parameters. The ends of the adjacent single segments are joined by straight lines, resulting in a flat reference polygon.

On the axis corresponding to a certain parameter, its value Pi is postponed. By connecting in similar fashion the ends of Pi of neighboring segments, we obtain the actual polygon. Each variant of the renovation project will have its own actual polygon on the convolution model.

The ratio of the areas of the actual and reference polygons characterizes the quality level of a renovation project, which is related to each other.

4. Discussion

In the domestic construction science, the considered flat infographic models is called "star" [16]. The models have been using in studies of multi-component and multi-parameter systems of the construction industry of the former USSR and Russia.

Subsequent studies confirmed the wide possibilities and unlimited range and application of "star" infographic models [26].

This class of models is widely used in the economy (where they are called the "Kiwiat diagram"), on transport (they are called "a portrait of object of interest"). The scope of application of these models is constantly expanding due to the simplicity and speed of their use in computer information technology. These models can be multi-layered as shown in figure 3, they are used for research objects that have a multifaceted characterization and application areas, for example, when reorganizing the territories of former industrial enterprises.

It is proposed to deduce the infographic models from a local application to a wide use.

The given models could be transformed into a package of R-language. R is an open source language and environment for statistical computing and graphics. It is preferred to presume of it of an environment within which analytical techniques are implemented. R can be extended via packages (tidyverse, ggplot2, dplyr, tidyr, purrr, readr roxygen2, testthat, devtools, etc.). There are many packages are available through the Internet sites. R is an integrated suite of software facilities for data manipulation, calculation and graphical display. One of the purposes of R-language is multifaceted data representation and visualization in complex systems as well [32].

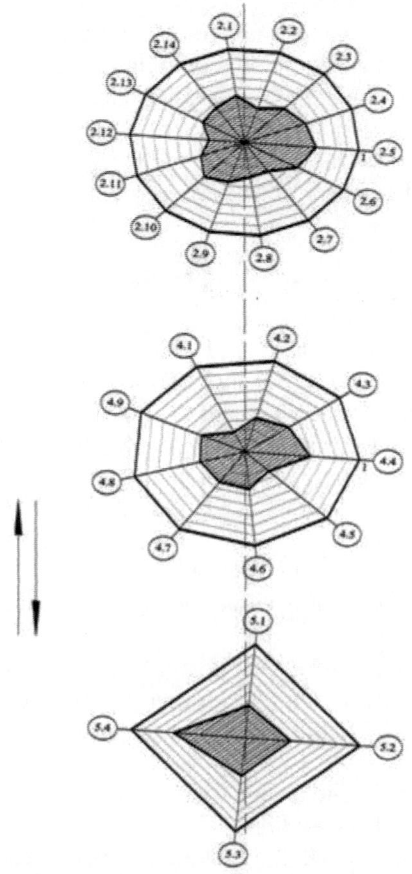

Figure 3. A multilayered model of the "convolution" of the parameters of the object of investigation when renovating and reorganizing the territory of the ZIL car factory in Moscow [25]

One of R's strengths is the ease with which providing compressed pivot data outlining the identified trends in the data set. In other words, R-language is a useful tool for modelling, visualization and analysis.

The complexity of the coverage of all aspects of renovation, as a multi-layered and multi-parametrical cyber-physical and socio-technical system, presupposes an examination of the significant diversity of individual characteristics, revealing the degree of their significance and interrelationships, requiring the development of new information technologies. The use of such technologies is able not only to interconnect individual characteristics, to perform diagnostics and monitoring changes in the quantitative values of individual parameters, but also to operate with integral assessments of the interaction of all the above-mentioned aspects of renovation. Domestic construction science pays serious attention on the creation and application of "folding" models for certain parameters and characteristics into integrated comprehensive assessments. It allows to operate with such data on different management levels when assessing the quality of processes and performance results. One of the varieties of such models is considered.

IOP Conf. Series: Earth and Environmental Science **177** (2018) 012026 doi:10.1088/1755-1315/177/1/012026

5. Conclusions

1. In order to provide comprehensive assessment of urban area renovation evaluate results and provide results of its implementation, as form of cyber-physical system activity, it is advisable to use a "star" multilayered infographic model, considered in the article. It represents the continuity of multi-component and multi-parameter systems studies have been developing in the construction industry of the former USSR and Russia.

2. The study proposes R-language as such future models development tool. It is proposed to transform flat and multilayer "star" infographic model into a package of R language in order to use the advantages of the language environment and put the models into common international level of data analysis for renovation of buildings and city blocks.

3. Also, the study has indicates to the fact that city environment began to be evaluated not only as a socio-technical, but also as a cyber-physical system.

References

[1] Altes W K K 2009 Taxing Land for Urban Containment: Reflections on a Dutch Debate *Land Use Policy* **26** 233–41

[2] Bots P, van Bueren E, Heuvelhof E, Hardoy I, Mitlin J E and Satterthwaite D 1992 *Environmental Problems in Third World Cities* (Earthscan, London)

[3] Cooke P, Heidenreich M and Braczyk H (eds.) 2004 Introduction: regional innovation systems – an evolutionary approach *Regional Innovation Systems* (Routledge, London)

[4] Duffy A 2009 Land use planning in Ireland – a life cycle energy analysis of the recent residential development in the Greater Dublin Area *The International Journal of Life Cycle Assessment* **14 (3)** 268–77

[5] Jaeger J A G and Schwick C 2014 Improving the measurement of urban sprawl: Weighted Urban Proliferation (WUP) and its application to Switzerland *Ecological Indicators* **(38)** 294–308

[6] Short J, Fleming S and Witt S 1986 *House Building, Planning and Community Action: The Production and Negotiation of the Built Environment* (London: Routledge & Kegan Paul)

[7] Kaiser E, Godschalk D and Chapin F S 1995 *Urban Land Use Planning* (University of Illinois Press. Urbana, IL.)

[8] Kamal-Chaoui L, Leman E and Fufei Z 2009 Urban Trends and Policy in China *OECD Regional Development Working Papers* 2009/1 (OECD Publishing)

[9] Kohler N and Lützkendorf T 2002 Integrated Life Cycle Analysis *Building Research and Information* **30(5)** 338–48

[10] Kohler N and Moffatt S 2003 Life-cycle analysis of the built environment *UNEP Industry and Environment* **April – September** 17–21

[11] Kunstler J H 2001 *The City in Mind: Notes on the Urban Condition* (Free Press, New York)

[12] Losev K Yu and Sinenko A 2017 Infography use to requirements specification for the design of the building *IOP Conference Series Earth and Environmental Science* **90(1)** 012197

[13] Millward H 2006 Urban Containment Strategies: A Case-Study Appraisal of Plans and Policies in Japanese, British and Canadian Cities, *Land Use Policy* **23** 473–85

[14] Povey M and Lloyd-Jones T 2000 Mixed value urban development: Mechanisms for sustaining the livelihoods and social capital of the urban poor *May 2000 ESF / NAERUS workshop*

[15] Seoul City Government 2009 *2030 Low-carbon Green Development Masterplan* (July 2, 2009, Seoul City Government, Korea)

[16] Chulkov V O 1989 *Geometric modeling in the complex documentation of engineering objects (infographics)* (Thesis abstract, Moscow, Publishing House of the DIA)

[17] Telichenko V I 1994 *Scientific and methodological basis for the design of flexible building technologies* (Thesis abstract, Moscow, Publishing House of the DIA)

[18] Sinenko S A 1992 *System engineering of designing of the organization of building manufacture* (Thesis abstract, Moscow, Publishing House of the DIA)

[19] Vlasov M Yu 1994 *Parametrization of infographic models in the automated documentation of engineering objects* (Thesis abstract, Moscow, Publishing House of the DIA)
[20] Yarovenko S M 1995 *Development of information technology of investment processes in construction* (Thesis abstract, Moscow, Publishing House of the DIA)
[21] Losev K Yu 2005 *Functional and structural automated design of comprehensive safety of low-rise residential construction objects* (Thesis abstract, Moscow, Publishing House of the DIA)
[22] Shchegol A E 1996 *System-engineering design of scientific support of construction* (Thesis abstract, Moscow, Publishing House of the DIA)
[23] Semechkin A E 1999 *Infographic methods for organizing the reconstruction of residential quarters* (Thesis abstract, Moscow, Publishing House of the DIA)
[24] Ginzburg A V 1999 *Automation of designing the organizational and technological reliability of the functioning of construction organizations* (Thesis abstract, Moscow, Publishing House of the DIA)
[25] Chulkov V O, Ghazaryan R K and Kuzina O N 2014 Basic cycle of reorganization. Innovations in the sectors of the national economy, as a factor in solving the social and economic problems of our time *International Scientific and Practical Conference* (Moscow, Dec. 5–6, INO-CRI EiUS) 82–94
[26] Volkov A A, Chulkov V O, Kazaryan R R, Fachratov M A and Kyzina O N 2015 Possibility of quantitative appraise components and guidance for constructional rearrangement of buildings attached to their confrence *Advanced Materials Research* **1065–69** 2585–88
[27] Tolstova K and Chulkov V 2017 Advantages of modern collapsible systems for exterior finishing of buildings in urban areas in Russia *MATEC Web of Conferences* **117** 00169
[28] Understanding industrial transformation: views from different diciplines 2006 *Environment and Policy* edited by Olsthoorn X, Wieczorek A J **44** (Springer, Netherlands)
[29] Czamanski D and Broitman D 2018 The life cycle of cities *Habitat International* **72 (2018)** 100–8
[30] Losev K 2017 The ratio of semantic information inside the life-cycle design stage of a construction object *Knowledge Sciences* **9 (2017)**
[31] Volkov A A 2017 Cybernetics of Construction Systems. Cyber-physical Construction Systems *Industrial ans Civil Engineering* **6** 4–7
[32] Wickham H and Grolemund G 2017 *R for Data Science: Import, Tidy, Transform, Visualize, and Model Data* (O'Reilly Media Inc., Canada)

IOP Conf. Series: Earth and Environmental Science **177** (2018) 012027 doi:10.1088/1755-1315/177/1/012027

Ensuring vibration safety of megalopolises from influence of the subways

A V Zamukhovskiy[1], N Shemer[2] and S V Privezentseva[3]

[1] Moscow Engineering Institute of Transport, 9b9 Obrazcova str., Moscow, 127994, Russia
[2] Representative in Russia companies GETZNER, Austria, Blyudents
[3] Moscow State University of Civil Engineering, 26 Yaroslavl highway, Moscow, 129337, Russia

e-mail: miit.ppx@inbox.ru, nikolaus.schemer@getzner.com, eledvendeesen@gmail.com

Abstract. This article is presented analyse of the latest technical methods and their realization for decrease in vibration impact of the subway on the housing estate and recommendations about decrease in vibrations from train service of the subway are provided in the houses located along Marshall Biryuzov Street from No. 11 to No. 34.

1. The first section in your paper

Development of the latest technologies, one of the priority directions in the town-planning sphere during creation of the steady environment and full-fledged life of society. In big agglomerations the main place is taken by transport infrastructure.

Today the Government of the Russian Federation allocates the key place to extra street rail transport connections, such as railroads and subway. In turn the last is the difficult mechanism, at the same time the fastest and with big passenger traffic. The developed length of lines of the subway in Moscow contains about 801.3 km of a way passing under buildings and constructions.

We seldom pay attention to the vibration influence arising in buildings where we should be, work and live. Rather we pay attention to noise, that is to the air fluctuations coming to us directly from a source, either the so-called secondary or structural noise generated by the fluctuating elements of buildings under the influence of the getting vibration (for example, a ware ring in buffet).

However impact of vibration on the person can have negative consequences even in that case when we don't feel vibration.

In this regard there are rather rigid norms regulating vibration impact on the person who is in the building [1].

It is also possible to note that importance of fight against vibration has found reflection and in number of normative documents: 19 normative documents while to noise– only 8 are devoted to problems of measurement and decrease in vibration.

Expansion of network of the subways within the existing building, construction of housing within a technical zone of the subways, increases in social inquiries to comfort of residential zones I demand ensuring sanitary standards on vibration effect, and sometimes and more strict requirements, in a residential zone of megalopolises.

Content from this work may be used under the terms of the Creative Commons Attribution 3.0 licence. Any further distribution of this work must maintain attribution to the author(s) and the title of the work, journal citation and DOI.
Published under licence by IOP Publishing Ltd

IOP Conf. Series: Earth and Environmental Science **177** (2018) 012027 doi:10.1088/1755-1315/177/1/012027

2. Another section of your paper

Any vibroprotection (or anti-vibration) construction provides dispersion of energy of vibration influence so that the energy which reached a securable object had the allowed level.

Any mechanical system can be considered as the system (set) of the masses connected among themselves by springs and dampers.

Each separate subsystem consisting of mass, a spring and a damper (figure 1), is called oscillator and is characterized by such parameter as natural frequency

$$f_0 = \frac{1}{2\pi}\sqrt{\frac{c}{m}}, \tag{1}$$

where c – dynamic rigidity of a spring and m – corresponding weight.

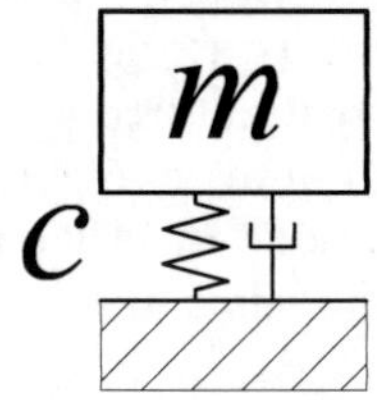

Figure 1. Elementary oscillator

Any oscillator works as the high pass filter, that is:

1) If the frequency of exciting influence is less than natural frequency, then this influence passes through oscillator practically without changes.

2) If exciting frequency is close or matches natural frequency, there is a resonance phenomenon which is characterized by sharp increase in a vibration amplitude.

3) The frequency of exciting influence is higher, the less there will be amplitude of forced oscillations. If the frequency of exciting influence is more, than √2 f_0, then speak about efficiency of vibroblanking as with further increase in frequency the vibration amplitude of system will be less than amplitude of exciting influence.

Follows from told that in order that through the vibroextinguishing system there did not pass influence with f frequency, natural frequency of the f0 system shall be at least twice lower.

The frequency is higher, the quicker energy of oscillations dissipates with distance in any environment (the high-frequency signal "whistle" dissipates to indiscernible level through several hundreds of meters, the low frequency signal "tyfon" is heard on several kilometers).

In the following subsection implementations vibroextinguishing (or anti-vibration) constructions are provided.

3. Another section of your paper

Methods of decrease in vibrations in buildings near tunnels of the subway can be divided into two big groups: the events held in a tunnel and the events held in the building or near him.

Treat the events held in the building or near it:
- a wall in soil;
- vibration insulation of the building in base level;
- floating floor;
- anti-vibration partitions.

The following belongs to the actions which are carried out in a tunnel:
- laying of the increased elasticity in knots of intermediate fastenings;
- the special vibroextinguishing fastenings;
- undo sleeper laying;
- undo ballast mats;
- weight spring system.

The specified actions are given in a type of the organizational chart in the figure 4 and given in decreasing order of their efficiency.

Actions of the first group, as a rule, will execute at construction of buildings near the existing subway lines though there are cases of vibroprotection of already existing buildings.

The actions providing decrease in vibration in the tunnel can be provided both at construction, and on already operating lines. The truth in the latter case won't exceed efficiency of decrease in vibration 15–20 dB.

Further the actions which are carried out in designs of a way of the subways are considered.

3.1. The vibroextinguishing fastenings

For the system of fastening the role of mass will be executed by the fluctuating part of a rail with the associated neobressorenny mass of crew carried to a wheel. A spring will be one or several elastomer gaskets. The rigidness of elastomer gaskets, as a rule, cannot be lower than 50 kN/mm. In case of neobressorenny mass of the cart of 7500 kg natural frequency of system can be estimated at 30 Hz. I.e. the efficiency of elastic fastenings can begin from 55–60 Hz.

At the same time, than soft gasket of a node of fastening, subjects is more turn of rail section in case of influence of side forces [2]. At the same time the spacer contracts nonuniformly and the efficiency of vibroblanking decreases (figure 2).

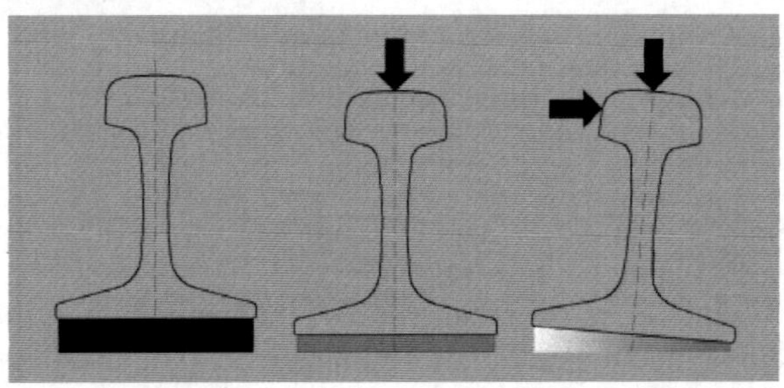

Figure 2. Operation of elastomer gaskets in a fastening node

Such decision was repeatedly realized in Moscow Metro: on a stage the October Field station – the Shchukinskaya station of the Krasnopresnensky line, from the PC 0113+00 to the PC 0116+00 rubber corrugated gaskets 10 and 20 mm thick under fastening like "subway" are laid down; on a stage the Krylatskoye station – the Strogino station of the Arbat and Pokrovsk line contact rail polyurethane spacers 14 mm thick (productions of the company Gettsner or NG-P) on the polimerbeton the korotyshakh of the KR and KRD type are laid down.

The efficiency of spacers on the Arbat and Pokrovsk line was 2–3 dB, and on the Krasnopresnensky line their efficiency so decreased that there was a need of their changeover.

3.2. Elastic laying in contact rail support (podshpalny laying)

Elastomer gaskets in contact rail support are most effective for constructions with cross ties (undo sleeper spacers). If contact rail support are not connected on the track width (units, lezhn), then there is the same restriction on rigidness of a spacer under the unit. And, the more unit height, the more will be influence of its turn on change of width of a track. For example, in case of service tests of the railroad switch on the Art. of Troparevo the turn of the units LVT in a lateral axis of a way of the plane is recorded in such a way that vertical relocation of an edge of the unit outside of a track is more than vertical relocation of a sole on a rail axis.

According to the vendor of the units LVT-HA their natural frequency makes 25 Hz, respectively the efficiency will begin approximately with 50 Hz.

In 1995–1996 in case of construction of a section of a section "Volga"– Maryino near the Lyublino station under Sovkhoznaya Street for lowering of vibrations under wooden cross ties laid down the special spacers consisting of two halves when superimposing which closed construction in the form of a plane parallelepiped with a cavity was at each other created inside. Such construction was applied in the Minsk subway in case of construction of the second line on a stage the station Kupalovskaya– the Nyamiha station in 1990–1991.

However the efficiency of this construction turned out so insignificant that several houses on Sovkhoznaya Street had to be settled.

3.3. Weight spring system

The weight spring system is realized in several options in Moscow Metro (two options near the Art. of Sviblovo with rubber elastic elements) and two options on St. "Exhibition" in the form of the massive plates leaning on cylindrical steel springs of Gerb. Efficiency of vibroprotection on St. "Exhibition" is confirmed with measurements of SES of Moscow.

Schematically the design "weight spring" is presented in the figure 3.

Lack of the weight spring system is its considerable construction height and pointed transfer of load of a tunneling.

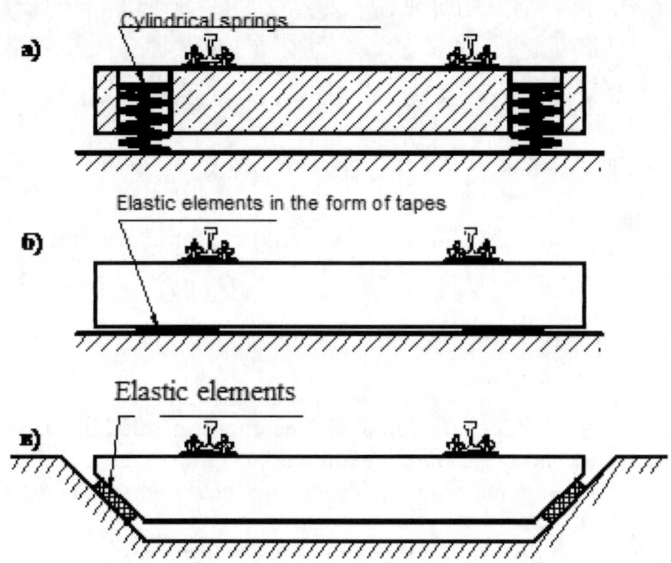

Figure 3. Designs of a way like "weight spring:
a) A traveling plate on twisted cylindrical springs (it is realized on St. Exhibition (now– St. Business center); b) A traveling plate on elastic elements in the form of tapes (it is realized on the project of Metrogiprotrans on a stage of St. A botanical garden– the Art. of Sviblovo); c) A traveling plate in the form of the return trapeze on the elastic elements working for compression and for shift (it is realized on the project of the engineer Velichko on a stage of St. A botanical garden– the Art. of Sviblovo)

Application instead of springs of plates (up to a continuous supporting) creates drainage system problems from space under plates. Besides, from the weight spring systems more effective are systems with steel springs as in this case static and dynamic rigidity are equal among themselves while in the best polyurethane they differ by 1.3 times, and in rubbers, especially synthetic more than twice.

3.4. Continuous supporting of rails (ERS, Embedded Rail System)

This construction represents a concrete plate with two trenches for laying down of rails. Under a rail lay down a continuous spacer which provides necessary rigidness in the vertical plane. The space on each side of a rail or is flooded elastic material, or filled with different options of spacers. The efficiency is reached at the expense of a continuous supporting (there is no sag between contact rail support) and low rigidness of a contact rail spacer. Side spacers provide spatial stability of a rail.

Though developers declare efficiency to 20 dB in the medium octave frequencies of 16 and 31.5 Hz, nevertheless vertical relocation continuously of a supported rail remain in the range of 1.5–2.5 mm as even the most elastic materials of side spacers will not be able to provide a bigger sag without friction about walls of a trench and as a result of wear.

This technical solution is applied in the subways of Spain and the Warsaw subway (figure 4).

Figure 4. Way of the ERS system in a tunnel of the subway

3.5. VGS-10 system

The elastic system of fixing of VGS is created and patented in 2003 in Russia by ABV group of companies. It for the first time in world practice allowed to reduce vertical rigidness of a contact rail support to 2.5–4 kN/mm, when maintaining horizontal rigidness and resistance to stealing at the level of normal contact rail support.

As showed calculations, and then and practical laboratory tests, VGS has the considerable efficiency in the frequency of 31.5 Hz that does it comparable only to the mass spring systems as any other type of contact rail support and rail fastenings does not provide efficiency in this frequency. Hereinafter in this point the efficiency is considered only in relation to frequency 31.5.

The previous (earlier applied) VGS 2 and VGS 5 models were created with the maximum adaptation to the existing technologies of laying down of a way, tools and rail fastenings that considerably restricted possibilities of growth of their efficiency. Nevertheless, they reduced vibrations on different sections on values from 8 to 21 dB that SES [2] is confirmed by the full-scale measurements taken by different organizations including (and generally). Separately it is necessary to mark that having installation potential on operating sections in night windows, VGS 2 allowed to take measurements most correctly in the same points before its installation. The majority of vibroprotective constructions have no such potential that forces to determine their efficiency by measurements on "similar sections" that is always not absolutely correct because of different soil conditions, depths of tunnels, existence of joints of rails, etc.

The best measured results (comparative measurements on different sections with wooden cross ties and fastening of VGS 5 showed efficiency 18 dB) are received in case of new construction, however and here often the quality of laying down left much to be desired.

VGS 10 became low-detail, threadless (it is unified with ARS-4, see a figure 5), with a broad range of regulation of a rail in the plan and in a profile. Special constructive decisions reduce loads of traveling concrete and anchor bolts, allowing to reduce their cost.

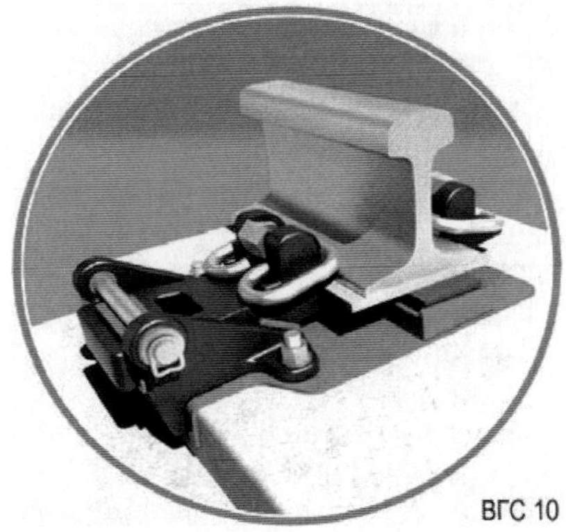

Figure 5. Fastening of VGS 10

The sleeperless design of a way has allowed to change essentially a rail supporting, having increased quantity of elastic contact rail support without changing the number of rail fastenings twice. Thus, the rail was evenly supported, having approached continuously supported rails in designs with the integrated rail (embedded rail).

Thanks to the high level of automation when laying with use of the 3D systems of navigation the probability of the construction marriage influencing efficiency is minimized. The efficiency of VGS 10 will be slightly higher, than at her predecessors. It will be reached at the expense of more evenly supported rail, and at the expense of the accuracy of laying of the rail without speakers on the level of contact rail support.

Besides it is necessary to consider that VGS 10 design (unlike the predecessors) can, in case of need, allow to increase vertical movement of a rail, for example, to 6 mm on those sites where it will be necessary. For this purpose it will be required to replace only material of elastic contact rail support. Installation of additional contact rail support directly on high-precision traveling concrete for providing a continuous supporting of a rail can become other option of the subsequent increase in efficiency.

Lack of VGS-10 is the lack of operating experience.

4. Another section of your paper

The approximate efficiency of the given actions, including by results of direct measurements before their carrying out is presented in table 1.

The data in table 2, values of efficiency, given in the table are received from reports of "Müller-BBM International"[3]; FBUZ "The Center of Hygiene and Epidemiology in the City of Moscow"[4], NO Ekogorod fund[5, 6] and other sources[7–9].

Table 1. Efficiency of actions for decrease in vibrations from a railway tunnel on nearby buildings

Action	Efficiency according to various sources	The efficiency confirmed (the place of realization)
System weight spring	up to 30 dB in the medium octave frequencies of 16 Hz and above	The measured values of vibration haven't exceeded settlement (the Art. Exhibition)*
Subballast mats	up to 25 dB in the medium octave frequencies of 31.5 Hz and above	8.3 dB (GABT)1)
Continuous bearing of a rail	up to 25 dB in the medium octave frequencies of 31.5 Hz and higher 2)	–
Podshpalny pokladka	up to 25 dB in the medium octave frequencies of 31.5 Hz and above	6–18 dB (23 on 1905 year St.) (VSP) 1)
LVT design (Low Vibration Track – a way of the lowered vibration)	5–10 dB is higher than 50 Gts3 of LVT HA)	according to JSC VNIIZhT

Action	Efficiency according to various sources	The efficiency confirmed (the place of realization)
The vibroextinguishing fastenings of VGS 02 and VGS 02MD	up to 12 dB in the octave frequencies of 31.5 Hz and above	11.3 dB (Museum of the Gun-on), 8.3 dB (Bolshoi Theatre) (VGS-02 and their modifications) 1)
VGS5	up to 8 dB in the medium octave frequency of 16 Hz and up to 14 dB– 31.5 gts above	up to 3 dB in the medium octave frequency of 16 Hz and up to 18 dB – 31.5 gts and higher 1)
VGS 10	up to 8 dB in the medium octave frequency of 16 Hz and up to 14 dB– 31.5 gts and higher 4)	–
Laying of the increased elasticity	up to 2 dB in the medium octave frequencies of 31.5 Hz and above	2 dB (Strogino)

Notes:
1) The efficiency is given on the measurements which are carried out on Moscow Metro [4–7];
2) According to the Edilon-Sedra company [8];
3) According to the Sonnevil AG company [9];
4) According to ABV group of companies [10].

Table 2. Comparison of characteristics and parameters of the vibroextinguishing systems

Design type	Efficiency, dB, in an medal octave strip, Gts		Need of concrete works	Approximate time for vibroprotection of the site during the work of 1 crew (2.8 km in single-line calculation), months	Need of polygon tests
	16	31,5			
1	2	3	4	5	6
VGS 02MD	0 (12)	12 (10)	no	6.5	no
LVT-M	-42) (16)	4 (18)	yes	17.5	yes
LVT-HA	-3)	-3)	yes	26.0	yes
VGS 5	3 (9)	18 (4)	yes	17.5	no
VGS 10	84) (4)	144) (8)	yes	10.0	yes

Notes:
1) The lack of efficiency, dB is given in brackets;
2) The negative value means growth of vibrations;
3) There are no data;
4) According to the producer.

4.1. Recommendations about decrease in vibrations from the subway

We will consider possible technical solutions on the example of a stage of St. "The October field" – the Art. Shchukinskaya.

From the provided analysis it is possible to draw the following conclusions and recommendations:
1) Ensuring the allowed vibration levels in houses down the street of the marshal Biryuzov requires decrease in vibrations by 12 dB in an medium octava strip of 16 Hz, on 23 dB in an medium octava strip of 31.5 Hz and on 22 dB in an medium octava strip of 63 Hz.
2) It is possible to apply the following designs to ensuring the required efficiency:
- VGS-10;
- continuously supporting rail;
- "weight spring".

However all these actions demand carrying out reconstruction of a way on the problem site.

IOP Conf. Series: Earth and Environmental Science **177** (2018) 012027 doi:10.1088/1755-1315/177/1/012027

For the purpose of cutting of costs for vibroprotective measures and reductions of terms of their realization it is possible to offer the following events:
1) Replacement of all defective cross ties on the site;
2) Fixing of all the shaken of cross ties for ensuring their dense contact with traveling concrete on the lower plate;
3) Replacement of all isolating joints on glue-bolt high-strength with metal-composite slips;
4) Replacement of intermediate fastenings like "subway" on fastenings of VGS2-50 MD (see table2).
5) Welding of all electro wire joints.

After performance of events of the specified actions repeated monitoring of a vibration situation in the buildings which are in a technical zone of the subway on the site of St is necessary. "The October field"– the Art. Shchukinskaya of the Tagansko-Krasnopresnenskaya Line in the volume specified in.

By results of monitoring to make the decision on further actions.

Follows from table 2 that the most effective solution for the operating line is installation of fastenings of VGS 10, however confirmation of their efficiency requires carrying out ground tests.

From the point of view of the smallest time for installation of vibroprotection installation of fastenings of VGS2-50MD is. In case of their installation carrying out concrete operations is not required. However before installation of fastenings of VGS2 it is necessary to revitalize sleepers economy completely: to replace unsuitable cross ties, to fix shaken cross ties for an exception of "keyboard" effect.

From table 2 it is visible that any of the proposed solutions does not provide 100% of lowering of the measured vibrations. Nevertheless it is necessary to consider that the shaken of cross ties ("keyboard" effect) and mechanical joints can make additional effect of elimination from 4 to 8 dB in the octave frequency of 16 Hz and above.

5. Conclusion

New technologies in the modern world allow to improve quality of life of the population, to do the world surrounding us safer and comfortable. Vibroprotection of a way – provides sustainable development of the cities and buildings. These measures will allow to level negative impacts from transport infrastructure on the housing estate and to improve quality of life of people

References

[1] CN 2.2.4/2.1.8.566-96 *Production vibration, vibration in locations of residential and public buildings*

[2] Grechanik A V, Zamukhovsky A V and Naumov B V 2011 The vibroextinguishing fastenings for the subway *Transport of the 21st century: researches, innovations, infrastructure: materials science.-technical conferention to the 55 anniversary of URGUPS* (Ural State University of Railway Transport, Yekaterinburg) **97**(180)

[3] 2010 Report on research "Reconstruction of the Bolshoi Theatre, Moscow. Detailing of measures of vibroprotection on the ways of the subway" No. W62 705/413. *Müller-BBM Interneshenel* on July 19

[4] Protocol of laboratory tests of levels of vibration No. P-654-14 of 30.10.2014 federal budgetary institution of health care "*Center of hygiene and epidemiology in city of Moscow*".

[5] 2015 The report on research *"Monitoring of a vibrational situation in the buildings which are in a technical zone of the subway on a section of St. "An October field" – the Art. Shchukinskaya of the Tagansko-Krasnopresnenskaya Line"* (Moskow: Ecological fund of development of the urban environment Ekogorod)

[6] 2012 The report on research "*Monitoring of a vibrational situation in the Underground rehearsal concert hall the Bolshoi Theatre of Russia with assessment of efficiency of the vibroprotective measures executed in tunnels of Sokolnichesky and Zamoskvoretsky lines of Moscow Metro*" (Moscow, Ecological fund of development of the urban environment Ekogorod)

IOP Conf. Series: Earth and Environmental Science **177** (2018) 012027 doi:10.1088/1755-1315/177/1/012027

[7] 2015 The report on research "*Assessment of impact of vibrations on traveling concrete and an obdelka of a tunnel on the Sokolnichesky line of Moscow Metro on sections of a ballastless upper structure of a way on the units LVT and with driving on wooden cross-pieces*" (Moscow, JSC VNIIZhT)
[8] 2018 Site edilonsedra.com, catalog Newsletter edilon, sedra edition April
[9] 2011 SYSTEM of the WAY of the UNDER VIBRATION (LVT). Report on vibroattenuation. Sonneville AG. Industrial zone 2 | 3225 Müntschemier | Switzerland Document No.: DPD078.GN
[10] Site abv1991.ru, prezintaition "Metro de Moscou. Les nouvelles structures et les technologies de la rparation rapide de la voie sur dalle (2012)"

Published under licence in Journal Title by IOP Publishing Ltd

Research physic-mechanical properties of composite materials on the base of crushed wood

A M Ibragimov[1] and A A Fedotov[2]

[1] Moscow state construction university, 26 Yaroslavl highway, Moscow, 129337, Russia
[2] Kostroma state university, 17 Dzerzhinsky str., Kostroma, 156005, Russia

e-mail: aafedotoff@yandex.ru

Abstract. Influence of traditional thermoset (urea – and phenol-formaldehyde) and thermoplastic alternative (recycled high density polyethylene and recycled polypropylene) binders on the performance of slab composite materials using as a filler of the special pieces of wood chips are considered. Results studies of physic-mechanical properties of considered materials are presented. Rational conditions production of plate materials based on recycled thermoplastic matrices are revealed. Comparative analysis of wood-polymer composite materials based on two types of adhesives have allowed to establish the possibility of obtaining competitive materials based on recycled thermoplastics with higher water resistance (in terms of swelling in thickness meet the requirements of GOST (Interstate standard) 32399-2013 for plates of type P7) and strength characteristics compared with materials based on traditional thermosetting adhesives. Using of thermoplastic binders indicates the feasibility of their use in production of particle boards due to the positive effect on the environment.
Keywords: composite materials, thermoplastic polymers, thermosetting binders, environmental friendliness, physical and mechanical properties.

1. Introduction

Currently, in domestic production of particleboards (composite materials based on synthetic binders and special pieces of wood chips) are widely used synthetic thermosetting binders based on urea-formaldehyde (UFO), urea-melamine-formaldehyde (UMFO) and phenol formaldehyde (PFO) oligomers. The problem of such boards is that one of the condensation products in the synthesis of oligomers is formaldehyde, a toxic gas of the second class of danger [1], which is a carcinogen, and presumably, a mutagen and an allergen. In a free state (due to incomplete condensation) it will be present in the binder and in finished product based on it.

The current domestic standard for particleboards (GOST 10632-2014) provides maximum permissible levels of formaldehyde in the board at the emission class E0,5 – not more than 4 mg, at the emission class E1, respectively, from 4 to 8 mg of formaldehyde, inclusive. If the production of boards with class of emission E1 on the basis of traditional thermosetting binder is applied in practice, the production of plates E0,5 very problematic. In this case, it is possible to use formaldehyde acceptors [2], however, it will not be possible to reduce its content to almost zero.

From other side, there is now a widespread use of polymeric (primarily polyethylene and polypropylene) packaging materials, which creates the problem of their disposal. It is not efficient to

Content from this work may be used under the terms of the Creative Commons Attribution 3.0 licence. Any further distribution of this work must maintain attribution to the author(s) and the title of the work, journal citation and DOI.
Published under licence by IOP Publishing Ltd

carry out the disposal of polymer waste, because the time of their decomposition is approximately from 50 to 100 years [3]. The burning causes harm to the environment (produce large quantities of toxic substances). An alternative would be sorting and recycling plastic wastes for use in various fields. In recent years, in Russia there are enterprises for processing of polymeric waste. In Kostroma, this set earned in 2016 (design capacity of 100 000 t/year).

Scope of primary and secondary polyolefins is quite wide. The polyolefins used for the production of wood-plastic composite materials [4, 5] mainly for construction purposes: decking, window and door profiles, window sills, siding, panels, stairs [6, 7]. The filler here is usually wood flour. Thermoplastics can be used to obtain slab composite materials, where the filler used the bark of the tree [8–10], agricultural wastes [11].

The work provides production of wood-polymer composite materials on the production technology of particleboards on hard metal pallets. As a binder, it is proposed to use secondary thermoplastic polyolefins, as a filler – a special cut wood chips used in the production of chipboards.

This study will differ from the traditional production of particleboards type of binder used, and from the production of wood-polymer composite materials – type of filler used and production technology.

The use of secondary polyolefins as binders will solve the problem of toxicity of particleboards and partly-the problem of disposal of polymer waste.

The aim of the study is a comparative analysis of physical and mechanical properties of wood-polymer composite materials based on secondary thermoplastic polyolefins and traditional thermosetting binders.

2. Methods of research

Influence of the specific pressure of pressing (varied from 3.0 to 6.7 MPa) was investigated by active single-factor experiment) on physical and mechanical properties of plate composite materials. As a binder used traditional thermosetting adhesives based on urea- (UFO) and phenol-formaldehyde oligomers (PFO) and alternative thermoplastic – granular secondary high-density polyethylene (SHDPE) and secondary granular polypropylene (SPP). In the preparation of adhesives based on urea-formaldehyde oligomer was additionally introduced hardener – ammonium chloride in the amount of 1 % by weight of the oligomer, the remaining adhesives were one-component. Thermosetting binders (UFO and PFO) were taken from existing production NAO «SVEZA Kostroma», Kostroma, Russia (GOST 14231-88 and GOST 20907-75, respectively), thermoplastic – with OOO «EcoTechnologies», Kostroma, Russia. The filler used was a special cut birch shavings for particleboards production enterprise, the NAO «SVEZA Kostroma» (average particles length 10.6 mm, width 0.7 mm, thickness 0.4 mm). Manufacture and testing of plate composite materials was carried out according to the technology of pallet pressing particleboards in a laboratory hydraulic press P100-400 on the base of Kostroma state university at the following constant factors:

- nominal thickness of board – 10 mm;
- specific duration of pressing – 1.0 min/mm;
- ratio: thermoplastic binder / filler is 45/55 %;
- consumption of thermosetting binder – 11% by weight of absolutely dry wood filler;
- temperature of pressing + 230 °C.

The levels of constant factors are determined on the basis of preliminary experiments, the results of which are partially presented in the already published works [12,13].

Physical and mechanical properties of particleboards were determined in accordance with the requirements of GOST 10633-78, GOST 10634-88 and GOST 10636-90. When conducting research on the determination of material properties the mean value of the indicator was calculated on the basis of data for the four duplicate experiments.

IOP Conf. Series: Earth and Environmental Science **177** (2018) 012028 doi:10.1088/1755-1315/177/1/012028

3. Results of research

After statistical processing, the following results of studies of the properties of the boards. For boards based on secondary thermoplastics values are presented in table 1, on the basis of traditional thermosetting binders – in table 2.

Table 1. Properties of particleboards on the basis of secondary thermoplastic binders.

Type of the binder	Specific pressure of pressing, MPa	Limit of tensile strength perpendicular to the plate, MPa	Swelling in thickness, %	Volumetric swelling, %	Water absorption by weight, %
SHDPE	3.0	0.58[a]	6.97	8.87	28.55
	3.7	0.60[a]	5.23	7.84	26.01
	4.4	0.70[a]	4.10	5.60	23.03
	5.2	0.59[a]	5.39	6.04	24.26
	5.9	0.55[a]	6.55	7.91	25.82
	6.7	0.53[a]	8.86	10.12	27.39
SPP	3.0	0.52[a]	3.64	4.68	17.96
	3.7	0.56[a]	3.27	4.26	14.17
	4.4	0.55[a]	3.08	3.30	11.82
	5.2	0.53[a]	3.03	3.43	11.94
	5.9	0.51[a]	3.24	4.10	12.89
	6.7	0.50[a]	6.21	7.17	15.43

[a] rupture occurred on the adhesive layer

Table 2. Properties of particleboards on the basis of traditional thermosetting binders.

Type of the binder	Specific pressure of pressing, MPa	Limit of tensile strength perpendicular to the plate, MPa	Swelling in thickness, %	Volumetric swelling, %	Water absorption by weight, %
UFO	3.0	0.22	62.33	64.78	88.60
	3.7	0.24	63.87	65.73	89.33
	4.4	0.26	65.86	67.13	91.58
	5.2	0.27	61.72	65.49	89.52
	5.9	0.28	57.16	59.09	87.71
	6.7	0.30	56.82	57.52	86.65
PFO	3.0	0.53	21.81	24.59	69.45
	3.7	0.56	16.69	19.03	62.66
	4.4	0.64	21.77	24.00	41.68
	5.2	0.61	22.89	25.21	60.20
	5.9	0.59	24.08	27.22	60.89
	6.7	0.57	24.58	27.48	61.44

Influence of specific pressure of pressing on physical and mechanical properties of particle boards shown in figures 1–4.

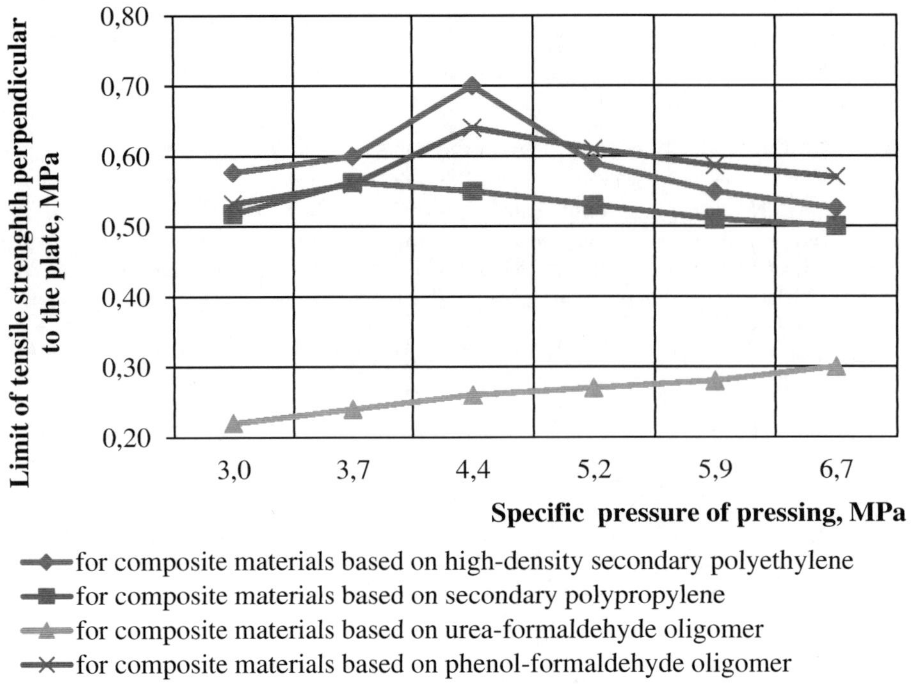

Figure 1. Influence specific pressure of pressing to the limit of tensile strength perpendicular to the plate

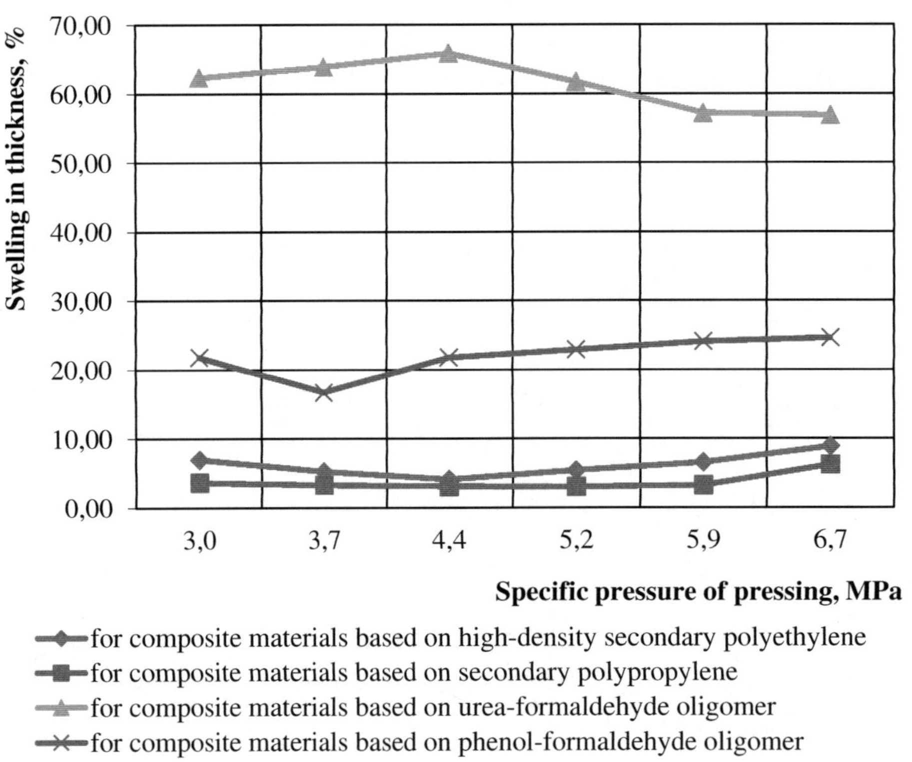

Figure 2. Influence specific pressure of pressing on the swelling in thickness

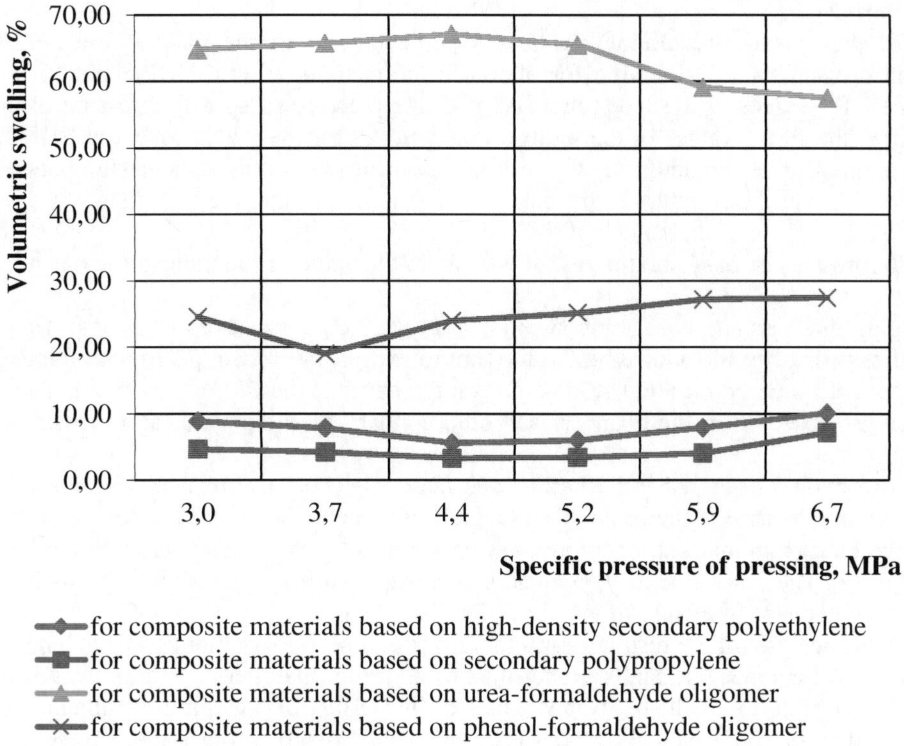

Figure 3. Influence specific pressure of pressing on the volumetric swelling

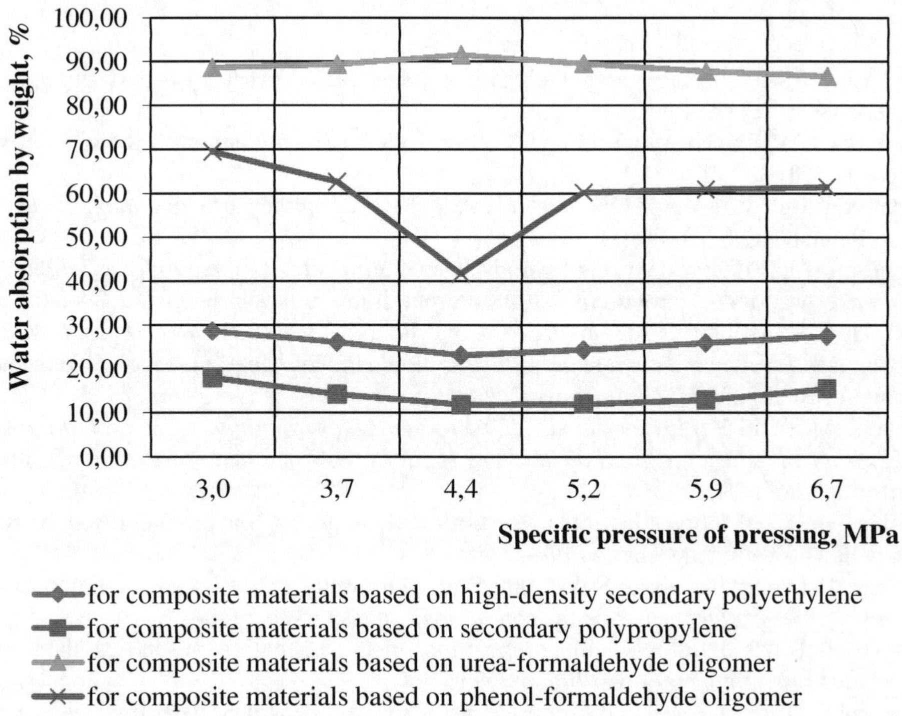

Figure 4. Influence specific pressure of pressing on the water absorption by weight

4. Conclusion

The limit of the tensile strength of the boards perpendicular to the plate is the best value at a compression pressure of 3.7–5.2 MPa for plates on the basis of PFO, SHDPE and SPP, for boards based on the UFO values of the index increase with the increase of specific pressure of pressing. At the same time, the actual values of the index will be higher for the plates based on SHDPE and SPP (since the rupture of the sample during the test did not occur due to higher adhesion between the glue and the wood filler in the structure of the sample). At the same time, the materials based on the UFO compared to all the others, the values of the indicator are significantly lower and range from 0.22 to 0.30 MPa (figure 1). The plates on the rest of the binder the value of the indicator is much higher (0.5–0.7 MPa).

Swelling in thickness and volumetric swelling (figures 2–3) composite materials based on recycled thermoplastics (primarily SPP) have best value (not to exceed 10 %) compared with the thermosetting binder (the boards is based on UFO – 56–67 %, on the basis of the PFO – 16–28 %). The best values of the index are observed in the boards based on secondary thermoplastics at a pressing pressure of 3.7–5.2 MPa.

Water absorption is minimal in SHDPE-based materials (no more than 29 %) and SPP (no more than 18 %). At the boards on the basis of UFO and PFO reaches 92 and 70 % respectively (figure 4). Values of the index minimum at a compression pressure of 4.4 MPa for boards on the basis of PFO, SHDPE and SPP. The best value of the indicator was observed from plates on the basis of the SHDPE and SPP at a compression pressure of 4.4–5.9 MPa.

Comparative analysis of the results shows that the use of secondary thermoplastic binders values of the studied parameters is significantly higher than using traditional thermosetting. Boards on the basis of secondary polyolefins will be water-resistant (in terms of swelling in thickness meet the requirements of GOST 32399-2013 for boards of type P7). Thus, by replacing the binder, it is possible to obtain environmentally friendly water-resistant materials with a slight change in the current technology of production of particleboards.

References

[1] GN 2.1.6.1338-03. *Maximum permissible concentration (MPC) of polluting substances in atmospheric air of populated areas*

[2] Leonovich A A, Shpakovski V G and Vojtova T N 2015 *Low-emission wood boards Furniture maker* **2(70)** 38–39

[3] Azarov V I, Burov A V and Obolenskaya A V 2010 *Chemistry of wood and synthetic polymers* (St. Petersburg: LAN' Press)

[4] Shcherbinina E 2016 Production of wood-plastic composites *LespromInform* **5** 128–33

[5] Abushenko A and Voskoboynikov I 2008 Eternal, liquid tree *Derewo.RU* **2** 78–84

[6] Glukhikh V V, Mukhin N M, Shkuro A E and Bryndin V G 2014 *Receipt and use of products from wood polymer composites with thermoplastic polymer matrices* (Ekaterinburg: Ural state forest engineering University Press)

[7] Tasekeev M S and Yeremeeva L M 2009 *Production of biopolymers as one of the solutions to the problems of ecology and agriculture* (Almaty, national center for scientific and technical information Press)

[8] Rudenko B D 2011 Investigation of operational factors in the production slabs of bark and recycled polyethylene *Forest journal* **4** 85–8

[9] Rudenko B D and Plotnikov S M 2009 Statistical study of the mode of flat pressing boards WPC *Actual problems of forestry complex* (Bryansk) issue **24** 146–9

[10] Rudenko B D 2010 Influence of composition on the formation properties slabs of bark and recycled polyethylene *Forest Bulletin* **4** 151–4

[11] Rangavar H, Taghiyar H R, Ghofrani M and Khojaste-Khosro S 2016 Improving physical and mechanical properties in particleboard by recycled polyethylene and canola residues *Sci. Technol.* **13** 857–64

[12] Pobirskij A M, Kotikov S A and Fedotov A A 2017 Study properties of wood-polymer composite materials based on secondary polypropylene *Int. ecol. Conf.* (Krasnodar, Kuban state agrarian University Press) 592–4
[13] Pobirskij A M, Kotikov S A and Fedotov A A 2017 Study of composite materials based on crushed wood using thermoplastic and thermosetting binders *Int. sci.-pract. Conf.* (Neftekamsk: World of science) 55–1

Published under licence in *Journal Title* by IOP Publishing Ltd

Water-Resistant Gypsum Binding Agents and Concretes Based Thereof as Promising Materials for Building Green

I Domanskaya[1,2], S Bednyagin[1] and H-B Fisher[3]

[1] Ural Federal University, 28 Mira str., Yekaterinburg, 620002, Russia
[2] Ural State University of Economics, 62 8 Marta str., Yekaterinburg, 620144, Russia
[3] Bauhaus University Weimar, 11 Coudray Road, Weimar, 99421, Germany

e-mail: i.k.domanskaya@urfu.ru

Abstract. For gypsum-based cementing materials, environmental advantages are highlighted over Portland cement, and major method for increasing their water resistance are described. By the example of gypsum-lime slag binders, a successful application of water-resistant gypsums is demonstrated in the severe climatic conditions of the Ural region. Phase composition analysis results are presented for gypsum-lime slag stone-like material after years of operation as part of the concrete walls of enclosures. It is found that durability of the material can be explained by the formation of stable compounds: calcium sulfate dihydrate, calcium carbonates and calcium hydrosilicates of the type CSH(B).

1. Introduction

Under conditions of globalization, the continuous advance of science and technology raises environmental problems that become truly global in scale. It therefore appears both timely and necessary to spread impartial environmental information among public at large. This will promote to the modern society the principles of Sustainable Development, which has been officially assumed by the United Nations Organization to be a World Development Strategy. An integral part of the Sustainable Development is Building Green concept, which abates negative environmental impacts and complies with relevant ecology standards. The Building Green normally relies on the following most important criteria: effective use of natural resources; minimization of a construction project's negative impact on the environment in the processes of construction and further operation; creation of microclimate comfortable for humans; application of green structural materials that are environmentally assessed throughout their life-cycle and others.

Currently, truly green (ecologically clean) structural materials, such as wood and natural stone, have limited application, usually in residential building projects of individual housing construction. State-of-the-art industrial construction relies mostly on using artificial stone-like materials that may be attributed to different groups: one group may be called kilned materials (ceramics, glass) to be manufactured using high-temperature treatment; and the other group is non-kilned materials (concretes, mortars) to be formed using the process of solidification of binding agents. Among non-kilned composite materials, a leading position is taken by Portland cement, which is used mostly for making concrete and steel concrete. In terms of technical and economic characteristics, cement concrete represents a state-of-the-art structural material of widest application, which contributes significantly into creating material basis for human environment of modern civilization [1].

Content from this work may be used under the terms of the Creative Commons Attribution 3.0 licence. Any further distribution of this work must maintain attribution to the author(s) and the title of the work, journal citation and DOI.
Published under licence by IOP Publishing Ltd

Non-kilned structural binding substance-based composites have certain advantages over natural and kilned stone-like materials: high manufacturability, economic efficiency, and a wide range of structural and technical performance characteristics. At the same time, these materials appear difficult for assessment in terms of their long-term behavior under specific operational conditions because they are characterized with prolonged processes of structure formation and may exhibit possible instability under the effect of external and internal factors. This concern is supported by a number of publications devoted to the issues of the materials durability and safety in operation. Many authors have reported about higher rates of corrosion processes in structures made of steel concretes, which become considerably costlier for repair and reconstruction. More frequent occurrences of construction accidents are observed involving sudden collapse of concrete structures, which as a rule date no more than 30–40 years back [2–4]. Although Portland cement has a long history of use nearly as long as 200 years, there are still unceasing discussions going on over the "driving force" of hydration processes in clinker minerals, over the roles played by surface phenomena occurring on the cement grain/water interface, and over internal structural strains that inevitably result in the drop of strength and in cyclic "saw-like" solidification behavior of cement stone [5].

Environmental compatibility of cement concretes also demands serious scrutiny, considering the resource-intensive and environmentally hazardous technology applied for manufacturing Portland cement, which reportedly produces 5 % of CO_2 emissions globally. These issues cause serious concern over the prospects for cement industry development; active search for alternative technologies is thus initiated; and synthesis research is intensified in creating and investigating alternative low-clinker, geopolymeric, gypsum-based water-resistant materials and other binder systems [6, 7].

2. Characterizing gypsum-based binding agents and methods for increasing their water resistance

Gypsum-based binding agents represent a kind of very few synthetic structural materials that can indeed be regarded eco-friendly. Raw materials used for making such binders are sedimentary rocks containing calcium sulfate dihydrate; possibly, as starting materials can also be used the chemical industrial wastes, such as phosphogypsum, borogypsum, fluorogypsum. The annealing temperature for the prepared starting material does not exceed 150–200 °C (compare with the annealing temperature for Portland cement at 1,500 °C), and the corresponding reaction byproduct is water:

$$CaSO4\cdot 2H2O = CaSO4\cdot 0.5H2O + 1.5H2O \quad (1)$$

When thus obtained gypsum-based binder is mixed with water, a spontaneous hydration reaction runs intensely, i.e., the reverse reaction occurs:

$$CaSO4\cdot 0{,}5H2O + 1.5H2O = CaSO4\cdot 2H2O \quad (2)$$

Hydration is a complicated physicochemical process that involves dissolving calcium sulfate hemihydrate, creating nucleation centers for solidification of a new phase, growing the nuclei and crystallization of calcium sulfate dihydrate. As the gypsum crystallites grow and interweave, the polycrystalline internal structure of synthetic stone is formed, with its macroscopic structure featured with high porosity (25–40 %) because of the great (up to 70 %) water consumption at the stage of mixing with water. Therefore, thus solidified gypsum binder is characterized with a high absorption of water and a low softening coefficient: in the water-saturated condition the strength of synthetic gypsum-based stone decreases by 40–50 %. Gypsum-based structures exhibit considerable creep under humidification and show low resistance to weather and, being such, are not to be recommended for operation in weather conditions.

A universal method for increasing water and weather resistance of non-kilned structural composite materials is decreasing their porosity. To this end, binder systems are introduced with small amounts of polymeric or plastizing additives to increase their density. For binder systems non-resistant to water, which include gypsum-based binders, a more drastic way of increasing weather resistance is to add up to 10–40 % of mineral components that, due to their intrinsic binding properties, solidify into water-resistant synthetic stone-like material. Such components are referred to as "hydraulic" components and thus produced binders are often called "mixed gypsums". There are two major directions known in the world to have been formed in the practical approach to solving the issue of increasing water-resistance of gypsum-based binders: one direction has been reported [8–10] to be aimed at using some amount of Portland cement as a hydraulic component; and the other direction [11–13] is to develop cement-free compositions based on lime and active mineral supplements.

It should be noted that the processes of hydration in mixed gypsum binders have thus far received insufficient research coverage. Under certain conditions, there is a danger of forming destructive phases (such as ettringite, thaumasite), which may result in increasing volume, cracking development and destruction of the solidified synthetic stone [14]. Such risks appear to be the reasons for insufficient use of water-resistant gypsum-based binders in construction, despite a multiple body of laboratory investigations and research publications available.

3. Industrial experience in manufacturing water-resistant gypsum-lime slag binders and concretes in the Ural Regionother section of your paper

In Russia, among water-resistant gypsum-based binders, most known are multi-component gypsum cementitious pozzolanic binders (GCPB) and gypsum-lime slag binders (GLSB), developed in the last century. Authorship rights for these binder systems are known to belong to Russian Institutions of Higher Learning: GCPB was developed in the Moscow Institute of Construction Engineering (now S.M. Kuibyshev's Moscow State Construction University); and GLSB system was developed in the Ural Polytechnical Institute (now B.N. Yeltsyn's Ural Federal University). The Moscow school of thought in science and engineering (A.F. Volzhensky, A.V. Ferronskaya, A.F. Korovyakov, et al) suggested using Portland cement and active mineral supplements (pozzolans) as the hydraulic component in the mixed gypsum-based binders. The scientists from the Ural Federal University (A.A. Antipin, L.I. Ryabokon, S.V. Bednyagin, et al) directed their attention to creating water-resistant gypsum-based compositions through the use of alkali and sulfate activation of blast-furnace acidic slags avoiding the use of cement.

In 1952, in the Ural Polytechnical Institute, a Gypsum Concrete Laboratory was established under the direction of A.A. Antipin, which developed GLSB compositions and studied GLSB properties, with further implementation of the developments into industrial technologies for manufacturing light concretes (GLSB concretes). At different stages of the lab's scientific activity, a number of various teaoms were involved in joint efforts: the teams included scientists, faculty members and students from eight departments of the Ural Polytechnical Institute and Sverdlovsk Architectural Institute; there were also members from medical sanitarian staff and from the staff of the specialized design and scientific research organization of the Sverdlovsk region, Ufa, Moscow and Moscow region [15]. Three experimental industrial plants were designed, which had over the period from 1960 to 1972 been used to manufacture GLSBs and GLSB-based wall blocks and sanitary units. The manufactured products were used to build over a hundred of experimental structures of various purposes: one and two-story residential houses, administration buildings, manufacturing facilities, agricultural buildings and structures. Monitoring the construction projects and laboratory tests have proved that increasing water-resistance of GLSB concrete as compared to commonly used gypsum concrete manifests itself not so much in increasing softening coefficient as in a considerable, over two orders of magnitude, decrease in the parameters of creep. The results were used as a basis for designing and constructing the-first-in-the-country gypsum works to manufacture GLSB blocks, with no similar works known abroad. The gypsum works was established in Krasnoufimsk, Sverdlovsk region, in 1976. The major gypsum concrete products were large-size wall blocks of 400 mm in unified thickness, 800 to 3,600 mm in length, and 800 to 1,600 mm in height. Thousands of buildings were constructed using the manufactured blocks: residential houses of various purposes, schools, kindergarten facilities, drug-stores, cattle-breeding facilities, fertilizer stores, granaries, garages, agricultural machinery service

IOP Conf. Series: Earth and Environmental Science **177** (2018) 012029 doi:10.1088/1755-1315/177/1/012029

shops, etc. The gypsum works stopped producing GLSB concretes in the 1990s as construction business experienced a slump in building volume.

For ten buildings erected in 1965–1985 of GLSB concretes on the territory of Sverdlovsk region, scientists from the Ural Federal University some years ago carried out a checkup which demonstrated that the buildings have remained in good condition and showed no sign of decay. The analysis results of the physical and mechanical properties of the wall structures have showed not only a high weather resistance of GLSB concrete but have also demonstrated considerable hardening of the material. At a density ranging between 1,200–1,700 kg/m^3 (depending on the type of filling) and at a designed strength of 5–10 MPa, the actual strength of the concrete was found to be 10–15 MPa [12].

4. Features of structure formation of gypsum-lime slag binders

The mechanisms underlying solidification of water-resistant gypsum-based binders is a complicated physicochemical process covering, on the one hand, hydration and crystallization of sulfate compounds to produce calcium sulfate dihydrate, and, on the other hand, formation of water-resistant crystalline and/or gel-like phases, such as calcium hydrosilicates and hydroaluminates running at the expense of reactions between the components of the hydraulic supplement. For GLSB concrete produced at the Krasnoufimsk gypsum works, such components were blast-furnace ground granulated slag having a basicity factor not exceeding 1 and burnt and ground lime.

To explore the features characterizing the prolonged structure formation of GLSB, the present study reports about studies carried out on small fragments of concrete taken from external uncovered surfaces of wall structures built in 1973 in Patrushi, Sverdlovsk region. Physical and chemical studies carried out on the selected samples were performed at the Laboratory of Structural Analysis Methods and Properties of Materials and Nanomaterials at the Ural Federal University, with some of the tests carried out at the Electron Microscopy Laboratory of the Bauhaus University (Weimar, Germany).

It is found that synthetic stone based on GLSB prolonged solidification is a rather dense and uniform structure comprising impregnated grains of slag (Figure 1, a). At a large magnification, it can be seen that the phase composition of stone material contains both well-crystallized structures and fragments of gel-like formations (Figure 1, b).

a b

Figure 1. Microstructure of GLSB after long years of hardening

X-ray phase analysis showed that the major crystalline phase identified with the reflections: 7.58; 4.26; 3.78; 3.05; 2.86; 2.67; 2.21; 2.07; 1.89; 1.62 is calcium sulfate dihydrate. It is likely that there is also low-base calcium hydrosilicates (3.04; 2.84; 1.88) and secondary calcium carbonate, which reflections are overlapped with the hydrate phases reflections (3.02, 2.27, 2.08, 1.87).

IOP Conf. Series: Earth and Environmental Science **177** (2018) 012029 doi:10.1088/1755-1315/177/1/012029

The presence of hydrosilicate and calcium carbonates in the composition of synthetic stone is supported with the results of differential thermal analysis. Endothermal effect at 160 °C confirms removal of chemically combined water from the hydration products of gypsum part of the binder; and exothermal effect at 369 °C is associated with the polymorphous transformation of dehydrated sulfate phases; the blurred endothermal effect in the range of 700–750 °C is due to the processes of decarbonizing of the secondary calcium carbonate formed as a result of lime carbonization in GLSB.

In the same temperature range, there seems to be dehydration of low-base crystalline and gel-like calcium hydrosilicates produced as a result of pozzolanic reactions between fine-grained slag component of GLSB and lime. In our opinion, it is crystallization of dehydrated hydrosilicates that may be associated with the exothermal effect observed in the temperature range of 830–860 ºC. From the temperature of phase transformation, the basicity of calcium hydrosilicates of CSH(B), which were fromed as part of GLSB composition, can be assumed to be equal to 1.36 (CaO/SiO_2 = 1.36) [16].

For the formation of calcium hydrosilicates as part of the mixed gypsum composition, the source of silicon oxide is ground granulated blast-furnace slag. Our investigations have shown that dissolving of the slag surface in the GSLB composition occurs non-uniformly (Figure 2).

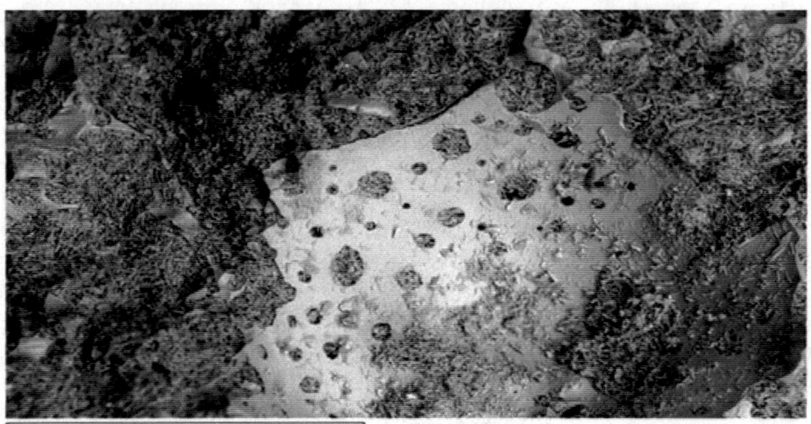

Figure 2. Particle of granulated blast furnace slag covered with products of pozzolanic reactions

The fragmentariness revealed in the “erosion” pattern of a slag particle, which is the source of silicate component for pozzolanic reactions, confirms the presence of microgranularity observed for the vitreous phase of granulated blast-furnace slags and supports the toponymic mechanism underlying the formation of calcium hydrosilicates [17].

Therefore, hardening of GLSB concretes revealed upon decades of their operation as part of the external wall structures has objective reasons and can be explained by slow processes of interaction between the vitreous phase of sag and calcium hydroxide resulting in the formation of both gel-like and crystalline calcium hydrosilicates.

5. Conclusions

Water-resistant gypsum-based binders can by no means be regarded as a fully adequate replacement for Portland cement; however, they meet the Building Green requirements for structural materials and show promise for low building structural applications. Successful production and application of GSLB for making concrete wall structures in the Urals in the 1960–1980s is a good evidence to that. After nearly fifty years of use in rather severe climatic conditions, the materials have revealed not only the absence of any signs of deterioration but even exhibited a 1.5–2.0 time increase in their structural strength. According to our studies, the reason for that is the stable phase composition of the synthetic gypsum-lime slag stone-like material without any destructive compounds. Along with well-crystallized calcium sulfate dehydrate, the material also contains some quantity of carbonate compounds and calcium hydrosilicates having both crystalline structure of CSH(B) type, with the basicity presumably equal to 1.36.

References

[1] Tamrazyan A G 2014 Concrete and reinforced concrete – look to the future *Vestnik MGSU* (Proceedings of Moscow State University of Civil Engineering) **4** 181–89
[2] Tang S W, Yao Y, Andrade C and Li Z J 2015 *Cement and Concrete Research* **78** A 143–54
[3] Yankovsky L V 2012 On the issue of durability of cement concrete in the transition to European standards *Building materials* **1** 16–8
[4] Hornbostela K, Larsena C K and Geikera M R 2013 Relationship between concrete resistivity and corrosion rate – A literature review *Cem and Concrete Composites* **39** 60–72
[5] Pshenichny G N 2017 Wheat about interaction, structure formation and hardening of cement systems *Innovation and expertise* **1** 149–63
[6] Huntzingera D N and Eatmonb T D 2009 *J of Cleaner Production* **17** 668–75
[7] Juengera M C G, Winnefeldb F, Provisc J L and Idekerd J H 2011 Advances in alternative cementitious binders *Cement and Concrete Research* **41** 1232–43
[8] Camarini G and De Milito J A 2011 *Construction and Building Materials* **25** 4121–25
[9] Butakova M D and Gorbunov S P 2016 *Proc. 2nd Int. Conf. on Industrial Engineering* (Chelyabinsk, Russia) **150** 1461
[10] Korovyakov V 2003 Gypsum binders and their use in construction *Russian chemical journal* **4** 18–25
[11] Cherevatova A, Fisher H-B, Wojtowicz E, Alekhin D and Sivashova S 2014 Application of mineral component in cement-free composite gypsum binder *Proc. XXI scientific readings on Science-intensive technologies and innovations* (Belgorod, Russia) **3** 418–22
[12] Ryabokon L, Bednyagin S and Domanskaya I 2016 *Building materials* **7** 21–4
[13] Kondratieva N, Barre M, Goutenoire F and Sanytsky M 2017 Study of modified gypsum binder *Construction and Building Materials* **149** 535–42
[14] Alksnis F F 1988 *Hardening and Destruction of Gypsum Cement Composite Materials* (Leningrad: Stroyizdat)
[15] Antipin A A 1959 *Application of Gypsum Concrete Blocks in Housing Construction* (Ufa: Bashknigoizdat)
[16] Kalousek G L 1976 The hydration Processes at early stages of hardening of cement *Proc. 6th Int. Cong. on the Chemistry of Cement* **2** 65–81
[17] Sarsenbaev B K, Sarsenbayev N B, Aubakirova T S, Abdramanova K S and Karymsakov S D 2013 Physicochemical processes of hydration and hardening of slag-alkali binders based on electrothermophosphor slags *Technique and technology of silicates* **20** 21–5

The reversible ventilation for administrative buildings

V Yu Kravchuk and A G Rymarov

National Research Moscow State University of Civil Engineering, 26 Yaroslavl highway, Moscow, 129337, Russia

e-mail: valerik_kravchuk@mail.ru, rymarov@yandex.ru

Abstract. The article describes the method of ventilation of administrative buildings premises with the use of reverse flow of supply and exhaust air on the same flues. A scheme of a reversible ventilation system and methods of designing this systems are presented, the conditions of its work are described.

1. Introduction

According to numerous surveys of administrative buildings with mechanical ventilation systems, the microclimate in their premises does not conform to standard. This situation occurs due to errors in the design, installation and operation of such systems. The number of office buildings is growing steadily in megapolises, hence the quantity of employees faced with the problems of poor air quality in the workplace is consistently increased. This very common issue compels the engineers to develop ventilation systems that use most of the natural forces. The efficiency of these systems is already confirmed by some researches in [1]. The reversible ventilation system belongs to such systems [2].

2. General information

The reversible ventilation is hybrid system (half mechanical and half natural) which combined the advantages of mechanical and natural systems. It has two operation modes:

1. Natural
2. Reverse

The change of operation modes is carried out by the valve group switching and activation of air supply device.

Figure 1a comprises the scheme of natural operation mode. In this mode outdoor air goes to premises via supply and exhaust devices. Further, it assimilates excess heat and moisture and is removed by air flues. In this case air movement is provided by natural forces only therefore the air supply device is inactivated and disconnected from the air flues. The natural mode may exist during the cold season of the year when the pressure difference is sufficient to provide the air exchange.

Figure 1b comprises the scheme of reverse operation mode. In this mode the air supply device takes the outdoor air and performs its pre-processing, subsequently moving it to the premises. Exhaust air is removed by supply and exhaust devices under the effect of excess pressure in rooms. The reverse mode may exist during the warm season of the year due to low natural pressure difference during this period. Hence the proposed ventilation system makes the most of the natural forces to provide the air exchange.

Content from this work may be used under the terms of the Creative Commons Attribution 3.0 licence. Any further distribution of this work must maintain attribution to the author(s) and the title of the work, journal citation and DOI.
Published under licence by IOP Publishing Ltd

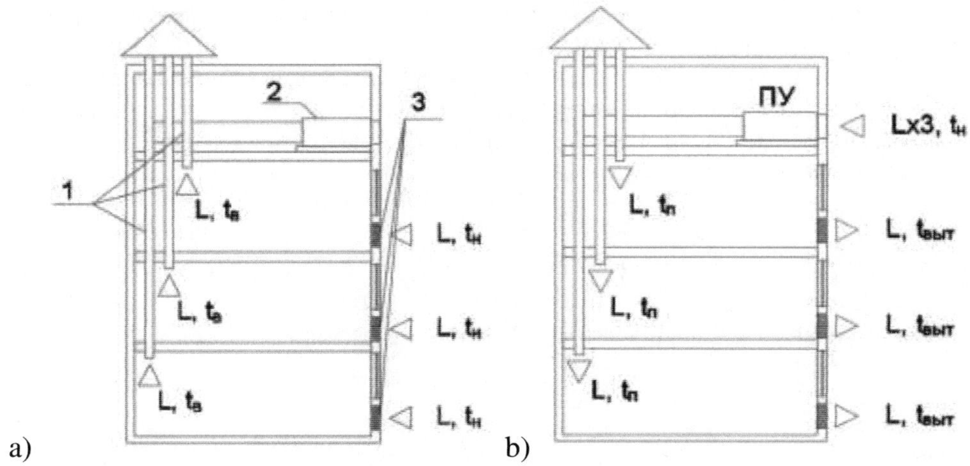

Figure 1. a) Scheme of natural operation mode.
1 – air flues, 2 – supply device, 3 – supply and exhaust devices.
b) Scheme of reverse operation mode

The application of reversal of air flows is not absolutely new idea. The investigations of this conception performed before in [3], however all of them discuss providing air exchange in a single or few number of premises and did not offer to use this principle centrally. The idea of air-permeability constructions was considered earlier [4] but not solved until the end.

The change of operating modes occurs due to the following circumstance: in natural operation mode the inlet air needs to be heated, therefore the supply and exhaust device is located directly behind the heating surface. When the outdoor temperature drops below +8 °C (in Russia), in the case when the heating system is switched off, the supply air begins to flow into the rooms with too low temperature. This situation violates the sanitary standards, therefore the operating mode of ventilation system is necessary to be changed to reverse. In this mode the inlet air is heated in air supply device and moves to premises with an acceptable temperature.

3. Supply exhaust devices

As part of the research, the author proposes a design of supply and exhaust device which corresponds to all necessary requirements. The proposed design of this device is presented on figure 2.

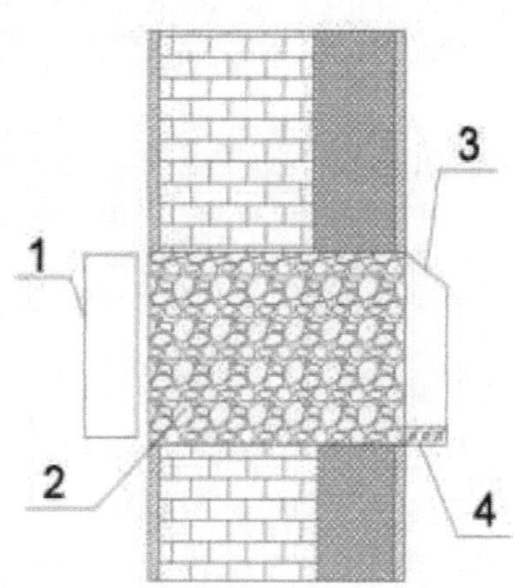

Figure 2. The location of supply/exhaust device in external wall.
1 – heat surface, 2 – breathable element, 3 – windbreak, 4 – regulating damper

Due to the relatively large area of the device, the air velocity in the supply jet is low. The low velocity of the supply jet does not cause overcooling of the water in the heating device and allows heating the air sufficiently. In order to fill the breathable element, the author proposes to apply large-porous media. These include backfilling of bloating clay or expanded polymers. In accordance to preliminary calculations, these materials provide the necessary hydraulic behavior during filtration. This can also be investigated from the papers [5].

Filtering through porous media is not a fully studied section of hydraulics. Because of low amount of information describing hydraulic mode filtering in large-pore filling, the author has assembled an experimental appliance for studying these modes. The equation of the exponential type and binomial type and experimental expressions for the filtration coefficients are known.

As part of the research, a simulation of filtration by equation 1 was performed.

$$\frac{dP}{dx} = C_{R1} * w_{fx} + C_{R2} * w_{fx}^2 \quad (1)$$

The coefficients C_{R1} and C_{R2} were determined experimentally. Comparison of experimental and simulation results is shown in the figure 3. The deviation of results was less than 5 %. This suggests that it is possible to apply filtration equations to simulate the flow in large-porous backfills. However, filtration coefficients should be determined experimentally.

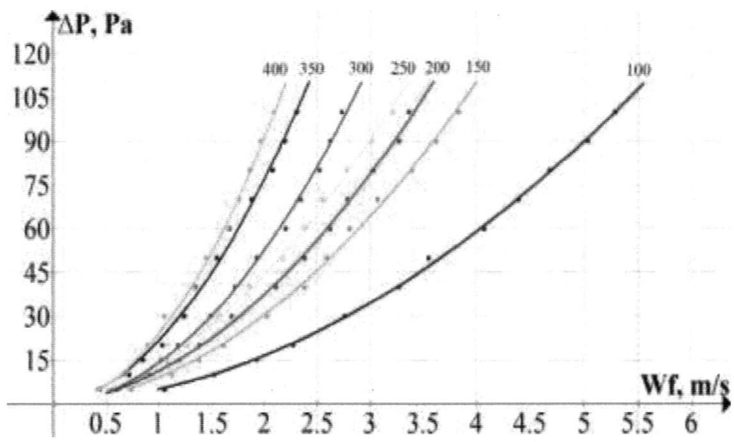

Figure 3. The experimental results (points) and simulation (lines) of air filtration in claydite filling. On the vertical axis, the pressure drop at the ends of the sample is marked. On the horizontal axis, the velocity of air flow in pores is marked. Labels indicate the thickness of simple, mm

For comparison, the author has performed modeling using the coefficients by expressions mentioned in other papers. In this case the convergence of simulation and experimental results was not found. It is known that the presence of any inhomogeneity in external walls violates thermal protection of a building, therefore it is necessary to consider the supply and exhaust device as a thermal conductivity insertion and apply the requirements of thermal protection norms to it.

4. The requirements of thermal protection

Russian standards describe two requirements to thermal protection. They includes hygienic and energy saving criterions [6]. The hygienic requirement limits the temperature on internal surface of external wall and the energy saving requirement limits the heat flux through external wall. The modern Russian method of accounting for thermal heterogeneity is developed by Gagarin and Kozlov [7]. In accordance with this method, in order to describe the specific heat fluxes through elements of external wall it is necessary to calculate temperature behavior of wall contour. The temperature of internal surface of a wall can be defined by the same simulation as well.

As part of the study the simulation of thermal behavior of external wall with supply and exhaust device was performed. In this simulation the filtration of cold air through porous filling was taking into account. The simulation was performed with varying values of the insertion height (from 100 to 600 mm) and the air velocity in the pores. The example of computation result is presented on figure 4. This simulation is allowed to calculate the specific heat fluxes through porous filling and apply the requirements of thermal protection norms to it. The dependence of specified heat flux on the air velocity in pores is presented on figure 5.

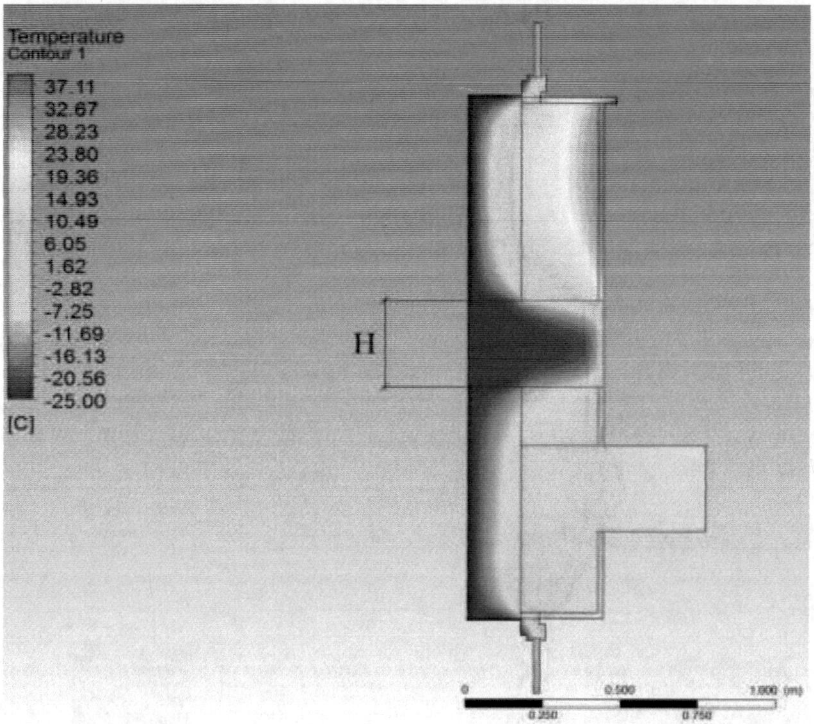

Figure 4. The results of thermal behavior simulation in external wall with breathable opening

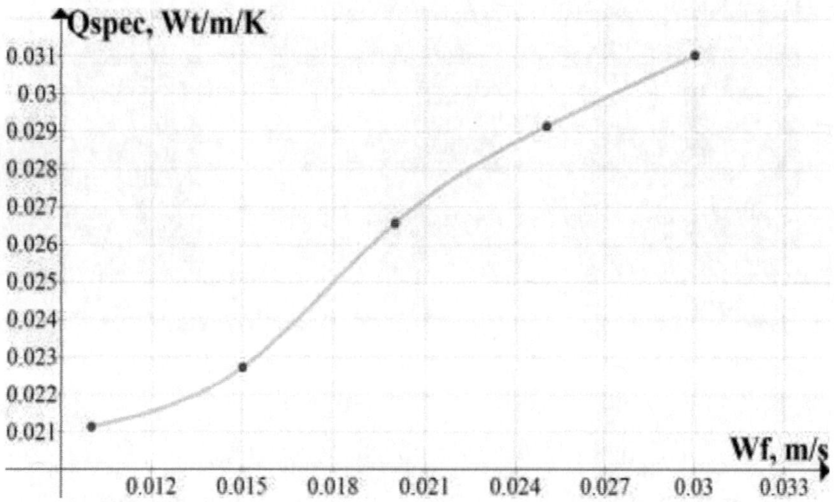

Figure 5. Depending specific heat flux with air velocity in pores

Here the dimension of specific heat flux is correct that proved in paper [7]. All calculations were performed in ANSYS CFX и Matlab computation systems. Summarizing the above, it is possible to propose an engineering technique that allows using breathable openings in external walls without violating the requirements of thermal protection. It can be formulated in the form of the following paragraphs:

1. The area of the supply and exhaust device is determined by the known flow rate and the velocity (from 3 to 30 mm/s) of the supply jet (the problem of non-isothermal filtration in filling at different velocities is considered in the article [8]). The height of device is determined by structural features of walls and should be as large as possible. The length of device is calculated by area and height values.
2. Such parameters as reduced thermal resistance of external wall and thickness of thermal insulation coat can be calculated using the method, described at [7] if height and length are known. This makes possible to execute energy saving criterion.
3. In order to take into account the hygienic criterion the temperature of internal surface of a wall should be calculated at known height and length. The air velocity in pores can be corrected providing that this temperature would not be too low.

5. Benefit cost analysis

As a part of the study the benefit cost analysis is performed. The reversible ventilation system was compared with usual supply and exhaust system. For example, the central-corridor administrative building with 6 floors was considered. In case of the usual supply and exhaust ventilation consists of 1 supply 2 exhaust systems. Total value of supply air rate is equal to 17700 m^3/h. The results of comparing are presented on diagram 1.

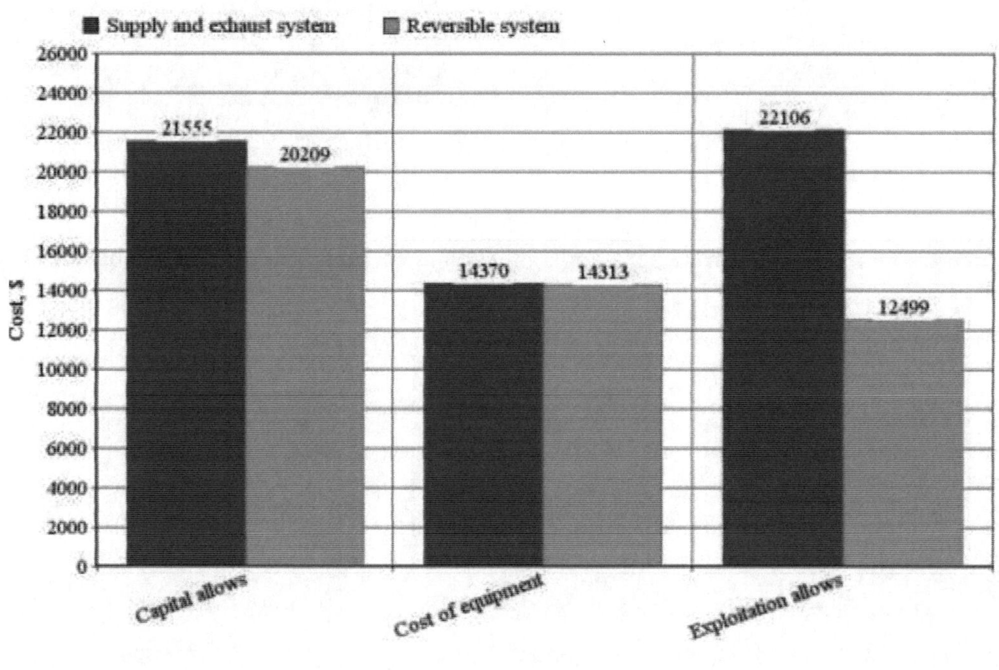

Figure 6. The results of benefit cost analysis

For calculating the value of capital and exploitation allocations in addition to the cost of equipment, the assembling, heat and electrical energy, repair activity and pay allocations was also considered. The results of cost analysis show the benefits in the effectiveness of reversible ventilation in comparison with usual ventilation was 25 %. The budget keeping arises from minor cost of equipment and also by inactivation of supply devices during cold season of the year.

References

[1] Artmann N 2008 *Cooling of the building structure by night-time ventilation* (Denmark: Departament of Civil Engineering Indoor Environmental Engineering Research Group)
[2] Engel P van den, Kemperman R and Doolard H 2012 Use the interaction of gravitational forces, the sun and wind for effective ventilation *REHVA*
[3] Kravchuk V Yu and Rymarov A G 2017 Structural features of reversible centralized ventilation system for administrative buildings Sanitary *Heat. Air-conditioning* **12** 86–8
[4] Hermann J W 1974 *Patent of ventilation system*
[5] Aristov YU I, Muhin V A and Mezencev I V 2004 *Patent of heat transfer regulation in ventilation system of office and residential premises and the regulation device*
[6] Malsagov A H 2007 *Patent of the climate-operation*
[7] Bogoslovsky V N, Titov V P, Parfentyeva N A and Medvedeva E V 1978 *Patent of ventilation device*
[8] Dong Chen 2017 Periodically reversible supply/exhaust ventilation strategy *Building and Environment* **124** 2590–97
[9] Aristov Yu I, Mezentsev I V and Mukhin V A 2006 A New approach to heat and moisture regeneration in the ventilation system of rooms. I laboratory prototype of the regenerator *Journal of engineering physics and thermophysics* **3** (79) 569–76
[10] Aristov Yu I, Mezentsev I V and Mukhin V A 2006 A New approach to heat and moisture regeneration in the ventilation system of rooms. II Prototype of the real device *Journal of engineering physics and thermophysics* **3** (79) 577–84
[11] Aristov Yu I, Mezentsev I V and Mukhin V A 2008 A New approach to regenerating heat and moisture in ventilation systems *Energy and buildings* **3** (40) 204–8
[12] Belyaev B C 1976 The using of porous media insertions in external walls. Thermophysical parameters and microclimate of large-panel buildings *Digest of research papers CRIED housing* (Moscow) **4** 63–71
[13] Valov V A 2011 The air-permeability walls in buildings Periodical journal of Siberian State Automobile and Highway Academy **13** 33–7
[14] Pastushkov P P 2013 Scientific approach to air-permeability walls *Roofing and insulation materials* **4** 13–4
[15] Cvyak A A 2004 *The development of external walls with changeable air-permeability* (Omsk)
[16] Maksimova M V 2000 *The prediction of external walls clodding* (Omsk)
[17] Dekhtyar R A, Sikovsky D Ph, Gorin V A and Mukhin V A 2002 Heat Transfer in a Packed Bed at Moderate Values of the Reynolds Number *High Temperature* **5** (40) 693–700
[18] Nizovtsev M I, Borodulin V Y and Letushko V N 2015 Regenerative heat exchanger with a periodic change in the airflow direction for room ventilation *Thermophysics and Aeromechanics* **6 (22)** 785–96
[19] Titov V P, Parfentyeva N A and Medvedeva E V 1981 The porous heat exchanger with fixed filling *Water supply and sanitary technic* **4** 27–8
[20] SP 50.13330.2012 *Thermal protection of buildings*
[21] Gagarin V G and Kozlov V V 2010 The theoretical preconditions of reduced thermal resistance calculation for external walls *Building materials* **12** (672) 4–12
[22] Kravchuk V Yu and Rymarov A G 2018 The use of through openings with a porous filling for ventilation *Sanitary. Heat. Air-conditioning* **2** 76–8

Published under license in Journal of Physics: Conference Series by IOP Publishing Ltd

IOP Conf. Series: Earth and Environmental Science **177** (2018) 012031 doi:10.1088/1755-1315/177/1/012031

Rationalization of water treatment as a top priority goal of sustainable development of residential areas and improvement of environmental quality

A Rumyantseva, M Berezyuk, L Abrzhina and I Barkin

Department of Environmental Economics, Ural Federal University named after the first President of Russia B. N. Yeltsin, 19 Mira str., Ekaterinburg, Russia

e-mail: alenarum@mail.ru

Abstract. The problem of searching for resource-saving and cost-effective technologies for cleaning rinsing water within the filter structures of waterworks remains relevant for many years. Most of contaminated wastewater of Sverdlovsk Region comes to the surface water bodies from Municipal Unitary Enterprise 'Vodokanal'. The solution of engineering tasks aimed at the environmental situation improvement in urban areas and water bodies protection from pollution subsequently requires continuous improvement of wastewater treatment methods. The paper examines the construction project of rinsing waters treatment unit of the filter station of Municipal Unitary Enterprise 'Vodokanal' with the formation of dehydrated sludge and provides the calculations of technical and economic parameters of the project. The project implementation is expected to result in a decrease of the man-caused environmental impact by eliminating the rinsing waters discharge, as well as by reducing the amount of water taken from the surface source into the technological process of potable water treatment.

1. Introduction

In the modern world, ecology is one of the basic components of the concept of favorable human habitat, along with its social and economic aspects. In the context of a global climate balance change, which has a significant impact on the redistribution of water resources, the public's attention is substantially focused on the problem of quality and usage of water resources.

According to the UN, potable water will play a decisive role on the Agenda of the Third Millennium. In 2000 the fresh water shortage within our planet, including agricultural and industrial needs, was estimated at 230 billion cu. m per year. By 2025 this number is expected to increase up to 1.3–2.0 trillion cu. m per year.

In terms of total freshwater resource amount, Russia has a leading position among the European countries. Russia is said to have more than 20 % of the world's freshwater resources in reserves (excluding glaciers and groundwater). Among the six countries with the largest river flow (Brazil, Russia, Canada, the United States of America, China, India), Russia is the second largest in the world after Brazil, and the third one in terms of water availability per capita (after Brazil and Canada) [1].

Environmental and water management safety is a guarantee not only for the industrial complex development of the country, but also for the development of the economy and for the population well-

Content from this work may be used under the terms of the Creative Commons Attribution 3.0 licence. Any further distribution of this work must maintain attribution to the author(s) and the title of the work, journal citation and DOI.
Published under licence by IOP Publishing Ltd

being. Absence of necessary and sufficient amount of good quality potable watery is already a serious problem nowadays.

2. Facts and figures of water management and wastewater disposal in the Russian Federation

Water management in Russia is carried out in an overwhelming degree due to the fresh water intake. In 2010, fresh water intake from water bodies was 72.7 billion cu. m; in 2011 – 68.7; in 2012 – 66.3; in 2013 – 65.1; in 2014 – 64.8; in 2015 – 62.2 and in 2016 – 63.0 billion cu. m.

Thus, there is a sustainable trend for the decrease of the indicator mentioned above, except for the last year. In 2015, the volume of contaminated wastewater discharged into the country's water bodies reduced to 14.4 billion cu. m, in other words it was 2.4 % less than in 2014. In 2016, this figure increased up to 14.7 billion cu. m, in other words it became 2.1 % more than a year earlier.

The share of polluted wastewater in the total amount of wastewater discharge in water bodies in 2010, 2014 and 2015 remained generally stable and its value is about one third. This, to some extent, indicates that the overall dynamics of the water usage and sewerage significantly influenced the change in the wastewater discharge under consideration. At the same time, it should be noted that, despite appreciable positive trends in the absolute change of the wastewater discharge, its relative share in the total volume of wastewater disposal in water sources has remained unchanged in recent years (its value is about one third).

According to Roshydromet (Russian meteorological service), in 2016, there were 2,990 cases of high pollution (HP) and extremely high pollution (EHP) registered within the territory of the Russian Federation. Extremely high levels of surface water pollution occurred in 638 cases at 136 water bodies, which is 6 % more than in 2015 (598 cases at 144 water bodies). High levels of pollution were observed in 2352 cases at 321 water bodies (in 2015 – 2423 cases at 331 water bodies). Over the past five years, the total amount of HP and EHP of surface waters remains approximately at the same level.

Over the past ten years, if compared with other subjects of the Russian Federation, Sverdlovsk region has the largest number of cases of HP and EHP [1].

3. Facts and figures of water management and wastewater disposal in Sverdlovsk region

The quality of the surface waters is largely formed under the influence of economic activity, primarily under the influence of industrial and domestic wastewater discharges. While the share of discharged polluted wastewater in the Russian Federation in 2016 is 34.2 % of the total volume of wastewater discharge [1], the same value in Sverdlovsk region is 75 % [2].

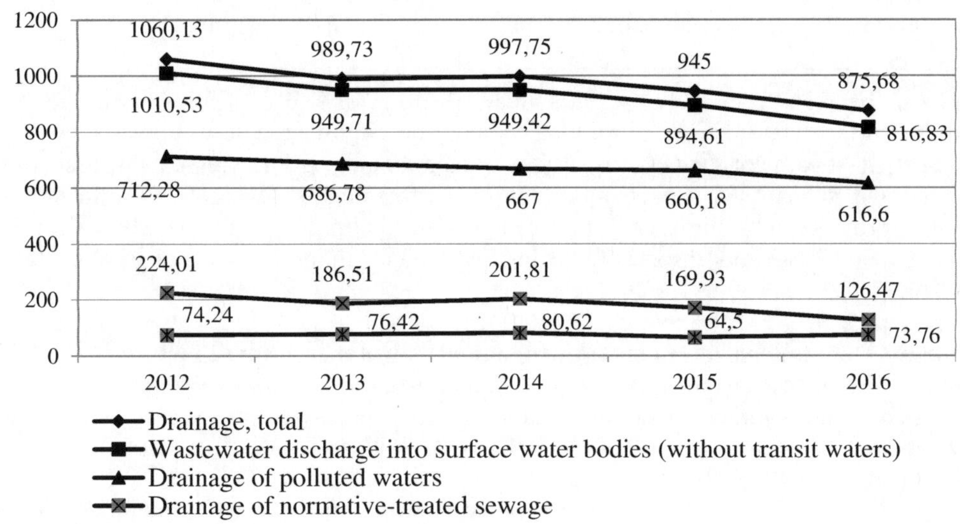

Figure 1. Dynamics of wastewater disposal in Sverdlovsk region (million cu. m)

The dynamics of wastewater disposal in Sverdlovsk region is shown on Fig. 1.

The greatest amount of contaminated wastewater enters the surface water sources of the region from a number of enterprises working in the spheres of:
communal services – 55 percent;
processing industries – 30 percent;
minerals extraction – 9 percent.

The list of enterprises which are the main sources of pollution of surface water bodies in Sverdlovsk region in 2015–2016 is given in table 1.

Table 1. List of enterprises which are among the main sources of pollution of surface water bodies in Sverdlovsk region in 2015–2016

No.	Name of the Enterprise	Wastewater discharged, total (mln cu. m)		Containated wasterwater discharged (mln cu. m)	
		2015	2016	2015д	2016
1.	Municipal Unitary Enterprise 'Vodokanal', Municipal Settlement 'Yekaterinburg City'	144.90	135.24	144.90	135.24
2.	OOO 'Vodokanal-NT', Nizhny Tagil, Gornouralsky Urban District	44.39	41.71	44.39	41.71
3.	Joint-stock company 'EVRAZ Nizhny Tagil Metallurgical Complex', Nizhny Tagil, Gornouralsky Urban District	29.20	33.43	29.20	33.43
4.	PAO 'Uralchimplast', Nizhny Tagil	31.97	28.95	31.97	28.95
5.	Joint-stock company 'Siberian-Urals Aluminium Company', Krasnoturinsk Urban District, Kamensk-Uralsky	25.66	27.40	25.66	27.40
6.	Pervouralsk Production Municipal Unitary Enterprise 'Vodokanal', Pervouralsk Urban District	24.15	20.77	24.11	20.73
7.	PAO 'VSMPO-AVISMA Corporation', Verkhnyanya Salda Urban District	20.25	20.22	20.21	20.18
8.	Municipal Unitary Enterprise 'Vodokanal', Novouralsk Urban District	19.88	16.66	19.88	16.66
9.	OAO 'Vysologorsky Ore Mining and Processing Industrial Complex', Urban District of Kushva, Nizhny Tagil	15.18	16.41	15.11	16.41
10.	Joint-stock company 'Vodokanal KU', Kamensk-Uralsky	16.75	16.03	16.75	16.03

As it can be seen from table 1, contaminated sewage of urban utilities is the main source of pollution of surface water bodies of regional cities. The largest volume of contaminated wastewater after biological treatment comes to surface water bodies from Municipal Unitary Enterprise 'Vodokanal' which has 22 % of the total discharge of contaminated sewage in Sverdlovsk region [2].

The data provided in accordance with the state statistical report of 2TP Vodkhoz for consecutive five years dynamics of water disposal by Municipal Unitary Enterprise 'Vodokanal' in Yekaterinburg is shown on Fig. 2.

4. Importance of modernization and improvement of water treatment facilities

Transition of Russia towards the technological standardization makes modernization and improvement of water treatment facilities important. Technological standardization process is regulated by the information and technical reference book ITRB 10-2015 'Waste Water Treatment Using Sewerage for Settlement and Urban Districts' [3].

Along with the changes in the system of wastewater disposal from municipal facilities and according to the Federal Law No. 7-FZ, newly adopted Federal Law No. 225-FZ has completely changed the technological regulation system of sewerage for settlements (VSSS) [4], which was previously reflected in Chapter 5 of Federal Law No. 416-FZ 'About water supply and Sewerage' [5]. Starting from 2019, the technological regulation system of wastewater treatment facilities will become completely different.

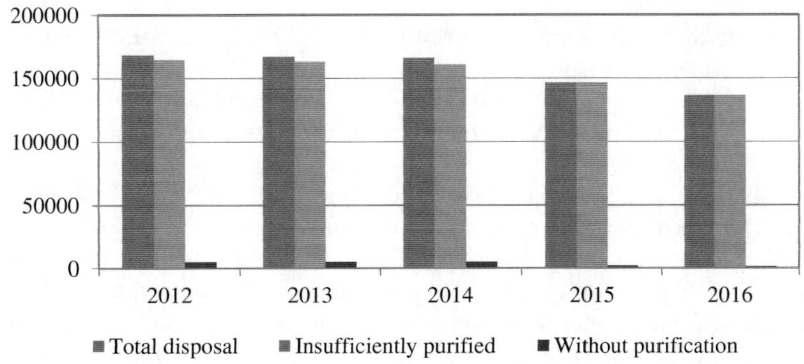

Figure 2. Dynamics of wastewater disposal of Municipal Unitary Enterprise 'Vodokanal'

The practical implementation of the system of technological regulation was initiated by the adoption of the Federal Law of 21.07.2014 No. 219-FZ [6]. Starting from 2019 this document will radically change the existing principle of standardization of the impact of production and economic activities on the environment, including wastewater discharge into surface water bodies: it provides the transition from water quality standards established for water objects for fishery usage towards regulation established on the basis of the best available technologies. The law considers the essential features of the municipal sewerage industry. It is very important that the law discloses the basic provisions of ITBR 10-2015 on linking the technological indicators of BAT with the categories of water bodies and the size of municipal wastewater treatment facilities. In addition, the law will seriously reduce the fees charged to the water companies for discharging pollutants [7, 8].

Thus, in order to improve the quality of discharged wastewater and to reduce its impact on water sources and the volume of discharges, it is necessary to build new wastewater treatment facilities or reconstruct and expand existing ones, build local treatment facilities, post-treatment units, and start using recycling and water reuse systems.

Despite the actual quality of surface water sources, clean potable water provision for the population is accompanied by the formation of contaminated rinsing waters and waste sediments, which also make a significant contribution to the pollution of water bodies and represent a problem that needs to be addressed. The amount of the rinsing waters is 7–10 % of the average amount of daily water consumption.

The sediment formed in the process of water purification, as a rule, is less dangerous for the environment and a human himself than, for example, the wastewater sediment is. At the same time, water sediment (especially formed during the purification of high values of water color and low-turbid waters) has higher values of the resistivity index of filtration and requires high costs for thickening and dehydration. Placement of water sediment in accumulation reservoirs or on dehydration sites leads to the expropriation of significant land areas that have been withdrawn from economic use for a long time, which is especially important for large cities. In addition, the hydroxide precipitate, in its original form, often poses a certain danger to the environment and the man, since the substances contained in it can, under certain conditions, be included in geochemical and biogeochemical cycles [9, 10].

The problem of water sediment is most often solved by its discharge into the lower lands, water bodies and streams, by placing it on silt areas (lagoons) with drying and subsequent export to dumps. Until recently, in many countries of the world, including Russia, the water sediment (with a single use of the coagulant) was often discharged in the form of sludge pulp into water bodies and streams, which

IOP Conf. Series: Earth and Environmental Science **177** (2018) 012031 doi:10.1088/1755-1315/177/1/012031

resulted in the supply of significant amounts of sedimentary material contaminated with products of hydrolysis of coagulants and various impurities contained in reagents and isolated from the source water. Negative ecological phenomena are also observed when placing water sediments at dehydration sites or in oxidation ponds.

The problem of finding resource-saving and cost-effective technologies for cleaning rinsing waters on the filter structures of waterworks is urgent.

This task is strategically important for the sustainable development of the urban environment and improving the quality of life of the population of the city of Yekaterinburg.

5. Feasibility study for the construction of a rinsing waters treatment and sludge dewatering unit at the main waterworks of MUE 'Vodokanal', Yekaterinburg

To solve this problem at the main waterworks of MUE 'Vodokanal', it is necessary to build a rinsing waters treatment and sludge dewatering unit. This workshop will allow not only to exclude the discharge of polluted rinsing waters into the environment, but also to substantially reduce the required intake of fresh water due to the re-use of rinsing waters, the main pollutant of which is suspended solids.

The aim of the project is to process the rinsing waters of the filter station of the Municipal Unitary Enterprise 'Vodokanal' along with the production of a dehydrated sediment in order to reduce the man-made influence on environment by eliminating the rinsing waters discharge, as well as by reducing the amount of water intake from the surface sources into the technological process of water purification.

The urgency of the project is explained by the high level of pollution of the reserve water source with its own production waste, a high payment for the negative impact on the environment, decrease in the quality of life and in the urban development.

Nowadays, various modern sludge dewatering technologies are used. Advantages of mechanical sludge dewatering include high productivity, reduction of the required area of the site due to elimination of waste areas, elimination of unpleasant odors, reduction of number of maintenance personnel, absence of climate factors influence on the dehydration process.

In the process of investigation, the authors considered three methods of mechanical sludge dewatering, among which are a chamber filter press, a band filter press and a decanter [9, 10].

A strip press filter will be used in this project due to the combination of low energy consumption, acceptable dehydration rates and low initial investment.

Realization of the investment program of the enterprise for 2014–2025 "Development of Water Supply and Sewerage Infrastructure" [11] is expected to result in the performance increase of the station from 120 to 300 thousand cu. m per day. This will lead to an increase in the total amount of discharged rinsing waters, and, as a consequence, to the increase in the amount of sediment.

Laboratory tests conducted at the plant in 2016 showed that a significant rate of water contamination was influenced by the dry residue and suspended solids. The average monthly suspended solids content in the discharged water is 67.3 g / cu. m, which is about 30 %. Regularly a filter station discharges about 353.73 tons of suspended solids per year. As a result of station modernization and reaching full working capacity, this figure will reach 880 tons per year.

Dewatered sediment can be used for disturbed lands remediation. The evaluation of the water treatment sludge in terms of its suitability for remediation was carried out in the Analytical Laboratory of the Ural State Mining University in accordance with the appropriate standards for remediation. The evaluation revealed that three main indicators (pH, dry residue and the amount of toxic salts in the water extract) show that the water treatment residue belongs to the potentially fertile group.

Sediments represent poor soluble multicomponent mixtures, the priority components of which are aluminum, iron, calcium, silicon, manganese. The solubility of sediments is negligible. Sediments are not a radiological hazard; they do not have any ecotoxicological properties. Based on this, they are classified as non-toxic, environmentally safe waste [9, 10 and 12]. The sediment corresponds to the requirements for soils of sanitary protection zones (land allocation for transmission lines).

Calculation of fees for negative impact on surface water bodies by the rinsing waters of the filter station was carried out in accordance with the tariffs set by the Russian Federation Government Decree No. 913 dated of September 13, 2016 [13]. In carrying out all the proposed measures, the absence of payment for negative impact can be considered as an additional economic effect. According to the conditions of the previous negative impact, its value is estimated for more than 800 million rubles per year for the maximum performance of the filter station.

The results of laboratory studies of the contamination of rinsing waters after the removal of suspended solids unequivocally confirm that this water is not advisable to be discharged, because it is most suitable for potable water. The biological oxygen demand (BOD_{20}) approximately coincides with the BOD of the water source. According to inorganic components, the same statement applies to the rigidity of water, to the content of dry matter, chlorides, sulfates, calcium and magnesium. Water contains an average of 1–2 g / cu. m of active chlorine, which corresponds to the level of microbiological indicators. Thus, up to 84 % of the generated water is returned to the water treatment process and can be used for re-treatment and further transportation to the consumer (Table 2).

Table 2. The amount of washing waters formation and process water usage

Capacity of filter station, thousands of cu. m per day	Amount of rinsing waters, thousands cu. m per day	Amount of water returned to the technological process, thousands cu. m per day
120	14.4	12.1
150	18.0	15.1
300	36.0	30.2

For the implementation of the project, it is necessary to purchase vehicles (dumper), equipment for the site (flotator, compressor, belt press filter, sludge pump, flotation water pump, metering pump) and to build the facility itself. The total amount of investments is estimated at 2,221,097 rubles. The sources of this project is investment program (investment component in the tariff for the potable water) and budgetary appropriations (finances of budget funds of various levels). Project implementation period is 12 years.

The project includes the calculations on the operating costs for sludge dewatering and rinsing waters treatment in accordance with the calculations on the increase in the performance of filter station; the cost of water treatment is also determined. The profit of the enterprise, which results from the use and recycling of rinsing waters is shown in Table 3.

Table 3. Profit from rinsing waters recycling

Period	Cubic capacity of water recycling, thous. of cu. m per year	Cost of the drinking water		Cost of the drinking water		Profit, thous. of RUB per year
		Rate, RUB.	thous. of RUB per year	Rate, RUB.	thous. of RUB per year	
2020	4362.48		37517.32		150941.81	113424.48
2021–2023	5453.10	8.60	46896.66	34.60	188677.26	141780.60
2024–2029	10906.20		93793.32		377354,52	283561.20

Table 4. The main indicators of economic efficiency of the project

No.	Indicator Description	Unit of measurement	Indicator values
1	Net profit	thous. of RUB	6 771 597.22
2	Net present value	thous. of RUB	729 037. 82
3	Pay-back period	years	4.84
4	Return on investment index	–	3.05
5	Return on discounted investment index		1.33
6	Internal rate of return	%	21.38

The main indicators of the economic efficiency of the project calculated with the 15 % discount rate are given in Table 4.

Analyzing the obtained indicators of the economic efficiency of the project, it can be conclude that the project for the organization of the sludge dewatering site at the municipal unitary enterprise 'Vodokanal' is cost-efficient and advisable for implementation.

6. Conclusion

The problem of safe sludge deposition and the use of rinsing waters purification technologies on filter stations is extremely relevant for Russia, where surface water is the main source of centralized water supply. In this regard, the implementation of the proposed project is important and cost-effective. The main ecological results of the project are:

1. exclusion of dumping of contaminated rinsing waters from MUE 'Vodokanal' facilities and its reuse in the technological process;
2. prevention of harm caused to water resources and soils;
3. utilization of the dehydrated sediment and obtaining of the final product;
4. reduction of the amount of fresh water intake from the surface water source.

Final decision on the choice of methods for rinsing waters treatment and options for the water sludge utilization should be taken after the thorough feasibility study of the various options and their ecological and hygienic significance. Anyway, to introduce the already developed methods of utilization, it is necessary to accelerate the construction of shops for mechanical sludge dehydration at filter stations.

Besides, there is need for further research of the features of the formation and properties of a water-filled sludge formed in various water supply stations in various design features, in different fresh waters, improvement of well-known methods and development of new ones for economically and environmentally justified utilization to introduce them into practice. This will not only significantly improve the operation of water treatment facilities but will also contribute to the implementation of the objectives of environmental protection, improve the quality of life of the urban population and help in sustainable development of urban areas.

References

[1] State report *On the State and Environmental Protection of the Russian Federation in 2016*
[2] State report *On the Condition and Environment Protection in the Sverdlovsk Region in 2016*
[3] ITRB 10-2015 *Waste water treatment using centralized drainage systems of urban settlements* (Information and technical reference book on best available technologies)
[4] Federal Law No. 225 of 29.07.2017 *On Amendments to the Federal Law "On Water Supply and Sewerage" and certain legislative acts of the Russian Federation*
[5] Federal Law No. 416 of 07.12.2011 *On Water Supply and Sewerage*
[6] Federal Law No. 219 of 21.07.2014 *On Amendments to the Federal Law "On Environmental Protection" and certain legislative acts of the Russian Federation*
[7] Budnitskiy D M and Daniovich D A 2017 Discharge procedure for rationing water utilizes will change dramatically in 2019 *Best Available Technologies of Water Supply and Sanitation* **4** 2–7
[8] Danilovich D A 2018 2017 in Review: a change in rationing the dischargers of water utilities and their clients *Ecology of Production* **1** 10–18
[9] Blagorazumova A M 2014 *Treatment and dehydration of urban sewage sludge: a study guide* (Saint Petersburg)
[10] Chitikela S R 2016 Conditioning and Mechanical Dewatering of Municipal Wastewater Biosolids-A Critical Review *16th Annual World Environmental and Water Resources Congress* (West Palm Beach, FL) 166–74
[11] Decree of the Government of the Sverdlovsk Region of 30.11.2015 No. 1274-RP *On amendment in the investment program of Yekaterinburg municipal unitary enterprise of*

water supply and Sewerage (MUE 'Vodokanal') Yekaterinburg City in the years 2014–2025 (development of infrastructure of water supply and Sewerage)

[12] Dyakov M S, Tsybina A V and Gruzdeva U V 2014 Technologies of processing and dewatering of sewage sludge: a retrospective and prospective ways of development *Bulletin of Perm National Research Polytechnic University. Applied Ecology. Urban Studies* **3(15)** 111–26

[13] Decree of the Russian Federation Government of 13.09.2016 No. 913 (edited of 09.12.2017) *On rates of payment for negative environmental impact and additional coefficients*

'Published under licence in *Journal Title* by IOP Publishing Ltd'.

Alternative energy as a tool for the development of small and medium-sized enterprises in Russia

L Teslyuk, A Boyarinov, N Dukmasova, I Plastinina and O Romanova

Department of Environmental Economics, Ural Federal University named after the first President of Russia B. N. Yeltsin, 19 Mira str., Ekaterinburg, Russia

e-mail: au.boyarinov@net-ustu.ru

Abstract. Nowadays, the level of development of small and medium-sized enterprises (SMEs) in Russia lags behind that of the states with advanced economy. Problems with connection to the power supply attributed to prohibitive costs and time-consuming procedures can be distinguished among many other reasons. The aim of this paper is to consider feasibility of applying alternative energy sources to the needs of small and medium-sized enterprises through the study of the Sverdlovsk Region. While analyzing local energy sources of the region with due regard to its geographic and climatic peculiarities, we have identified the most promising sources of energy that require state support for their efficient use. Estimation of economic efficiency of the implemented project of alternative energy supply at a small enterprise proves feasibility of using a stand-alone solar panel based energy supply system.
Keywords: small enterprises, energy supply, alternative energy sources, solar panels, economic efficiency.

1. Introduction

In democratic states with advanced economy small enterprises are the backbone of their national economy. Their contribution to GDP illustrates the welfare of the population. They generate jobs for the able-bodied population and form the so called middle class whose core is represented by economically independent and socially active citizens. Small enterprises in the OECD member states are a powerful force for scientific and technological progress. The majority of innovation projects, patents and certificates obtained in different spheres, from medicine to IT, accrue to small and micro-sized enterprises.

The attitude towards small enterprises in Russia is slightly different. Since the existence of the USSR and Gosplan (State Planning Committee of the USSR) national economy has been showing the trend towards production consolidation irrespective of a specific industrial sector. After the collapse of the USSR, the situation changed in favour of small enterprises, but not dramatically. The share of large companies and public sector in GDP has been rising steadily. Small enterprises in our country causes problems to many entities related to entrepreneurship activities or their regulation. State government bodies must spend enormous amounts of money to monitor the activities of dozens of small companies with minimum annual turnover, manage taxes, etc. Large companies due to their clumsiness and internal bureaucracy cannot compete fairly and honestly with small enterprises at local markets and have to beg local authorities, which they fused with in the 1990s, for help. Due to public awareness formed by domestic media and the experience of the first years following the collapse of

Content from this work may be used under the terms of the Creative Commons Attribution 3.0 licence. Any further distribution of this work must maintain attribution to the author(s) and the title of the work, journal citation and DOI.
Published under licence by IOP Publishing Ltd

the USSR, the population has quite a negative attitude towards the entrepreneurs considering them as people who became rich following the wrong path and who have criminal background.

The situation is aggravated by problems which, as it seems, should not occur in a market economy. SMEs have difficulties with connection to resources, such as gas distribution network, electrical grid, water supply lines, and sewerage network even *on a fee-for-service basis*. Prohibitive costs and time-consuming procedure required for connection are not always viable in case of entrepreneurs whose activity suggests low power consumption.

This paper examines feasibility of using alternative energy sources by small enterprises located far away from the centralized power supply networks, in the geographical and climatic conditions of the Middle Urals. Among renewable resources available in this region, the following can be identified: solar energy, wind power, small hydro power, and biomass.

2. Challenges for SMEs in Russia

Small business in the states with advanced economy plays a key role in the economy, providing fair competition, supplying new commodities and services to the market, and covering the needs of large enterprises. Development of small enterprises in the states with advanced economy is more rapid, as national authorities elaborate a lot of programs to support them and grant privileges considering this sector very important, which influences micro- and macroeconomic performance, i. e. GDP, loan interest rates, economic competitive ability, employment of population, etc. (Table 1) [1].

Table 1. Key figures reflecting the role of small business in different countries (2015)

Value	Russia	USA	Japan	Germany	England	Singapore
Share of small enterprises in GPD of the country, %	20	52	51.6	57	52	53
Share of small enterprises in total employment, %	26	50.1	69.5	69.3	55.5	62.3
Share of small enterprises with regard to total number of enterprises, %	98.5	97.6	99.2	99.3	99.1	98.5

For example, in Germany subsidies granted to small enterprises account for about 4 billion euros annually (in comparison, in Russia in 2017 it was 7.5 billion rubles or, on the average weighted exchange rate of the Russian rouble to euro equal to 70 RUB/EUR, 107 million euros). Two committees deal with the problems of small businesses in the U.S. Congress. Their aim is to support small business at a government level. Japan, with a particularly high number of small enterprises, distinguishes those of them that cannot develop on their own under market economy conditions without any state financial aid. States with advanced economy support small business financially and give privileges, particularly tax privileges [2].

In the developed countries, the main problems related to government interference into business are the complexity of the tax system and government overregulation of certain business spheres. However, if in Russia bureaucracy and excessive activity of inspection authorities are the main reasons hindering the development of small businesses, in the advanced economies it happens primarily due to economic and external factors. Small businesses in the West are very sensitive to contraction of domestic demand. The second largest problem affecting western small enterprises is a tremendous amount of pirated items coming from Asian countries. Fraudulent use of brand names and low quality of pirated goods inflict substantial damage to reputation of American and European trademarks. The third challenge for small enterprises in the developed countries is high standards of living assured not only by their economic strength, but also by government regulation. The OECD member states provide high minimum wage level; contributions to medical insurance funds and pension contributions are mandatory. This makes western companies spend substantial funds for job compensation, which increases the production cost of goods and decreases their competitive ability [3].

In Russia, due to insignificant influence of small businesses on GDP, analysts, experts and state officials, as a rule, distinguish and describe *small and medium-sized enterprises (SMEs)* rather than small business to increase the effect of their analytical efforts, development forecasts, and elaboration of SMEs' support programs. The total number of operating SMEs in Russia as of 2017 was about 5.5 million (according to data from the Federal State Statistics Service).

Federal support includes federal SME support programs, initiated by the Ministry of Economic Development of the Russian Federation, Ministry of Agriculture of the Russian Federation and other ministries and agencies. The following government agencies were established: *Federal Corporation for the Development of Small and Medium Enterprises* (*SME Corporation*, AO), *Russian Bank for Small and Medium Enterprises Support* (*SME Bank*). *SME Corporation* is responsible for monitoring support granted to SMEs as per Federal law No. 209-FZ dated July 24, 2007 *On the SMEs' development in the Russian Federation*. *SME Bank* is responsible for government financial aid program (bank loans, microloans, receipt of assets under the finance lease agreement, accounts receivable financing) [4].

However, despite SME support on the federal, regional and municipal levels, small and medium-sized enterprises face a lot of problems. As per the data provided for by the analysts of the Center for Macroeconomic Analysis and Short-Term Forecasting [4], 3,227 companies have gone bankrupt in Russia in different spheres since the beginning of 2017 due to the following reasons: lack of stability, inflation and economic disenfranchisement, weakening of the Russian currency, and increase in rates of the Bank of Russia. Other experts declare that SME problems are attributed to the critical number of different inspectors who strangle business by their inspections. SMEs find it difficult to get government financing funds, face problems during tenders and auctions pertaining to state procurement. These problems are fundamentally different from the difficulties faced by SMEs in the USA, Japan, and EU countries.

In the 2000s, SMEs, in our opinion, encountered a very unexpected problem that substantially inhibits SME development and remains a problem even today, i. e. arbitrary behaviour of such natural monopolies as *RAO Unified Energy System of Russia* (after reformation of energy industry *Federal Grid Company of the Unified Energy System of Russia* and regional grid companies), *Gasprom* (PAO). Even large industrial companies had difficulties in connection to electrical grid and gas supply network, not to mention SMEs.

All the regions in Russia find themselves in a similar situation. Key economic indicators show that the Sverdlovsk Region is one of the most developed major industrial territorial entities of the Russian Federation (according to data from the Federal State Statistics Service). The Government of the Sverdlovsk Region also makes efforts to support SMEs. Subprogram *Development of small and medium-sized entrepreneurship* has been approved and is being implemented (within the framework of the 2013–2020 State Program of the Russian Federation *Economic Development and Innovative Economy*). The analysis of different ratings and research works pertaining to the analysis of entrepreneurial and investment climate in the regions of the Russian Federation showed that the Sverdlovsk Region is not the leader in this sphere and holds middle positions among the territorial entities of the country [5]. The Sverdlovsk Region is one of the regions with a very low level of SME state support, i. e. only 293.4 rub of federal financing per capita. As many entrepreneurs remark, the most time-consuming procedure when connecting to the power supply system in the Sverdlovsk Region is elaboration and approval of the project for external electricity and gas supply.

3. Analysis of alternative energy sources in the Sverdlovsk Region

In the Urals, in general, and in the Sverdlovsk Region in particular, the main sources of electrical energy for large companies, small and medium-sized enterprises, and for housing and community amenities are heat power, hydropower, and nuclear power (*the Beloyarsk Nuclear Power Station* with the fast neutron reactors).

Renewable sources of energy, such as solar energy, wind power, and biomass also find their application, though it happens very rarely. This is due to several reasons. First of all, low specific power necessitates large sizes of electric power stations, which leads to the increase in capital investment in comparison with conventional power stations. Second, climate variability influences certain kinds of energy, for example, solar energy and wind power depend on the season and meteorological conditions. Third, the obtained energy has to be immediately used (which is not always efficient) or accumulated. Accumulation of energy in large quantities is an unsolved technical challenge. Finally, procurement and maintenance of special equipment incur high costs, which makes such sources of energy uncompetitive in comparison with conventional sources.

In view of the reasons stated above, it is clear that large companies do not see any benefits in application of renewable sources of energy on a large scale, but it is different in case of SMEs. In the Urals alternative sources of energy are often adopted in SMEs in the Chelyabinsk and Kurgan Regions. The following can be identified among renewable resources available in the Sverdlovsk Region: solar energy, wind power, small hydro power, and biomass [6, 7].

3.1. Solar energy

In 2014 *RealSolar* performed research pertaining to the amount of solar energy in different regions of Russia. The data on daily solar radiation is given in Table 2 for several Russian cities.

Table 2. Daily solar radiation, kWh/m^2, horizontal platform

City	Jan	Feb	March	Apr	May	June	July	Aug	Sept	Oct	Nov	Dec	Average per year
St. Petersburg	0.35	1.08	2.36	3.98	5.46	5.78	5.61	4.31	2.6	1.23	0.5	0.2	2.8
Moscow	0.5	0.94	2.63	3.07	4.69	5.44	5.51	4.26	2.34	1.08	0.56	0.36	2.63
Kazan	0.68	1.44	2.82	4.29	5.52	5.93	5.72	4.49	2.86	1.51	0.83	0.54	3.06
Rostov-na-Donu	1.27	2.09	2.98	4.09	5.53	5.76	5.86	5.17	3.85	2.38	1.31	1	3.45
Nizhny Novgorod	0.64	1.45	2.75	3.95	5.34	5.6	5.5	4.27	2.69	1.45	0.75	0.45	2.91
Yekaterinburg	0.64	1.5	2.94	4.11	5.11	5.72	5.22	4.06	2.56	1.36	0.72	0.44	2.87
Novosibirsk	0.69	1.37	3.02	4.08	5.05	5.48	5.01	4.29	2.93	1.44	0.8	0.62	2.91
Khabarovsk	1.64	2.72	4.11	4.61	5.39	5.86	5.42	4.53	3.81	2.56	1.72	1.28	3.64

The research showed that solar energy in the Sverdlovsk Region has small potential. However, as domestic and foreign experience shows [8–10], even such a moderate irradiation makes it possible to use solar panels in separate remote objects with the aim of cost reduction of diesel power generation (for 1 kWh) and building of a stand-alone power supply system with low power consumption.

In our opinion, the climate in the Urals makes it feasible to apply solar energy to power SMEs if there are no other sources of energy.

3.2. Wind power

Annual average wind speed is a very important indicator for application of wind power plants.

The Sverdlovsk Region is characterized by quite a nonuniform distribution of wind current within its territory. As per the data provided by *Meteoblue*, annual average wind speed in the south and east of the Sverdlovsk Region reaches 1–2 m/sec, while in the west and in its central part, 2–4 m/sec, and in the north, 3–6 m/sec. The highest annual average wind speed (5.5–10 m/sec) can be found at the tops of the Ural mountains' spurs (Blagodat, Kachkanar, Magnitnaya, etc.) and in the North Sosvinsky height where annual average wind speed reaches 6–12 m/sec [10].

Application of wind power plants to power large companies in the Sverdlovsk Region is not efficient. First of all, wind is unstable, it can rise and slacken. Second, the speed of wind in the Sverdlovsk Region is quite low. Third, wind turbines of good quality are very expensive. In the fourth place, wind power plants are characterized by high level of acoustic pollution. Finally, they cause interference to TV and different communication systems. However, wind turbines can be used as a source of electrical energy together with solar panels, diesel-, petrol- and gas generators for country houses, auxiliary premises and farm enterprises, small houses, small-scale producers in remote regions where there is no central power supply network or the parameters of the available electric power are not appropriate for the consumer.

3.3. Biopower

There is quite a long list of possible sources of raw materials for biopower, for example, oil crops, sugar and starch crops, their waste products, animal fat, timber, wood wastes, peat, biodegradable solid municipal wastes, sewage sludge, manure, etc. But peat, wood wastes and wastes from farm enterprises are the most promising sources from this list for the Sverdlovsk Region. Nevertheless, these resources remain untapped, and the development of this trend in the near future is unlikely.

3.4. Mini hydro power plants

There are a lot of river water resources in the Sverdlovsk Region. There are 18,414 rivers with total length of 68,000 km, besides 135 water-collecting areas, more than 1,200 ponds, and 2,500 lakes (according to data from the Ministry of Natural Resources and Environment of the Sverdlovsk Region).

Mini hydro power plant is one of the most promising ways for the development of the alternative energy. The companies located in remote and hard-to-reach areas, far away from power lines can use small hydro power plants. They can operate using current energy of small rivers and even creeks. Modern plants are easy to operate and fully automatic. They can operate both standalone, and as part of an electrical grid. What is more important, the quality of current produced by small hydro power plants complies with the requirements of state standards (GOST) with regard to voltage and frequency.

All the enumerated alternative sources of energy are currently important for SMEs. The choice of this or that source for a SME is determined by its economic efficiency in comparison with other sources, including connection to public networks despite high capital investment.

Several examples of successful implementation of alternative sources of energy application despite the difficulties of connection to public electric power and (or) gas distribution networks can be found in the Middle Urals. One of such examples is described below.

4. Stand-alone power supply of a small enterprise

An enterprise registered in Yekaterinburg as a legal entity X is engaged in distribution of tools and components of technical production goods, including those of own production, within the territory of the Urals and Russia. In compliance with the criteria stipulated in regulations of the Russian Federation, this enterprise falls within the category of small business. Nowadays, the main territory of this enterprise includes production areas and warehouses, but also there are two additional sites for shipping of the goods with convenient approaching roads.

One of the sites that was launched in 2013 is 30 km away from Yekaterinburg. When the site was under construction, the entrepreneurs faced a serious problem, i.e. absence of power supply. The total length of power lines required for connection to the centralized power supply was 1.7 km, which necessitated capital investment equal to 1,500,000 RUB (or 47,100 EUR). The owners of the enterprise found such a connection economically inefficient due to high nonrecurring expenditures and time-consuming procedure of connection for a relatively low power consumption. Eventually, they chose a project of alternative power supply which was successfully implemented.

Warehouses, an office building and an accommodation building are located within the premises of this site.

The choice of the room lighting devices, household and office equipment was based on their power consumption bearing in mind the capacity of installed solar panels, i.e. 3 kWh/24 h or 1,095 kWh/year. LED lighting devices were installed in the office, while mercury lamps mounted in the warehouses. To save electric power consumption, PCs were replaced with laptops, and ink-jet multifunction printer was used. Radio connection was adjusted through a channel with 5 Mbit/s. Expenses associated with electric power in the accommodation building included lighting, a fridge, a

TV set, and a phone. The rooms were heated with radiant convector gas heaters; LPG delivered in balloons was used as fuel. This site was less operated during winters.

To generate electric power, a stand-alone solar panel based energy supply system was used. The equipment was installed by the in-house personnel (Figure 1 and 2).

Calculations of capital investments for installation of this system is given in Table 3. Funding of the project was carried out by the enterprise itself.

Table 3. Capital investment

Items for capital investment	Sum, RUB	Sum, EUR[a]
1. Solar panels, 200 W, 24 V, 9 pcs	87,000	2,730
2. MPPT controller for solar panels	20,000	620
3. *Docker* accumulator batteries 12 V, 200 A/h, 3 pcs	17,000	530
4. Inverter, 2,500 W	8,500	270
5. Installation materials, commutation and other equipment	10,000	310
6. *Hitachi* generator, 2 kW (for emergency recharging)	17,000	530
7. LED lighting	12,000	380
8. *Alpine* radiant convector gas heater, 4.5 kW (heating of the office and accommodation building), 2 pcs	12,000	380
Total	183,500	5,760

[a] Official average ruble-euro exchange rate of 2013: 31.85 RUB/EUR.

The stand-alone energy supply system demonstrated stable operation for more than two years. There were practically no failures, and if any, it was mainly attributed to human errors. The system did not require activation of an emergency petrol generator for 9-10 months a year. The least efficient months were October, November, and December due to low luminous flux and short length of light days. With low recharging of the solar panels during the day and with a 24-hour operation of the accommodation building, accumulator capacity was reduced to zero within a week time.

Figure 1. Solar panels. General view

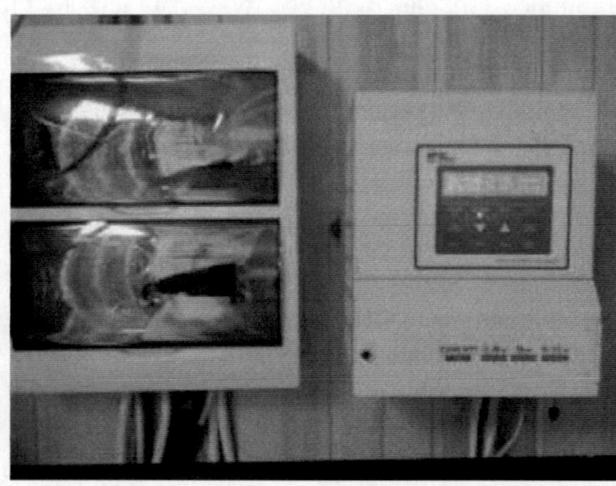

Figure 2. MPPT controller for solar panels

To evaluate economic efficiency of application of a stand-alone solar panel based energy supply system, we can compare expenses incurred for three variants of generation of power that can be divided into two kinds, i. e. electric power (for lighting, communication, operation of the household and office equipment) and heating energy. We consider the following as sources of energy:

Variant 1: the system on the basis of solar panels and gas convectors.

Variant 2: *Hitachi* generator (fueled with petrol) and gas convectors.

Variant 3: centralized power supply network. This variant implies heating of the rooms with electricity.

Calculations of operating energy costs for three variants of energy generation is given in Table 4.

Table 4. Calculation of energy costs.

Parameter	Variant 1	Variant 2	Variant 3
1. Capital investment, RUB/EUR[a]	183,500/5,761	116,000/3,642	1,500,000 /47,096
2. Annual energy and fuel consumption, including			
2.1. electric power, kWh	1,095		7,420
2.2. petrol, l		1,003	
2.2. LPG, l	794	794	
3. Annual operating costs, RUB, including	33,992	59,593	191,552
3.1. electric power			41,552
3.2. petrol		33,701	
3.3. gas	14,292	14,292	
3.4. depreciation	19,700	11,600	150,000
4. Present Value[b], RUB	-213,744	-290,240	-1,447,499
5. Prime cost of 1 kWh, RUB/EUR[a]	16.89/0.53	37.29/1.17	25.82/0.81

[a] Official average ruble-euro exchange rate of 2013: 31.85 RUB/EUR.

[b] This parameter takes into consideration capital investment and operating cash outflow. It is calculated for a 10-year period with a discount rate of 17 %.

Our analysis showed that the first variant, i.e. application of a stand-alone solar panel based energy supply system, is the most cost efficient for the enterprise. However, the comparison of the costs of 1 kWh shows that solar energy is more expensive, limited and unstable source of electric power in the Middle Urals.

High energy costs in case of a centralized power supply network are attributed to high expenses for connection to the electrical grid, which is an obstacle for the small business development both in the Urals, and within the territory of Russia. Even if technological connection of power receiving appliances of electrical energy consumers to the available power lines is possible (generally, not further than 300 m), nonrecurrent costs amount for 750 thousand rubles. Therefore, it is very important to consider means of small business support that can be provided by state government with regard to power supply. Application of a stand-alone solar panel based system is efficient as a backup electric power supply for communication systems, alarms, heating pumps, and the server of the enterprise.

5. Conclusion

1. Small and medium-sized enterprises are the backbone of the economic growth in the developed states, and account for more than half of the gross domestic product. According to the plans of the Government of the Russian Federation, this figure will be reached in Russia by 2040.

2. SMEs throughout the world require special treatment and support. Global experience show that despite state support the number of new small businesses registered every year equals approximately the number of the companies shut down every year. Without state support existing imbalance with regard to companies-bankrupts and companies that successfully continue their activity would dramatically change in favor of the former.

3. Besides conventional problems related to corruption, bureaucracy, and frequent changes of legal framework, SMEs in Russia face specific challenges, such as connection to public networks (electric power supply, gas and water supply, sewerage).

4. One of the solutions of these specific problems can be application of alternative sources of energy. Feasibility of applying some alternative sources has been analyzed through the example of the Sverdlovsk Region. The most efficient kinds of alternative energy appropriate for this region have been identified.

5. Calculations for adoption of alternative sources of energy have been given for a specific enterprise located in the Sverdlovsk Region, including a respective feasibility report.

References

[1] Salyakhova E 2016 *Novainfo: Economic sciences* **57-4** 200–7

[2] Mills K G and McCarthy B 2014 *The State of Small Business Lending: Credit Access during the Recovery and How Technology may change the Game* (Cambridg: Harvard Business School)

[3] Lee S and Denslow D 2004 Fall/Winter. *J. of Small Business Strategy* **15** **(2)** 77–89

[4] Ushakova A 2017 *Stat support of Small and Medium Enterprises in the Russion Federation: Challenges an Opportunities in the Republic of Tatarstan. Thesis* (Helsinki: Helsinki Metropolia University of Applied Sciences)

[5] Tolmachev D and Ulanova E 2015 *J. Economy of Region* **1** 115–31

[6] Rukavishnikova I, Rumyantseva A and Plastinina I 2016 Environmental component in economic efficiency of renewable energy projects *WIT Transactions on Ecology and the Environment* **205** 203–12

[7] Velkin V 2014 Research efficiency of complex systems with spread spectrum renewable energy for electric power supply decentralized objects in Russia *WIT Transactions on Ecology and the Environment* **190** 437–43

[8] Paino S, Hoang H and Hukkalainen M 2017 Energy and emission analyses of solar assisted local energy solution with seasonal heat storage in a Finnish case district *J. Renewable Energy* **107** 147–55

[9] Saikku L, Tainio P, Hilden M, Antikainen R, Leskinen P and Koskela S 2017 Diffusion of solar electricity in the network of private actors as a strategic experiment to mitigate climate change *J. of Cleaner Production* **142** **(4)** 2730–40

[10] Velkin V and Shcheklein S 2017 Influence of RES Integrated Systems on Energy Supply Improvement and Risks *J. of Problemy Ekorozwoju* **12(1)** 123–29

IOP Conf. Series: Earth and Environmental Science **177** (2018) 012033 doi:10.1088/1755-1315/177/1/012033

Sustainability of the urban environment in the construction of city complexes

Ya Silin, A Kokovikhin, E Ogorodnikova and A Plakhin

Urals State University of Economics, 620014, Yekaterinburg, Russia

e-mail: cmb_8@mail.ru

Abstract. The article contains the results of a study aimed to assess the impact of the multifunctional city centers (MCC) construction in various spatial models of cities in terms of providing parameters for the sustainable development of the urban environment. The authors used the methodology of international standards for the assessment of "habitat stability", which includes a three-component analysis of the MUC impact on the city space. The received results allow to draw a conclusion about the best parameters of sustainability development of the city environment at MUC creation within the framework of multicore city planning.

1. Introduction

The sustainable development of the urban environment is due to the formation of a rational spatial structure of the city. The complexity of the interposition of various urban elements poses the task of preserving a favorable urban environment in the process of transforming planning-related structures. For example, the reconstruction of the business center requires a corresponding increase in the capacity of the transport infrastructure elements and creation of recreational zones. The rationale for investment projects for the construction of the multifunctional city centers (MCC) forms the actual task of multi-criteria analysis of the impact of these centers on the urban environment. With a compact planning layout of the MCC, a comprehensive solution to the problems of reducing the excessive load on the infrastructure, environmental and transport environment of the city is possible. The solution of this problem is possible with the introduction into the practice of spatial planning of the city of a single methodology for assessing the parameters of sustainable development of the urban environment in the construction of major facilities.

The purpose of this study is to assess the impact of the MCC construction in various spatial models of cities from the perspective of sustainable urban development parameters.

2. Methods of research

The study is based on classical models of the city spatial structure, including zones for various purposes. In the basic models, the interconnected areas of residential development, industrial, communal, transport and recreational zones are allocated.

The primary is the model of Burgess's concentric zones (1925) [1, 2], which describes the location of urban zones in the form of concentric circles (Fig. 1). The construction of a MCC in this model is made in the historical center of the city and, as a rule, replaces and complements the historical building.

Content from this work may be used under the terms of the Creative Commons Attribution 3.0 licence. Any further distribution of this work must maintain attribution to the author(s) and the title of the work, journal citation and DOI.
Published under licence by IOP Publishing Ltd

The planning solution presented in Fig. 1 corresponds to the applicable concept of the MCC construction in most Russian megacities such as Ekaterinburg, Novosibirsk, Nizhny Novgorod. The main characteristic of this model is the remoteness of residential quarters from the central district and the corresponding increase in the load during the pendulum movement of MCC employees to and from the place of residence.

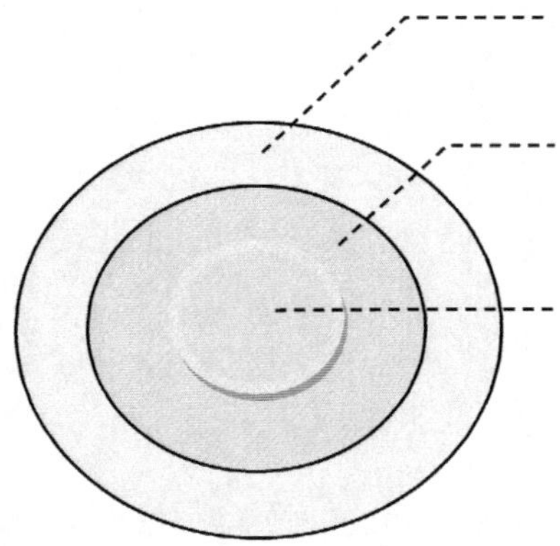

Figure 1. Placement of the MCC within the Concentric Zones Model

A similar arrangement of multifunctional city centers is presented in the sectoral model of H. Hoyt (1939, 1946) [3,4], which complements the previous model of spatial development of the city by the main transport corridors (Fig. 2).

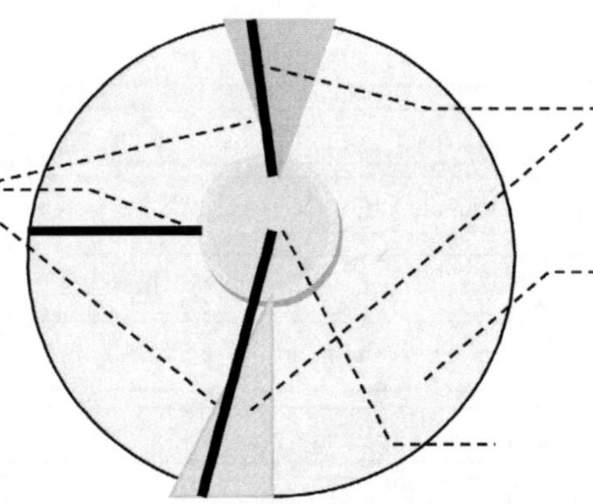

Figure 2. Placement of the MCC within the Sector Model

The planning solution presented in Fig. 2 corresponds to the concept that was used when placing Moscow-CITY and Lakhta Center in St. Petersburg. The proximity of the MCC to the transport axles makes it possible to neutralize the negative effects of the pendulum migration; however, such a placement model increases the ecological load on the city center and makes it necessary to include large recreational zones in the layout of the district.

The fundamentally different arrangement of multifunctional urban centers corresponds to the multi-core model of Harris and Ulman (1941) [5]. This model characterizes the spatial development of the largest cities with evolving suburban zones, forming independent business districts (Fig. 3).

The planning solution presented in Fig. 3 corresponds to the conceptused in the construction of the Défense area in Paris and the Dockland area in London.

Solving the scientific problem of developing a methodology for studying the MCC impact on the sustainable development of the city, it should be noted that a wide range of urban indicators is used in world practice.

IOP Conf. Series: Earth and Environmental Science **177** (2018) 012033 doi:10.1088/1755-1315/177/1/012033

In particular, we can cite the UNHabitat housing indicators system that establishes a rating system for assessing the sustainability of people's habitats that meets the goals of the current generation in meeting their needs in a comfortable living environment and performing public functions through the use of residential and public buildings without reducing the level of such opportunities for future generations (Trutnev, 2008).

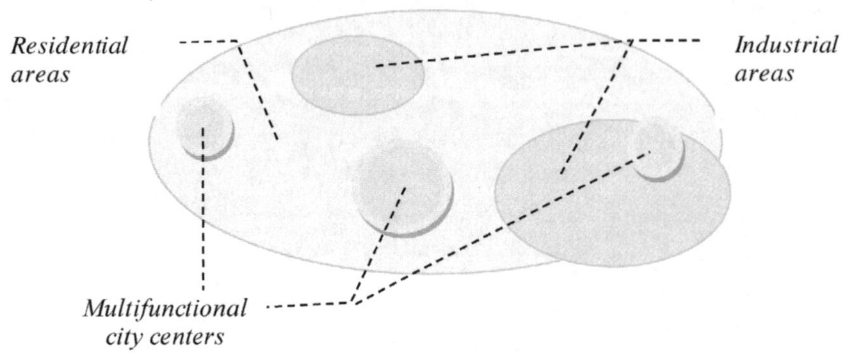

Figure 3. Placement of the MCC within the Multi-core Model

In Russia also has been developed a set of standards for real estate valuation, which introduced the notion of "habitat stability", identical in its meaning to the concept of «sustainability in building», adopted in international standards (ISO) [8,9]. This methodology is taken as a basis by the authors of the article, the evaluation parameters according to which are presented in Fig. 4.

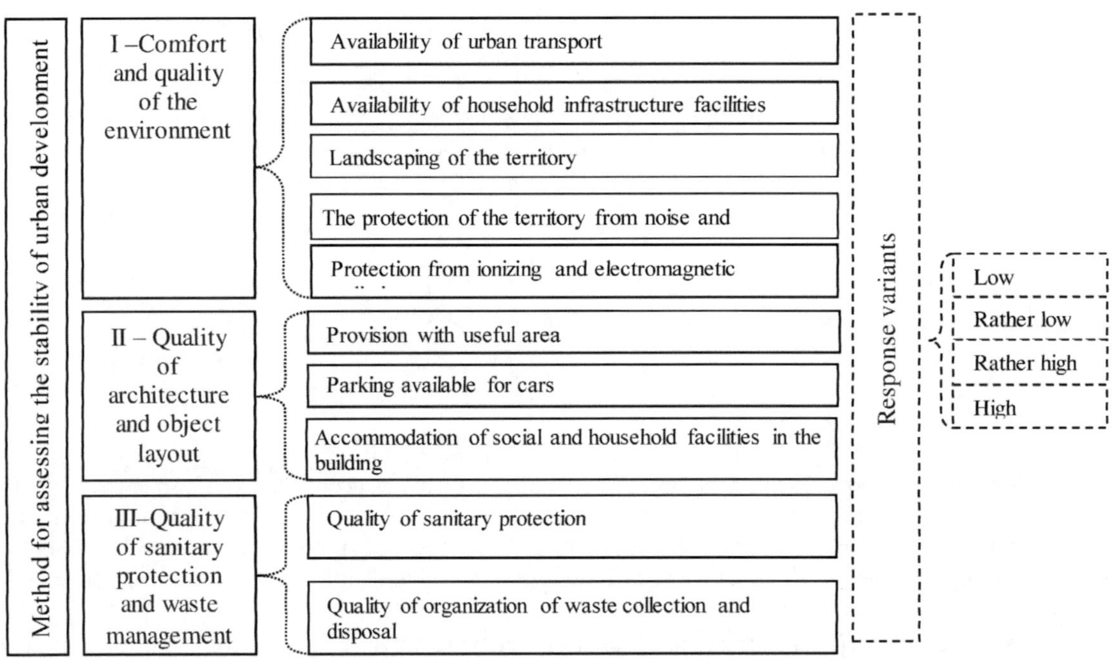

Figure 4. Methodology for assessing the impact of the MCC construction on the Urban Development Stability

The authors believe that the implementation of the construction projects of the largest Multifunctional city centers within the framework of urban zoning models has different effects on the parameters of sustainable urban development. The analysis was carried out in the context of the following groups of indicators:

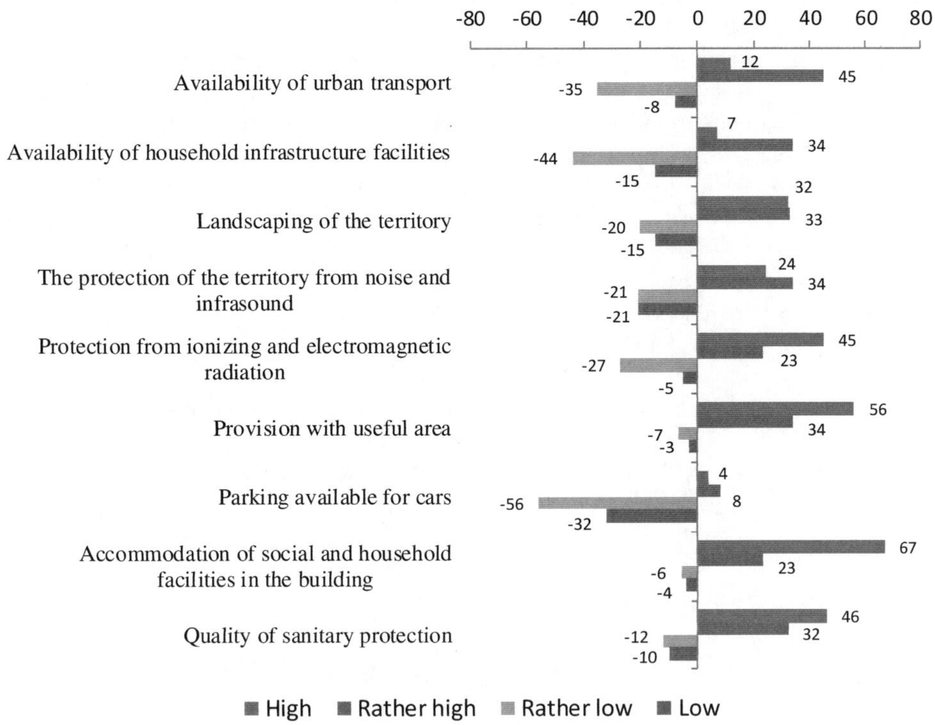

Figure 5. Assessment of the impact of the MCC construction on the urban development stability within the Sectoral Model (based on the Moscow-CITY example), %

Firstly, it is necessary to assess the comfort and quality of the external environment when placing the MCC, such an assessment is made in the context of the provision of transport, household infrastructure, and the possibility of external negative impact.

Secondly, the object itself is evaluated from the point of view of utility for the urban environment.

Thirdly, the quality of sanitary protection and waste recycling created by the input of the MCC operation is assessed.

Interpretation of the results obtained using the above assessment methodology was carried out in accordance with four possible options: high, rather high, rather low and low value of the indicator. Conclusion on the favorable impact on the urban development sustainability during the construction of the facility is carried out in the case of obtaining a "high" and "rather high" estimates for all three groups of parameters.

Conclusion on the adverse impact on the urban development sustainability in the MCC construction was carried out in the case of obtaining a "Low" and "rather low" estimates for the three groups of parameters.

In general, the conclusion about the negative impact on the urban environment development sustainability will also have an intermediate result when, in the aggregate assessment, there are values on the one hand as "high" and "rather high", and "low" and "rather low". In this case, it is possible to recommend a change in the project for the MCC construction with the aim of changing the negative assessments, for example, including additional parking spaces or expanding the recreational zone in the construction project.

3. Results

To assess the impact of the MCC construction on the urban development stability in the framework of sectoral and multicore urban planning models there were used the materials of the substantiation reports for the construction of the Moscow-CITY (2009) and UNHabitat (2007) [7] provided by the consulting company RBC-Real Estate. The results of the study can be seen in Fig. 5.

As can be seen in the figure, the assessment of the impact of the MCC construction has multidirectional consequences for the urban development sustainability. The most negative consequences of the MCC introduction in the sectoral model of the city planning will be caused by the lack of parking spaces and the availability of external objects of the household infrastructure. If the second drawback can be resolved by placing the necessary facilities on the territory of the complex itself, the solution of the first problem requires additional investment in the construction of parking lots.

Next, in Fig. 6, we present the assessment results of the impact of the MCC construction on the urban environment stability in the framework of the multi-core model.

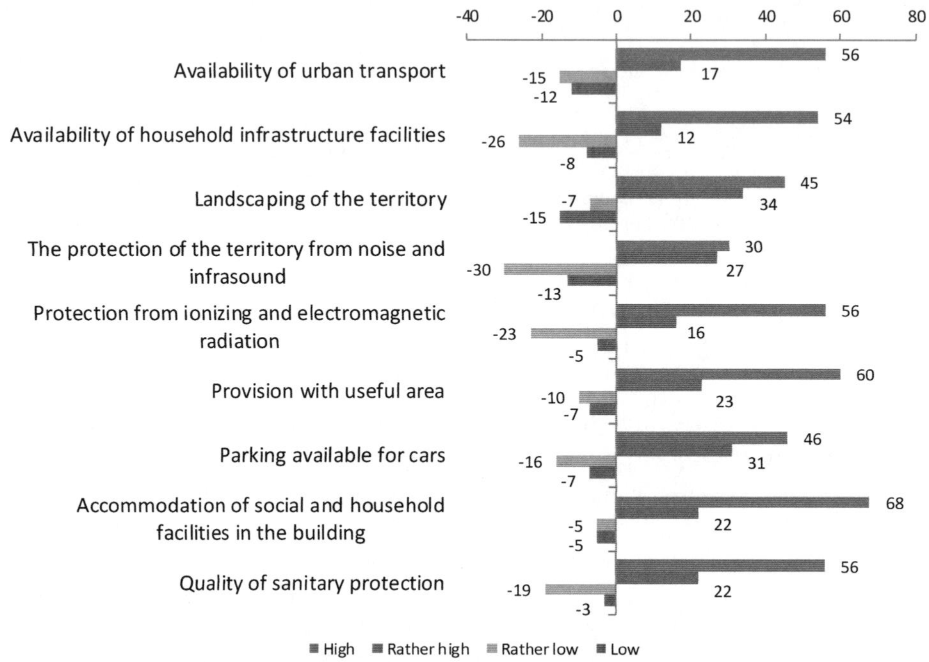

Figure 6. Assessment of the impact of the MCC construction on the urban development stability within the multi-core model, %

As can be seen in the figure, the respondents in general are positively assessing the MCC construction in terms of the urban environment stability. Rather, the low value of the indicator is observed in terms of the protection of the territory from noise and infrasound, the solution of this problem is possible with the installation of special soundproof shields.

To assess the impact of the MCC construction on the urban development stability in the framework of the concentric zones model, an empirical study was conducted. The method used by the authors was a questionnaire conducted during the implementation of research work on the topic «The Justification of the investment project for the construction of the MCC "Ekaterinburg-CITY"». Questioning on the block of questions «I – Comfort and quality of the environment» was conducted among residents of the streets of Boris Yeltsin, Chelyuskintsev, Marshal Zhukov, a total of 155 questionnaires. Questioning by blocks «II – Quality of architecture and object layout» and «III – Quality of sanitary protection and waste management» was carried out among the specialists and designers, on the whole 7 questionnaires were received. The results of the study can be seen in Fig. 7.

The survey showed that, in general, the respondents made a negative assessment of the consequences of the MCC construction on the urban environment development sustainability in the framework of the concentric zones model.

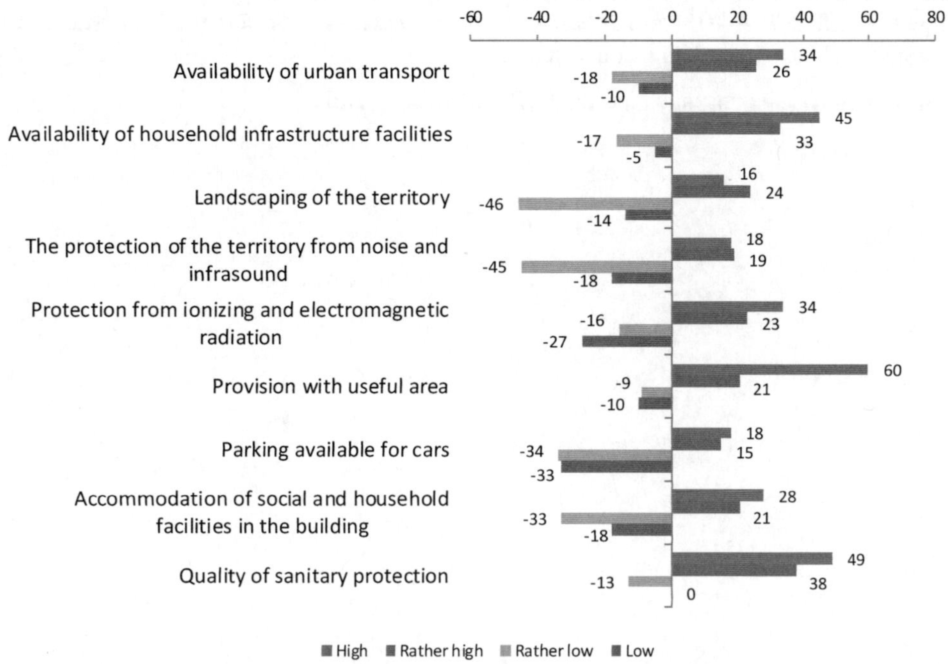

Figure 7. Assessment of the impact of the MCC construction on the urban development stability within the concentric zones model, %

4. Conclusion

The study shows that from the position of sustainable development of the urban environment the most promising is the construction of multi-functional city centers within the multi-core urban planning model. This result confirms the thesis about the future direction of cities spatial development, cited in the works of Harris, Ulman and their followers. The creation of multifunctional city centers is expedient to carry out unloading the center of the city and redirecting traffic flows to newly formed centers.

References

[1] Park R, Burgess E and McKenzie R 1925 *The Growth of the City: An Introduction to a Research Project* 35–41

[2] Cheshire P, Nathan M and Overman H 2014 *Urban Economics and Urban Policy* (Cheltenham, *U.K.: Edward Elgar Publishing)*

[3] Hoyt H 1939 *The structure and growth of residential neighborhoods in American cities* (Chicago, IL: Chicago University Press)

[4] Hoyt H 1946 Forces of urban centralization and decentralization *American Journal of Sociology* **6** 843–52

[5] Ullman E 1941 A Theory of Location for Cities *American Journal of Sociology* **XLVI** 853–64

[6] Kokovikhin A, Ogorodnikova E, Williams D and Plakhin A 2017 Factors of the institutional environment in the assessment of the investment climate of the municipal entity by an entrepreneur *The Economy of the Region* **13 (1)** 80–92

[7] Trutnev E 2008 *Urban regulation: Fundamentals of urban planning in the conditions of the real estate market* (Moscow Fund "Institute of Urban Economics")
[8] ISO/TS 21929-1: 2006 *Sustainability in building construction* – Sustainability indicators – Part 1: Framework for development of indicators for buildings
[9] STONOSTROI 2.35.4-2011 *Greenbuilding. Buildings and civil construction.* Rating estimation of Sustainability in building construction

Published under license in Journal Title **by IOP Publishing Ltd**

University-to-School Environmental Projects for Sustainable Development: A Case of Ural Federal University

D Sandler[1], M Volkova[1] and D Kochetkov[2, 3]

[1] Ural Federal University, Ekaterinburg, Russia
[2] Peoples'Friendship University of Russia, Moscow, Russia
[3] Institute of Economics, Ural Branch of the Russian Academy of Sciences, Ekaterinburg, Russia

e-mail: kochetkovdm@hotmail.com

Abstract. Sustainable development is a worldwide recognized social and political goal; it is discussed both in academic and political discourse. According to the authors, the formation of a new way of thinking will help to achieve this goal. A lot of research is related to sustainable development in higher education. However, mental models are formed even more effectively at school age. The paper was written in the context of Russia, where the subject of sustainable development in education is extremely poorly developed. The case of Ural Federal University was presented. The University has been working for several years on the creation of a device for the purification of sewage industrial water in the framework of an initiative student group. Recently, schoolchildren have joined this work. Such projects have been called university-to-school projects. Successful solution of inventive tasks contributes to the formation of mental models. The case has been analyzed in terms of institutionalism. The authors argued the primacy of mental institutions over normative in the process of sustainable society construction.

1. Introduction

There are two polar points of view on what is happening to the climate of our planet. But the proponents of the theory of global warming, and their opponents agree that humanity contributes to climate change. The growth of the world's population and technological progress requires more and more energy, which can only be obtained by burning fuel. As a result, technogenic pollution of the planet only grows from year to year. As a result, air quality in large cities does not meet the standards, which leads to an increase in the incidence of respiratory, cardiovascular, and allergenic diseases. This leads to annual financial losses comparable to the GDP of a small country.

Another problem is the quality of water. The amount of water in nature is almost identical. For use in industry, agriculture and everyday life, mostly fresh water is suitable. Discharge of contaminated water leads to a decrease in the quality of drinking water, thermal pollution of water bodies, and flooding of territories. At present, the demand for water has reached such a level that in many places on the planet, including Europe, there is an acute problem of lack of fresh water. Scientists warn that in one or two generations the majority of the world's population will lack fresh water [1].

Anthropogenic impact on the biosphere of our planet is so global that governments of the world have to cooperate in the field of ecology. But here, everything is not so clear. The desire to reduce

Content from this work may be used under the terms of the Creative Commons Attribution 3.0 licence. Any further distribution of this work must maintain attribution to the author(s) and the title of the work, journal citation and DOI.
Published under licence by IOP Publishing Ltd

emissions of carbon dioxide into the atmosphere led to heated debates around the Paris Climate Agreement. Someone sees in it a concern to reduce the anthropogenic impact on the planet; someone (e.g., Great Britain) uses a climate agenda to protect its market. There is an opinion that the global markets have a problem of carbon protectionism, which can lead to sanctions pressure on the national economies [2]. In this context, the US withdrawal from the Paris Agreement is simply protection of the national business.

If you take the problem more broadly, the whole history of the struggle for sustainable development is in fact, the history of the struggle of the population for the environment, and for entrepreneurs, it is for profits. And, if large plants and enterprises can afford pollution control facilities, then for middle and small businesses, it is often too expensive. Thus, the choice between profit and environment among the majority of entrepreneurs in the world is obvious. Almost the only deterrent for "ecologically irresponsible" is the regulatory role of the government in the area of environment protection, imposed penalties, sanctions, etc. Of course, there are entrepreneurs for whom taking care of the environment is not an empty shell, but most often profit comes first, and environmental expenditures with whatever remaining funds the firm has. The most environmentally conscious are sufficiently young people who grew up and received education at a time when environmental problems became relevant and discussed. As practice shows, if a person received an environmental education since childhood, a businessman and an ecologist can make a compromise in his mind.

However, in Russia, as in most emerging countries, there is practically no formal educational policy in the field of sustainable development. In this context, initiative projects are of particular importance. In our work, we analyze the experience of the partnership between Ural Federal University and schools in the field of sustainable education. We call such projects university-to-school projects. The rest of the work is organized as follows: in the next section we briefly described the background of sustainable education, then we analyzed the case of Ural Federal University; in the Discussion and Conclusions section we tried to interpret the results and formulate the insights.

2. Background

Sustainable development was originally conceived as a political concept in the late 1980s. The first definition of sustainable development belonged to the Brundtland Commission, which defined it as a development in which the needs of present generations are met without compromising the ability of future generations to meet their own needs [3]. Thus, at its earliest meaning, sustainable development was associated with the maintenance of the environment quality [4]. Over time, the concept expanded. A more modern interpretation includes environmental, as well as economic and social dimension [5–7]. The UN Resolution [8] introduced the vision of sustainable development: *We envisage a world free of poverty, hunger, disease and want, where all life can thrive. We envisage a world free of fear and violence. A world with universal literacy. A world with equitable and universal access to quality education at all levels, to health care and social protection, where physical, mental and social well-being are assured. A world where we reaffirm our commitments regarding the human right to safe drinking water and sanitation and where there is improved hygiene; and where food is sufficient, safe, affordable and nutritious. A world where human habitats are safe, resilient and sustainable and where there is universal access to affordable, reliable and sustainable energy.*

The vision emphasizes the role of education in building the future. Speaking about sustainable education, the authors of the article do not mean gathering recyclable materials, feeding birds in winter and participation in ecological events. Of course, these are essential components, but not the most important ones. Several studies showed that the growth of professional knowledge of scientific, technological, and socio-economic nature directly affects the possibility of the transition of the territory to an innovative development path [9]. Thus, in order for a "greening" of production to take place, there must be an increase in the general culture of people in this area. The most important result of the intellectualization of the population of the territory is the emergence of an intellectual and innovative space. A new type of behavior is required for its functioning, not only for individuals but for the social and economic systems as a whole.

Intellectualization of the population and the various innovations arising on its basis cannot come into being in a short time. Any society needs a relatively long period of time for its maturation, which first includes the accumulation of knowledge, and then turning it into a way of thinking for the bulk of the population. Intellectual loners can only offer original ideas, describe them, and even make prototypes, but only collective efforts of the majority of skilled working force can transfer the ideas into mass production.

As a source of knowledge accumulation and generation, secondary and higher education moves to the forefront in the development of sustainable society. Relevant skills and competencies are essential for implementing the paradigm of sustainable development [10]. Competence-oriented programs have been recognized as a critical factor in the transition of education to the path of sustainable development [11]. This concept has been supported by numerous empirical evidence in the form of best practices [12, 13]. Nevertheless, most discussions about the formation of competencies relevant to sustainable development are conducted in the context of higher education [14–16].

We believe that sustainable development requires not only skills and competencies but also the formation of stable patterns of thinking. And here secondary education comes to the forefront because the formation of mental models at an early age is much more effective. We all study at school from 7 to 16 years. In the junior classes, natural science lessons are usually conducted, and then during the course, the students are given an idea of the environment, the basics of ecological knowledge. There are Olympiads of ecological direction for pupils of schools. In some schools, the children, under the guidance of teachers or supervisors, write reports or conduct feasibility studies (often analysis of data from the Internet) on environmental topics. After entering the university, unfortunately, school work is lost. This is due to another organization of studies so that at the graduating departments (apart from specialized ones) sustainable development or ecology is not the main subject. As a result, a person has acquired some skills in the field, but a way of thinking has not been formed. In such a society it will be indecent to litter, it is young people who are quicker to get used to throwing out garbage separately, but there is hardly any innovative breakthrough in the field of sustainable education right now.

3. A Case of Ural Federal University

The authors of the article believe that it is necessary to use school environmental practices from the first days of study at the university. It is essential to involve students in solving practical problems. Those, who consciously engaged in environmental problems at school, take this work willingly. Only the task should not be narrow, but rather large, including knowledge from different areas. As an example, we can cite the disposal and treatment of polluted waters of small CHP and thermal power plants. The main sources of hot water supply and heating are also the main sources of pollution of the living area due to its location. Losses of CHP and TPPs with heat supply to consumers increase depending on the distance. Therefore, historically, many of them are in the urban environment where it is simply impossible to place cooling ponds, sludge accumulators, etc. At the same time, one can use a variety of heat exchangers, scrubbers, etc.

As a result of technological processes, a large amount of contaminated water is emitted during steam condensation at any CHP and TPP. An even larger amount (10–15 times more than has been discharged) is needed to ensure that condensation and cooling of the discharged water occurred in the end. The most conservative estimates show that even a small CHP plant incurs significant losses as a result, and if you add the cost of water treatment and the cost of its disposal, the amount looks impressive. Besides, CHP and TPP use water for washing boilers and other mechanisms. Thus, if this water is returned to circulation for reuse, then several problems can be solved simultaneously: reducing water consumption, reducing heat pollution, obtaining an additional product, which will make the work more attractive in the eyes of potential consumers and investors.

Such work has been conducted in Ural Federal University since 2008 under supervision of M. Volkova, exclusively by the students of volunteers who are not engaged in specialized departments, in their spare time. It should be noted that the problem involves a fairly large range of issues: from the creation of a purification technique with the help of developed biofilters to the study of the issues of complete utilization of filters. In different years, the team of students changed, and the range of tasks varied. Each of the students can choose the direction by which he will work. For example, the construction of a prototype is closer to the builders, the utilization and production are for physicists, the specialists in the field of biochemistry conduct a study of optimal conditions for the operation of biofilters, etc. Students publish their work; as a result, ecological thinking has been formed, what is more important. Students start to think and think a little differently.

The team of which we are speaking is quite unique. Its composition changes regularly, all students are of technical specialties, the basic principle is voluntariness. The work in the research team starts from the first to the second year, there is no compensation and funding. At the same time, our student team is working on a very topical issue. The head of the team wants to show that the protection of the environment is by no means such a costly expense item, as it is commonly believed. Teams of similar-minded people wishing to improve the ecological situation are quite capable of solving environmental problems. The need for such work, in the authors' opinion, is shaped by the educational component, in addition to the environmental one.

Thus, there was a desire to attract students to scientific work, in order to broaden their horizons and develop their creative abilities. The involvement of students in solving practical environmental problems proved to be very effective. Students' engagement from the first years of study empowers a system of continuous environmental education. Creativity enables personal freedom, unites similar-minded people, and distracts from reality. We are convinced that such teams are an excellent opportunity for the integration or inclusion of students with disabilities. Students with disabilities can realize their potential at creative discussions, when studying or processing experimental materials, etc. If the state of health does not allow them to participate personally, then the modern means of communication allows these children to communicate remotely. For people with disabilities, much is done in the sphere of creating an accessible environment, but we are sure that the most important thing is the sense of necessity and usefulness to others. It really can be achieved through the work in creative teams. It enables communication with others on an equal footing, a realization of one's thoughts and projects, and in the end, raises self-esteem.

What attracts students in a job that is not paid and been done in spare time? According to the results of unstructured interviews with program participants, this is:

- A possibility to develop creative ideas;
- An opportunity to participate in the scientific conferences. Maybe in the future, the student will not become a scientist, but an experience of public presentations is always useful. The first presentations are usually very "stammering", "stuttering" and not the most memorable for the audience. After a year or two, they are already self-confident speakers, quietly communicating with the audience. Perhaps, the only thing that remains for students to remember about their work is certificates, presentations, and publications.

On the basis of the data obtained, the team of researchers developed a schematic diagram of the device, which has such advantages as compactness and wastelessness, can be easily adapted to particular features of the production process. Thus, from the original idea, thanks to the initiative of students and proposals to explore this or that aspect, the project divided into three parallel studies:

- Application of purified water (for irrigation and plant cultivation);
- Studying the possibility of obtaining biogas from biomass grown during the purification of condensate;
- Development and improvement of the machine itself (patent documentation is being prepared).

There are quite a lot of issues that require closer examination and study in the future. Our team participated several times in the competitions to receive funding from UMNIK1, obtained good

[1] The program of the Fund for the Promotion of Innovation, which finances youth projects with scientific novelty. URL: http://umnik.fasie.ru

reviews. Therefore, there is a hope that sooner or later, but we will have funding, and all the projects, temporarily postponed, will be brought to an end. Fortunately, there are many creative children in our country, and they have a desire to contribute to the improvement of the environment. And most importantly, such people, whatever they do in their lives, will be able to rally around themselves like-minded people. There are interesting studies and real proposals, and this despite the fact that everything is done only on enthusiasm. Simply, the students are given an opportunity to realize their idea and do something interesting and necessary. Not all of them are enrolled in post-graduate or master studies in the environmental field. Most often, for each of them remains a shelf with their published works in memory of the work. For someone it is more, for someone it is less, it is not so important. The main thing is that students have a clearer understanding of the environment as a science. Some of the students continued their education (and the articles helped with the admission to the post-graduate or master's program), someone found a job.

Nevertheless, you can start this work even earlier. Beginning with grades 7–8th, a sufficient stock of knowledge accumulates among schoolchildren, as well as their desires to apply them in practical tasks. A striking example is the *Ural Project Session* held for gifted schoolchildren by Ural Federal University. To participate in the Session, the schoolchildren passed a contest, to which pupils from the 7th to the 11th grade were admitted. The work was conducted in several directions, one of which was *Intellectual Energy Systems*. Within the framework of the Session, the project *Purification of technical water from power plants using simple algae* was developed. The team consisted of 5 people (2 students of 11th grade, 2–10th, and 1–8th). The students not only perfectly understood the problem, but they also developed and assembled a workable prototype.

The device consists of three filters (Figure 1). Contaminated water entering the pipe into the first filter (far left) is subjected to primary purification from coarse contamination. In the filters 2 and 3, there are algae, due to which the polluted water is saturated with oxygen and changes the hardness. Indications of temperature, light, and hardness are automatically controlled by sensors on the control panel. In the case that the sensor shows a decrease in the hardness to a predetermined value, the valve automatically opens, and the purified water returns to the system or goes for technical purposes (on the stand it is a tap). If the water does not sufficiently reduce the hardness, which is controlled by the solenoid, then it returns to a new cleaning circle. Thus, a closed cycle enables the water to be purified to a predetermined value.

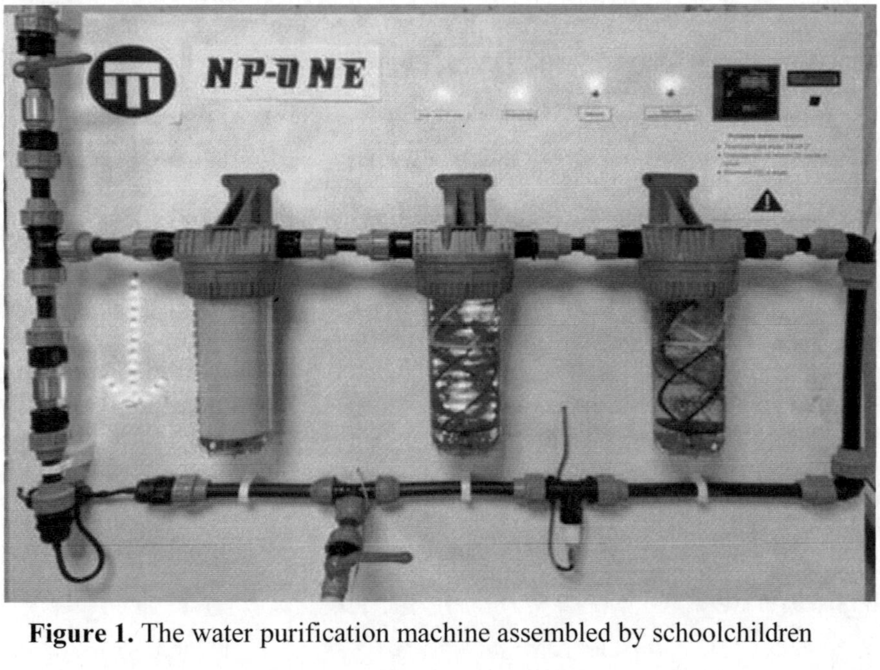

Figure 1. The water purification machine assembled by schoolchildren

IOP Conf. Series: Earth and Environmental Science **177** (2018) 012034 doi:10.1088/1755-1315/177/1/012034

The idea was so interesting for the schoolchildren that even after the end of Session the team did not fall apart. The guys themselves created the group NP Team in the social media and now actively participate in grant competitions. There is intent to continue the development of a comprehensive method for cleaning contaminated water from the CHP plants. The only deterrent is the final grade of the school this year for two of the team members.

Now the team is preparing to participate in the Ural Project Session in the educational center Sirius (Sochi). 100 students, who have successfully passed the competitive selection, will attend the Session. Students of public, municipal and private educational organizations located on the territory of the Russian Federation in the regions of the Urals and Western Siberia are eligible to take part in the competition. The purpose of the Ural Project Shift is identification, development, and support of gifted students in the field of design and research activities. The tasks to be performed during the Session are as follows:

- Activation of creative, cognitive, and intellectual initiative of students who showed interest and inclination to study mathematics and earth sciences;
- Identification and support of students who are prone to research and development activities;
- Generalization and development of the best practices in the studying of mathematics, physics, chemistry and biology in the upper grades, namely, preparation for Olympiads, development of research and educational projects, organization of extracurricular work of students;

Engagement of scientists, specialists from research institutes, higher education institutions to work with students.

4. Discussion and Conclusions

Of course, sometimes the students do not have enough knowledge, but enthusiasm and the desire to apply their knowledge in something important led to this result. Probably not all of the team will associate their lives with the environment in the future, but the more such programs are available, the faster the generation will grow with a sustainable way of thinking.

It is not justified to consider that the work of schoolchildren and students is insignificant and has no practical significance. Of course, there are a lot of ways to reduce emissions at CHP and TPP. Practically all methods of reducing the environmental load are quite energy-consuming, expensive, and often cumbersome. All of them are developed in the research institute, by a large team. Pupils and students do not have such a financial opportunity, so they use more democratic materials in their studies. The result is a fairly inexpensive and more attractive product. Thus, solving a specific applied problem it is possible to form an active position in relation to sustainable development in the younger generation.

Let's look at the problem from the point of view of institutional theory. We understand institutions as a set of formal norms, informal constraints, and coercive mechanisms [17]. In a broader sense, institutions can be divided into 4 groups:

1. Normative – norms, rules, customs, standards, conventions, contracts, etc. [17];
2. Functional – status functions and routines [18, 19];
3. Structural – organizational forms and models of transactions [20];
4. Mental – collective representations, beliefs, stereotypes, values, cognitive schemes, etc. [21].

Normative institutions are among the most rigid formal institutions that rely on violence as a coercive mechanism. As a rule, it is the government that relies on this type of institutions. Routines are repetitive, normal and predictable procedures for solving similar problems. The theory of structural institutes was developed for the firm level, but this model operates at the national level as well. It determines not only the structure itself but also the model of interaction of various elements of the system. Finally, mental institutions or mental models imply a similar perception and interpretation of reality. The core of mental institutions is values that are an integral part of the culture of the individual and society as a whole. At the same time values are divided into absolute, which the individual always follows, and relative, which are conducted only when it is profitable.

Functional and structural institutions are derived from normative and mental; so, we can define the latter as primary institutions. According to the authors, the importance of mental institutions is greater. The adoption of regulatory institutions is determined by the transaction costs of accession and evasion: if the transaction costs of evasion are lower than the costs of compliance, a person will choose not to comply with the norm. In the case of mental institutions, the transaction costs of evasion are extremely high, because they are supported by both the social and the internal mechanism of moral evaluation. The costs of changing mental institutions are also very high. Therefore, we can conclude that only people with stable mental models of sustainable development can guarantee a sustainable future. This does not negate the role of normative institutions in the transitional period because the formation of mental institutions takes a long time (at least one or two generations). During this period, normative institutions create hothouse conditions for the development of mental institutions.

Acknowledgments
The publication has been prepared with the support of the «RUDN University Program 5-100». The authors wish to thank Russian Foundation for Basic Research for financial support of this research (project 17-22-07001 "The complex algorithm of culture-based regeneration of old industrial cities").

References

[1] Connor S 2013 Lack of fresh water could hit half the world's population by 2050 *Independent*
[2] Wood T and Edis T 2011 Pain of carbon protectionism *Aust. Financ. Rev.* **59**
[3] 1987 *World Commission on Environment and Development (WCED) Our Common Future*
[4] Bell S and Morse S 2008 *Sustainability Indicators: Measuring the Immeasurable*? (2nd ed.; London: Earthscan)
[5] Shriberg M 2004 Assessing Sustainability: Criteria, Tools, and Implications *In Higher Education and the Challenge of Sustainability* 71–86
[6] Kuhlman T and Farrington J 2010 What is sustainability? *Sustainability* **2** 3436–48
[7] Farley H and Smith Z 2013 *Sustainability: if It's Everything, Is it Nothing?* (New York and London: Routledge)
[8] 2015 United Nations Transforming our world: the 2030 Agenda for Sustainable Development. *Gen. Assem. 70 Sess. 2015* **16301** 1–35
[9] Bochko V 2015 Intellektual'noye obespecheniye innovatsionnogo razvitiya ekonomiki [Intellectual support of innovative development of the economy] *In Innovatsionnoye razvitiye ekonomiki: tendentsii i perspektivy [Innovative economic development: trends and prospects]* 14–26
[10] Jacobi P, Toledo R and Grandisoli E 2016 Education, sustainability and social learning *Brazilian J. Sci. Technol.* **3** 3
[11] Dlouhá J, Glavič P and Barton A 2017 Higher education in Central European countries – Critical factors for sustainability transition *J. Clean. Prod.* **151** 670–84
[12] 2015 UE4SD *Leading Practice Publication: Professional Development of University Educators on Education for Sustainable Development in European Countries* (Prague)
[13] Mulà I, Tilbury D, Ryan A, Mader M, Dlouhá J, Mader C, Benayas J, Dlouhý J and Alba D 2017 Catalysing Change in Higher Education for Sustainable Development *Int. J. Sustain. High. Educ.* **18** 798–820
[14] Wiek A, Withycombe L and Redman C 2011 Key competencies in sustainability: a reference framework for academic program development *Sustain. Sci.* **6** 203–218
[15] Rieckmann M 2012 Future-oriented higher education: Which key competencies should be fostered through university teaching and learning? *Futures* **44** 127–135
[16] Lambrechts W, Mulà I, Ceulemans K, Molderez I and Gaeremynck V 2013 The integration of competences for sustainable development in higher education: an analysis of bachelor programs in management *J. Clean. Prod.* **48** 65–73

[17] North D 1990 *Institutions, institutional change, and economic performance* (Cambridge and New York: Cambridge University Press)
[18] Nelson R 1994 Routines *In The Elgarcompanion to institutional and evolutionary economics* **2** Hodgson G, Samuels W and Tool M Eds (Aldershot: Edward Elgar) 249–53
[19] Searle J 1995 *The construction of social reality* (New York: Free Press)
[20] Scott W 1995 *Institutions and organizations. Ideas, interests and identities* (Thousand Oaks: Sage)
[21] Denzau A and North D 1994 Shared mental models: Ideologies and institutions *Kyklos* **47** 3–31

IOP Conf. Series: Earth and Environmental Science **177** (2018) 012035 doi:10.1088/1755-1315/177/1/012035

The principles of sustainable development of architectural space of the Temernik River embankment in Rostov-on-Don

Ju V Gorgorova, M G Sarkisyants and M A Sotnikova

Don State Technical University, 1 Gagarin square, Rostov-on-Don, 344000, Russia

E-mail: yu.gorgorova@yandex.ru , maryland_23@mail.ru, 230719911@inbox.ru

Abstract. The problem of sustainable development of embankments is multifaceted and complex. The article formulates the principles of sustainable development of the architectural environment of the embankment of the Temernik River in Rostov-on-Don. The authors of the article suggest a systematic approach to the organization of the environment in the project to create the embankment of the Temernik River to solve environmental problems in Rostov-on-Don.

1. Introduction

Embankments in Russia appear recreation carrying capacity, which is not fully used. There is no functional development of a territory and waterside improvement, by that having negative impact on city-planning on the whole. A coastline in Russia is 65 thousand kilometers long. Embankments are space-occupying complexes, which occupy urban areas, related to urban development and river- and marine areas. Any site development in coastal territories is of short duration because of man-made and natural factors. Being the objects, which have geographical, ecological, economic and social components, coastal territories are to be in-depth studied in order to ensure sustainable development of their capacity [1].

Urban recreation zones of embankments have very important significance. It is arrangement of urban recreation zones, favorable microclimate, and fall in atmospheric temperature that is especially important for hot districts. The general problems for many cities are as follows: intensive urbanization process and development of coastal territories; expansion of transport infrastructure, loss of natural factors and landscape diversity of coastal stripes and water areas; levelling of architectural and landscape unicity, water levelling; loss of continuity in space development of urban environment; inaccessibility and neglected state of coastal territories; undeveloped territorial and functional resources, insufficient quantity of public spaces, recreation, tourism, physical training and sport, education, creative industries; loss of historic traditions, undervaluation of modern opportunities to bring water areas to urban life; lack of strategies and innovative methods of complex transformation and development of coastal territories within city-planning [1–4].

2. The existing state of the Temernik River in the city of Rostov-on-Don

The conducted researches show, that except for the above-mentioned problems there are serious ecological problems in Rostov-on-Don. As the result of surface water bodies monitoring, carried by Rostov antiplague institute, V. Cholerae O1 was found in the body of the Temernik River, near botanic garden in 2014. This water body was admitted dangerous, as a reservoir and infection

Content from this work may be used under the terms of the Creative Commons Attribution 3.0 licence. Any further distribution of this work must maintain attribution to the author(s) and the title of the work, journal citation and DOI.
Published under licence by IOP Publishing Ltd

transmission factor. That is why special preventive activities are needed in order to stop circulation of toxicogenic clones in the water body. The results of long-term biological testing data in the body of the Temernik River confirm that the toxic level of the river is referred to the 4–5th toxicity class, which is ranked as extra- and extremely toxic; and ecological status as poly- and hyper-toxic [5–7].

Silting of the body of the Temernik River significantly reduced its drainage property, by that contaminating groundwater in the vast territory and making groundwater useless for public water supply and domestic water supply. Discharge capacity of the body results in urban area underflooding in low rainfall flood with flood flow of 40–50 meters per second. In more heavy rainfall flood there is real damage of surplus activated sludge to contaminate the Don River. It might cause ecological disaster in the Gulf of Taganrog of Azov Sea [5–7].

3. Results of experimental design

Taking the above-mentioned into consideration, there are enhancement approaches in the Temernik River basin, integration of coastal territories in a functional planning urban structure and arrangement of park and recreation zone offered. In order to integrate coastal territories in a functional planning urban structure, a complex approach considering city-planning, ecological, economic and social aspects is required:

Analysis of the territory to assess current ecological state, to identify functional and specific connections; development of city-planning with regard to ecological and functional factors of the territory.

Consolidation of separated territories. Renewal of man-made frame in natural frame. Arrangement of a linear park alongside the Temernik River, local parklands, green corridors, which create a natural frame, in accordance with the approved general town-planning scheme.

Linear development. Arrangement of a single network of rivers and channels in the territory of Rostov-on-Don (arrangement of city water frame) with regard to the most problematic zones, by that contributing to widening and depth deeping of a river-bed, to its natural purification by means of flow rate increase and to ecological stability increasing. Arrangement of a single system of green planning, local parklands, green corridors, which connect the city and interact with line- and central structure (green city frame).

Implementation of control and monitoring systems, qualitative redistribution of resources in the city in order to achieve balance of nature, to prevent ecological disasters and emergencies related to man-made and natural phenomena. A drainage system: drainage of territories with high ground water table, ground water releasing with holes and arrangement of artificial channels (subterranean, surface channels) in order to minimize flooding and earth flow, to strengthen soil bearing capacity, to decrease contamination of water, lying a river-bed in underground reinforced-concrete channel (northern burial-ground). Due to such water transferring there will be possible to arrange various water bodies in existing parks and public gardens.

Development of local waste treatment plants for reclamation and filtration of ground water before flowing to the river.

With regard to analyzed schemes, approved general plan of water supply and sewerage in the city it is offered to lay holding chambers and suction hoses with shafts, with provided system of sluices, which are connected with storm-water drainage and municipal sewerage to control level of surface water, floodwater and waste water. The above-mentioned measures let surface water draining from streets and ensure discharge of sewerage systems in case of emergency (atmospheric condensation, floodwater, etc.). It shall be realized by means of using sluices, which let drain surface water to holding chambers, from where water shall be gradually (over a particular period of time) discharged to a central collector. It shall help to avoid flooding of residential areas, floods, flood flows and such consequences as resulted from a high level of atmospheric condensation in summer of 2016 in Rostov-on-Don. Such system shall also help to maintain a single water-level in the Temernik River regardless of environment conditions (falls, floods, etc.). The project provides sluices and suction hoses along the full length of the river at a pitch of 2 km.

It is also offered to process sludge deposits into agricultural fertilizer.

It is offered to use a biological module for effective and ecological water reclamation in the river.

Arrangement of linear connections with engineering and transport function. Arrangement and replenishment of transport infrastructure. Introduction of new pedestrian and recreational traffic with residential areas (transport- and communication frame of the city).

Arrangement of a lifting rale line, which let connect several urban planning areas: Voroshilovskiy, Oktaybrskiy and Leninskiy, providing citizens with a possibility to move quickly and comfortably. There shall be also a possibility to have convenient access to the main city facilities: the main railway station and the stadium 'Rostov-Arena'. A route of a lifting rale line is planned to be along the river-bed. It shall not disturb existing buildings and shall not demolish buildings and structures.

It is offered to lay an inland canal by partially using a river-bed. An inland canal is a large synthetic streamflow, which is intended for navigation of tourist boats. A process of natural river purification is quite long and complicated. The offered approach shall decide a problem of river-bed purification. There shall be water crafts, which purify the bottom of sludge deposits and clump of hazardous substances. There shall be also special activities organized to normalize biogenic substance in the river. The important fact is that an inland canal building shall not affect all areas, but leave the most plant life along the river intact, by that saving it.

Arrangement of an inland, shipping canal and a lifting rale line create transport routes, which help to unclog traffic arteries of the city and it becomes possible to organize touristic routes and circular tours in significant places, which coincide with cultural and historical frame of the city. One of the suggested tours goes through Temernitskiy ancient settlement, where there is a monument of architectural hereitage – Surb-Khach.

Arrangement of a bicycle track covered with solar batteries, which provide the park with power supply. Bicycle paths in the embankment are organized in such a way as to provide continuity of bicycle moving in the urban environment and therewith there are no obstructions to pedestrian flow. The bicycle paths are considered to be covered mainly with bright colors in order to help bicyclers to orientate better without motion speed changing.

The conditions for people with limited mobility moving are provided. Arrangement of access ramps and stairlifts with special coverage for people with limited mobility.

Embankments are crossing points of pedestrian, transportation and bicycle flows. One of the main aims was to separate and structure them. Pedestrian flow is closer to water than the other ones.

The project uses a principle of spatial continuity that is sequential and interrelated organization of coastal territories as a continuous and maximally extended linear facility in conjunction with re-arrangement and development of adjacent areas. All offered activities are aimed at deciding not only ecological, technical and real problems of the Temernik River and its improvement, but at including this territory to city-living as well. When implementing this concept we shall get the following results: the city is connected by pedestrian green corridors, a linear park of 18 km long and new parklands, which shall significantly improve air control of the city and develop microclimate in the territory of the whole river-bed.

Organization of contract public spaces with service function and public entertainment. After analyzing location of existing educational institutions it is offered to place five eco-stations, which include biotrons, amphitheatres (within the frameworks of the eco-education program), aquarium, water treatments system and open-space accessible roof on the designed site. On the low-level of the building a water-living zoo is located; it is possible to see one floor up through it, by that uniting spaces. Sports areas are also considered (sport frame of the city). It shall contribute to healthy lifestyle, gain in health and psychoemotional state of people. This territory shall be a municipal recreational area and a cultural center of Rostov-on-Don. New traffic arteries might be used as tourist trails. Based on the recommended river control system a system of surface water control system in the whole city area is offered. Providing the city with process water.

Considering the fact, that the seabed of Rostov Sea is anti-drainage it is impossible to drive piles or any other structures. The most appropriate structural solution is to use high-impact cables. It shall help

IOP Conf. Series: Earth and Environmental Science **177** (2018) 012035 doi:10.1088/1755-1315/177/1/012035

to use larger space and to redistribute streams of people, to increase variants of resting-places, to connect left and right river-banks.

Consequently, the proposed concept of development of architectural space on the Temernik River embankment in Rostov-on-Don shall help not only to improve a coastal territory, but also to solve many important problems in the city, for example, to improve ecological state of air basin and water basin, to take sporting and cultural behavior to a new level, to extend recreational areas, to solve transportation problem and to increase a quality of life on the whole.

The project to create Temernik embankment in the city of Rostov-on-Don.
The main architect of the project M. G. Sarkisyants

Except for the above-mentioned architectural, city-planning and engineering principles of sustainable development of the architectural space of the Temernik River embankment in Rostov-on-Don it is offered to use modern architectural and artful approaches to arrangement of embankment space, as follows:

To avoid using monumental spaces on the embankment and to arrange comfortable areas for resting and communicating, which are close to nature. It is achieved by using natural environment elements. For example, by combining natural and artificial forms in pitched work. To create idea of transfusion of natural and artificial environment.

When developing architectural solutions of the embankment, special attention is given to contact of a man with water and to arrangement of barrier-free environment. It is mainly related to visual contact without any visual barriers. There is no fencing, but if required natural elements instead of rock or reinforced concrete materials are used to determine boundaries [3, 8, 9].

A modular system is used for building infrastructure on the embankment. A multilevel park, raising above water, is designed within the territory of Rostov Sea [3, 8, 9].

Terraced planning of the multilevel embankment shall add theatrical elements, where urban environment is similar to a stage. An amphitheater, which might be used as cinema-hall or theater, is considered on the embankment. Moreover, body of water, floating stage and space above water might be used as a stage for special events [8, 9].

A proposed linear park is unique, as located along 18-kilometers water body, being within the city limits. Embankment is a linear space; for it not to look the same throughout, different measures are taken to avoid space monotony. For example, combination of transit promenades and leisure destinations, applying geoplastics and different planting. Color is used as one of the main spatial references along the embankment. A promenade of the embankment is considered to be paved with bright colors, combined throughout. It is considered to arrange a surface-mounted pedestrian mall and terraced areas, located along the park, at a level of +8.000 m. To avoid space monotony significant facilities are arranged. Such facilities shall become symbols of a place. They help to organize space and create cultural continuity: preservation and restoration of historical values, carrying traditions and including historical sites to modern urban environment.

It is proposed to use expanded materials, which prevent from water evaporation and contamination of water, in the project. Natural materials, natural fabric and colors, which are traditional in this region, are used to underline perception of the architectural space as natural surroundings [8, 9].

Compositional arrangement of plantations, visual reference means, hard landscaping is used as aesthetic harmonization aimed at improvement of man visual perception of the embankment qualities. It is planned to plant different sorts of trees by that improving air basin in Rostov-on-Don. It is considered to arrange art spaces, view points, leisure areas, entertainment and leisure centers in the park.

Conservation and restoration of landscape, arrangement of wilderness areas within water protection zone and water area. The project considers conservation of ecosystem and minimum changing of coasting in the process of embankment arrangement. The paths are raised on abutment pieces by that saving natural circulation of atmospheric condensation. A partial load system is applied, when arranging plots and paths on foundation made of gabion [8, 9].

Green covered floating islands, which contribute to conservation of natural flora and fauna, shall be organized on the river. Specially selected plants and water grass are screening large volume of river water, by that purifying it of dirt and harmful elements.

Special solar protection devices, transformable shields, tents and other shelters, pergolas, roofed-in paths are considered as overheat protection. Planting, including wall gardening performs the same function. Sheltering with the designed building and elements, for example, roof projection is considered, when locating zones.

The project of city public space of the Temernik River embankment is elaborated in accordance with the requests of different subcultures and city communities. The space should be dynamically changing and adapting to changes in lifestyle of people, cultural requirements and development of IT technologies, etc.

The project takes seasonal dynamics, year-round adaptability, i.e architectural, landscape and functional organization based on year-round use, development of possibilities for various and optimal use in summer and winter periods into consideration [10, 11].

Water shall perform some functions of the embankment. It makes possible to provide more territory for the embankment, for rest and walks. The project considers arrangement of floating restaurants, floating landing stage and floating modules for swimming and slow fleet station. It is considered also to use waterborne transport as alternative to surface transport.

4. Discussion

Modern architectural concepts and engineering solutions shall emphasize the coastal zone by that making various impression and new perception of a place. Energy conservation, improvement of energy efficiency of buildings and structures in the coastal territories and use of alternative energy plants are prospective.

The Temernik River embankment and the linear park shall become modern architectural, cultural, leisure and multifunctional elements of the city, keeping historical traditions. Moreover, the park is supplied with bicycle tracks, covered with solar batteries, providing the park with power supply and decreasing operating costs.

5. Conclusion

A proposed model of complex development of the Temernik River embankment in Rostov-on-Don is based on the principles, which determine system approach to arrangement of environment. It is necessary to use interconnection of methods and approaches to development of coastal territories for harmonic inclusion of coastal territories to modern structure of the city and city life, sustainable

IOP Conf. Series: Earth and Environmental Science **177** (2018) 012035 doi:10.1088/1755-1315/177/1/012035

development of urban environment. The above-mentioned principles of sustainable development of architectural space of the Temernik River embankment in Rostov-on-Don shall let prevent imminent ecological disaster, purify the river-basin and organize a recreational zone, which conforms to ecological requirements.

References

[1] Bobryshev D V 2011 Regularities of the functional-planning organization of the valley complex of a major river as a factor of sustainable development on the example of the Irkutsk agglomeration *Vestnik of the Irkutsk State Technical University* **7** 22–8

[2] Kramer D A, Neruda M and Tikhonova I O 2012 European experience of revitalization of small rivers *Journal of Biology, Ecology, Natural Science, Earth Sciences* **2**

[3] Nefedov V A 2002 *Landscaping and environmental sustainability* (St. Petersburg: Polygraphist)

[4] Litvinov D V *Gradoecological principles of development of coastal zones: the example of large cities of the Volga Region* (St. Petersburg)

[5] Yezhova M I, Arkhangelskaya I V et al 2015 *The results of monitoring cholera of water objects of the environment of the rostov region in 2014* (FKUZ Rostov-on-Don Anti-plague Institute of Rospotrebnadzor, Rostov-on-Don)

[6] Ignatova N A 2009 *Evaluation of the toxicity of waters and bottom sediments of anthropogenically polluted ecosystems by the method of biotesting (on the example of the Lower Don basin)* (Rostov-on-Don)

[7] Nikanorov A M, Horuzhaya T A, Minina L I and Mironova T V 2009 Influence of a megacity on the water quality of a large river (on the example of Rostov-on-Don) *Vestnik of the Southern Scientific Center* **5** (4) 45–51

[8] Gorgorova Yu V and Sarkisyants M G 2017 Modern trends in the design of the architectural environment of urban embankments *In the collection: Construction and Architecture-2017. School of Architecture, Design and Arts Materials of the scientific-practical conference. Ministry of Education and Science of the Russian Federation* (Don State Technical University, Academy of Construction and Architecture, Rostov-on-Don) 90–5

[9] Gorgorova, Yu V 2017 The Humanization of the Urban Environment in Conditions of Multi-Story Development (in the example of the Canary Wharf business district in London) *Inženernyj vestnik Dona (Rus)* 4

[10] Gorgorova Yu V 2017 Application of modifiable art objects and interactive elements to enhance the attractiveness of the public space, *Science, education and experimental design at the Moscow Architectural Institute: Abstracts of the reports of the international scientific-practical conference of faculty, young scientists and students* (Moskow: Moscow Architectural Institute) **2** 396–7

[11] Gorgorova Yu V 2014 Seasonal transformation of urban space as one of the ways to increase its attractiveness *Scientific Review* **11-3** 693–6

Published under licence in *Journal Title* by IOP Publishing Ltd

IOP Conf. Series: Earth and Environmental Science **177** (2018) 012036 doi:10.1088/1755-1315/177/1/012036

Assessment of the level of impact and degree of environmental safety of industrial facilities in the urban environment

V Bespalov[1] and E Kotlyarova[2]

[1] Department of Environmental Engineering, Don State Technical University, 1 Gagarin square, Rostov-on-Don, 344000, Russia
[2] Department of Urban Planning and Building Design, Don State Technical University, 1 Gagarin square, Rostov-on-Don, 344000, Russia

e-mail: ekkot.arch@gmail.com

Abstract. The problem of a comprehensive assessment of the level of impact and the degree of ensuring the ecological security of built-up areas has not been fully resolved. For its solution, we suggest a methodical approach based on the use of three coefficients for assessing the quality of the state of the environment and the effectiveness of measures to ensure environmental safety in urban areas, in particular, the coefficients of environmental hazards, economic losses and social discomfort. Its use will allow the maximum objective selection of a complex of ecologically efficient and economical architectural and planning, organizational, technological and special engineering and environmental measures in the design of new and operation of existing industrial facilities. The results received at the completed stage of the research can be used as a basis for further development aimed at creating a methodological basis for the assessment, selection, calculation, design and implementation of a complex of ecologically efficient and economical architectural, planning, organizational, technological and special engineering and environmental measures to ensure environmental safety built-up areas in general.

1. Introduction

Every year the problem of ensuring the ecological security of built-up areas affects more and more areas of human activity. The continuation of intensive anthropogenic activities without highly effective measures to ensure environmental safety becomes very damaging to the environment and contradicts the basics of the theory of sustainable development [1–2]. The removal of the most production facilities beyond the central part of urban settlements does not completely solve the problem of reducing the negative impact of these facilities on the environment. In addition, the tendency of increasing the number and size of agglomerations, along with an increase in the rate of de-urbanization, leads to a rapprochement of the residential areas under construction with new industrial areas [3].

The consequences of various impacts on the sustainability of the urban ecosystem can be divided into explicit, directly affecting the safety of the population, and hidden, which can include the results of social and economic changes reflected in the statistics or even in the value of real estate. The explicit consequences include the depletion of the fertile layer of urban vegetation, the

Content from this work may be used under the terms of the Creative Commons Attribution 3.0 licence. Any further distribution of this work must maintain attribution to the author(s) and the title of the work, journal citation and DOI.
Published under licence by IOP Publishing Ltd

disturbance of the temperature and aeration modes in the soil, and as a consequence, the complex conditions for the growth and development of vegetation, the scarcity of fauna, and the pollution of the air basin and the aquatic environment as a result of anthropogenic activity. In our opinion, the growth of certain types of diseases in various groups of the population, depending on the nature of the prevailing pollution, and also the decrease in the attractiveness of certain capital construction projects located in an unfavorable zone should be considered as hidden.

In this regard, one of the important directions for overcoming the ecological crisis and solving the problem of ensuring the ecological safety of built-up areas is the development of criteria assessing the impact of industrial facilities on the urban environment, with subsequent development on their basis of a methodology for selecting, designing and implementing a set of highly efficient and economical architectural, planning, technological and special engineering-ecological measures aimed at overcoming the negative effects of industrialization and increasing environmental safety.

2. Materials and Methods

Functional zoning of almost every settlement can be characterized as follows:
- residential areas, which includes both the existing building of the central (historical) part, and the zone of individual, as a rule, single-family private housing construction and the territory of new micro districts with multi-storied buildings;
- industrial zones, which include operating enterprises, warehouses, bases, etc.;
- zones of engineering and transport infrastructures;
- public and business zones used for placement of objects of trade, education, business, health, etc.;
- recreational zones or a natural complex, dividing into virgin and fallow lands, gardens, parks, boulevards and public gardens, and also forest parks [4].

In our scientific work, the main attention should be paid to the objects of industry in the urban environment. As it is known, their functioning is aimed at solving such important economic tasks of municipal entities as creating new engineering and transport infrastructure facilities, increasing the taxable base and revenues of the consolidated budget of the subject of the federation, increasing the rate of industrial growth, and creating new jobs, in turn, entails an increase in the standard of living of the population and an improvement in the investment climate. Thus, the geography of location of industrial objects plays a big role in building the territorial-economic system of a particular region from the point of view of economic geography.

In its turn, on the territory of each industrial site it is possible to perform a separate division of the territory into the following areas:
- industrial and technological, within which the main production and technological buildings and facilities, warehouses, engineering infrastructure facilities, power engineering, garages, cargo areas, railway access roads, berths and other transportation facilities, as well as supporting and maintenance facilities are located;
- social, which can include all buildings and facilities used to provide the required sanitary and hygienic conditions at the enterprise;
- recreational area within which employee recreation areas are located and green plantations of various types and structures (trees, bushes, annual and perennial decorative plants, etc.) grow and perform ecological, environmental and sanitary-hygienic and aesthetic functions.

The impact of a complex of sources of environmental pollution in the areas of production facilities in the urban environment leads to economic damage from air pollution D_{air}, water sources pollution D_{water}, land (soil) pollution D_{soil} and pollution by physical fields D_{phf}.

Herewith, three types of economic costs of production facilities as objects of environment are contrasted with the economic damage to the environment:
- expenditures on environmental measures aimed at preventing the impact of emissions, waste, discharges, etc. on the environment (modernization of production processes, compliance with environmental safety of manufactured products, the costs of designing environmental protection

measures for environment, investments aimed at increasing the ecologization of production, etc.) $V_{extrime}$;
- ecological payments of the object of anthropogenic impact (payment for discharges, emissions, removal and utilization of waste, use of natural resources, etc.) C_{ost};
- economic benefit (economic effect) for the urban settlement from the operation of environmental management facilities located on the investigated territory of the industrial zone (payments, taxes, contributions and other payments to the local budget, investment commitments aimed at the development of the social sphere, rents, etc.) E_{ef}.

The costs of industrial facilities aimed at preventing the impact of emissions, discharges, waste and the impact of physical fields on the environment and the health of the population of the adjacent territory of the industrial zone, including the sanitary protection zone $V_{extrime}$, consist of the measures costs to protect the air basin, water bodies, soil cover and vegetation, as well as measures to reduce the level of pollution by physical fields.

Assessment of the state of industrial facilities in the urban environment requires a complex accounting of the three main parameters groups: ecological, economic and social, determining the overall quality of the territory examined as a whole. Taking into account that the values of economic, environmental and social characteristics vary depending on the functional area of the examined urban area, in determining the integrated assessment, it is also necessary to take into account the inequalities of the indicators included in it. The approach suggested by us allows, taking into account various territorial characteristics with different dimensions, to obtain the resulting dimensionless coefficients of ecological danger, economic losses and social discomfort, respectively [5].

The factor of ecological danger of industrial objects in the urban environment directly characterizes the state of the environment in the area of their location. Experts use the parameters characterizing the level of pollution and the effectiveness of air protection, the degree of use and effectiveness of protection water bodies, the degree of use and effectiveness of protection the soil cover, as well as parameters characterizing the level of impact as the main parameters describing the quality of the environment physical fields (noise, vibration, electromagnetic and ionizing radiation).

As the main characteristics of the state of the environment, we took the amount of emissions into the air basin, discharges into water bodies, as well as wastes generated as a result of the functioning of the examined object and the actual values of pollution levels in the area where the object is located.

Based on these characteristics, we obtained dependencies for determining the relative levels of pollution of the main components of the environment:

$$V_j = \sum_{i=1}^{n} \frac{Q_{ap}}{Q_{ev}} \tag{1}$$

where Q_{ap} – the actual pollution level, mg / m^3 (mg/kg, dBA, V/m, microR/hour); Q_{ev} – value of permissible level of pollution, mg/m^3 (mg/kg, dBA, V/m, mcR/h).

A regulatory framework for assessing the statistical characteristics of the state of the environment is a hierarchically ordered system of general documents that regulate environmental factors both at the international and at the national level. In particular, in the Russian Federation they include: international legal acts, provisions of the Constitution of the Russian Federation, federal laws of the Russian Federation, decrees and orders of the Government of the Russian Federation, decrees of the President of the Russian Federation, legal acts of ministries and departments, as well as laws and regulations of regions of the federation and local government bodies.

When using the characteristics of the state of the environment, it is possible to determine the distribution of concentrations of harmful substances, as well as the levels of pollution by physical fields in the examined area. In our opinion, this is not enough to calculate the socio-ecological and economic assessment of the condition of the territories of industrial zones of large cities and the adjacent elements of urban infrastructure.

To combine the pollution coefficients of the components of the environment V_j into a criterion of total pollution V_{total}, we used the following formula:

$$V_{total} = (V_{air} + 1)^{D_v^{air}} \cdot (V_{water} + 1)^{D_v^{water}} \cdot (V_{soil} + 1)^{D_v^{soil}} \cdot (V_{phf} + 1)^{D_v^{phf}} - 1 \quad (2)$$

where D_v^j – the weighting factor of the corresponding component of the environment.

To calculate integral coefficients that take into account several components of the environment, it is necessary to apply the significance coefficients of the partial criteria B_{3j} that characterize the dynamics and influence of individual components of the overall assessment. In our approach, we offer to calculate the weighting factors in accordance with the values of economic damage caused by man-made objects of the corresponding environmental component j:

$$D_j = \frac{D_j}{\sum_j D_j} \quad (3)$$

Formula (2), in our opinion, is characterized by such advantages as
- ease of use in practical calculations;
- taking into account the weight factors of the main components of the environment;
- sufficiently high accuracy, since the opposite values of the factors that it includes are taken into account, which does not happen when using the linear method calculation of the average arithmetic sum and the ball method.

The use of the weighting factor D_v^j highlights the fact that some components of the environment are more significant than others, because at the present time they are mostly negatively affected, and therefore they need to be taken into account to a greater extent.

The method we developed is based on the measurements data of the state of various areas. The pollution coefficients of the main components of the environment have been determined in comparison with the normative (maximum permissible and background) values, which are in turn determined on the basis of hygienic and epidemiological conclusions.

The impact of a complex of sources of environmental pollution on the territories of industrial facilities in the urban environment leads to the emergence of primary effects that determine the state of the environment at a particular time. Our methodological approach to the determination of the coefficient of environmental hazard contains an assessment of the primary consequences of anthropogenic impact, which is based on accurate data from measurements of the states of different areas.

Thus, the coefficient of ecological danger of the state of the environment of industrial facilities in the urban environment C_{ecol} is calculated on the basis of the characteristics of the state of the components of the environment and the consequences of human activities, leading to negative changes in this state.

The essence of the economic losses factor in assessing the production facilities of the urban environment and the adjacent areas is in the characterization of economic benefits (economic effect), as well as economic damage to the environment and the health of the population of the adjacent territory, obtained as a result of the functioning of the man-made object.

The amount of economic benefit from the operation of each facility is affected by the volume of taxes, deductions, payments and other payments to the local budget, which determine the benefits of the object for the economy of the city and the region as a whole.

To determine the economic effect obtained from the functioning of a man-made object in this methodology, the following are used:
- profit of the environment object, directly affecting the amount of payments, taxes and deductions that come to the local budget from a specific enterprise;

IOP Conf. Series: Earth and Environmental Science **177** (2018) 012036 doi:10.1088/1755-1315/177/1/012036

- financial effect that determines the rent, the amount of taxes and other revenues to the local budget;
- the cost of production, which determines the profit of the environment object.

The values mentioned above are interchangeable – one can be used to calculate the other two. To determine the economic effect of $E_{ef.}$, which is included in the formula for calculating the economic losses factor, we offered to use one of these quantities, depending on the conditions. In addition, we took into account rent for land and premises. On the one hand, the final profit of the enterprise depends on it, and, on the other hand, it supplies the local budget.

Also, economic benefits from the operation of the facility include investment commitments aimed at developing the social sphere, the infrastructure of the region, payments related to the transfer of land from one category to another.

Significant financial resources from payments for negative impact on the environment can be accumulated in the local budget by municipal and regional authorities and subsequently used for environmental protection.

The general principles of assessing economic damage imply that the damages of industrial facilities include real damage to the environment and the health of the population of the adjacent territory. Calculation of the amount of damage in monetary terms from various types of violations of environmental legislation was carried out by us on the basis of the principles of compensation for damages set out in the Civil Code of the Russian Federation and the Federal Law "On Environmental Protection".

Estimation of damages from pollution is understood as the calculation in monetary terms of economic and non-economic losses caused by deterioration of buildings and structures [5], corrosion of metals, disruptions in technological processes, an increase in the incidence rate and a decrease in the working capacity of the population adjacent to the industrial zone, and other phenomena caused by chemical, physical and biological pollution of the environment. Finance in this case is not only an economic characteristic, but also the magnitude of social and economic damage. Financial (economic) assessment is possible only in the final values, but social damage can be endless with irretrievable losses of the basic life values of a person [6].

Calculation of annual values of economic damage from air pollution can be carried out according to the formula [6]:

$$Da = \gamma t \cdot \sigma(\lambda) \cdot f(\lambda) \cdot (\Sigma Ai(\lambda) \cdot mit), \tag{4}$$

Results where γ_t – monetary unit of emission, rub./cond.t; $\sigma(\lambda)$ – coefficient, by means of which regional features of the territory affected by harmful influence are taken into account; $f(\lambda)$·- correction including the dispersion of impurities in the atmosphere; $A_i(\lambda)$ – the coefficient of reduction of the impurity of the i species to the mono-pollutant, cond. t/t; m_{it} – volume of release of the i type of contaminant admixture, t/year.

Economic assessment of damage to water bodies is carried out according to the formula [6].

$$Dв = \rho t \cdot \beta(\lambda) \cdot (\Sigma Di(\lambda) \cdot Vit), \tag{5}$$

where ρ_t – monetary valuation of the discharge unit, rub./cond.t; $\beta(\lambda)$ – coefficient allowing to take into account the features of the reservoir, which is subject to harmful influence; $D_i(\lambda)$·– the coefficient of reduction of the impurity of the i species to the mono pollutant; V_{it} – volume of discharge of the i type of impurity pollutant, t/year.

The economic assessment of the amount of damage from the degradation of the environment afterwards by contamination with harmful substances and littering of the soil cover on the territory of the industrial zone is determined by the formula [*6*].

$$Dn = Nc \cdot S \cdot Ce \cdot Cenv, \tag{6}$$

where N_c – rate of land value, thousand rubles/ha; S – the area of the soil cover degraded, contaminated with harmful substances or littered in the reporting period, ha; C_e – coefficient

IOP Conf. Series: Earth and Environmental Science **177** (2018) 012036 doi:10.1088/1755-1315/177/1/012036

of ecological status and ecological significance of the territory; $C_{env.}$– coefficient for the territories under protection.

To calculate the economic evaluation of the damage caused by a high level of sound pressure to the health of the population, one can use formula [6].

$$Ds.pressure = B \cdot g \cdot c \cdot S / (\rho \cdot N \cdot 1002), \quad (7)$$

where B – average output per worker per year, rubles/person; g – percent of city-forming group, %; c – a coefficient that takes into account the decline in labor productivity, %; S – area of discomfort to the maximum permissible level (60 dBA), ha; ρ – density of production fund, m^2/ha; N – norm of total space per employee (15 m^2/person).

Economic evaluation of the damage from the vibration effects on the environment can be determined by means of the following formula [6].

$$Dvibr. = \gamma \cdot \Sigma\, zi \cdot vi \cdot ri \cdot [\Sigma(Bj - B_H)], \quad (8)$$

where B_j (B_{Hj}) – the actual (standard) vibration level of the j frequency level; n – number of buildings in the examined territory; z_i – coefficient of significance of the i building (it can be determined by the result of an expert evaluation or repairing costs); v_i – coefficient reflecting the rate of destruction of the i building (depending on the age of the building, the state of the structures, as well as used in the construction and reconstruction of materials); r_i – distance to the building; γ – the cost estimation for the reference influence option.

Since the results of calculating the economic losses factor, the status of the industrial zone in question should be attributed to one of the following:
- economic damage is greater than economic benefits from economic activity. This option is mostly unfavorable, because this territory "works" for economic indicators. It is necessary to recommend measures to improve the existing situation;
- economic damage and economic benefits from the operation of environmental management facilities are equal. The issues of environment protection and the dynamics of the economic development of the territory are of equal interest;
- amount of economic damage is less than the value of economic benefits from economic activity. This option is most favorable.

That is why the coefficient of economic losses of industrial sites in the urban environment of Racon was determined on the basis of parameters of economic damage to the environment, the costs of environmental protection measures, and non-environmental investments of environmental management objects into the local budget.

The essence of the social discomfort coefficient is the characterization of the health status of the working and living population, living conditions in the area of the industrial facilities of the urban environment and the surrounding urban infrastructure, including the sanitary protection zone.

The manuals on health statistics contain information on establishing the relationship between levels of diseases and mortality of the population and various characteristics of the state of the environment. The development of effective recreational activities is impossible without knowledge of this relationship. The most common reason for conducting such studies is the need to establish the causes of diseases associated with environmental pollution, while specialists apply methods to study the patterns of the spread of non-infectious etiology among the population adjacent to the industrial zone, including the sanitary protection zone, based on the use of statistical indicators.

The characteristics used to determine the social discomfort of urban industrial areas and adjacent urban infrastructure elements include the comfort and well-being of the examined areas, the emergence of public goods, the degradation of the environment and the loss of aesthetic properties of landscapes [7], as well as data on the social benefits of the adjacent to the industrial zone of the territory population, including the sanitary protection zone [8], employed in hazardous areas of production and the results of statistical analysis of existing information on the health of the population of the territory adjacent to the industrial zone, including the sanitary protection zone, based on the analysis of occupational injuries and occupational diseases.

Every case of disability to work entails losses – direct and indirect. Their performance depends on the production performance of the enterprise. Direct losses result from losses of the enterprise and economic damage. These losses are related to the economic activities of the enterprise and affect production performance. Indirect losses are payments made by the trade union organization and social insurance in case of disability.

Occupational diseases and incidents of injuries can have different consequences for the restoration of work capacity and the ability to perform labor functions. In diseases or injuries that cause a decline in labor productivity, a transfer to a work required an easier, lesser qualification that is not associated with a harmful production factor is required. For diseases or injuries that lead to a decrease in productivity and disability, a disability pension is required. Disability with a fatal outcome leads to the payment of benefits to the family of the victim.

Loss of ability to work for each of the above cases can be determined:

$$L=O+S+C, \tag{9}$$

where O – loss of organization; S – social costs; C – conditional losses.

On the other hand, the economic loss of working capacity can be divided into the following groups: losses caused by injuries L_{tr}; losses caused by microtrauma L_{mtr}; losses caused by occupational diseases L_O, and costs of compensation due to the presence of harmful production factors L_C:

$$L=LTR+LMTR+LO+LC. \tag{10}$$

The total amount of the conditional production losses in case of injury C_{tr} is compiled from the loss of production in connection with the absence of the injured worker at work C_{tr}^{1}, and the cost of nonproduced products on idle equipment as a result of injury C_{tr}^{2}.

$$Ctr = Ctr1+ Ctr^2. \tag{11}$$

The economic losses of the organization O_{tr} in this case are the following: from the cost of the spoiled materials; cost of damaged equipment, tools and destroyed structures O_{tr}^{2}; cost of repair works O_{tr}^{3}; payments on sick leave O_{tr}^{4}; surcharges to earnings in case of a decrease in professional capacity for work O_{tr}^{5}; payment of pensions for a disability or fatal outcome O_{tr}^{6}; costs for re-qualification of the victim and training of new workers O_{tr}^{7}.

Thus,

$$Otr= \Sigma\, Otri. \tag{12}$$

The social costs S_{tr}, which are covered from public consumption funds, are formed from the cost of outpatient treatment S_{tr}^{1}, clinical treatment S_{tr}^{2}, payments of pensions by the means of social insurance to disabled people with injuries S_{tr}^{3}, expenses for sanatorium treatment S_{tr}^{4}:

$$Str= \Sigma\, Stri. \tag{13}$$

Microtrauma, which does not lead to loss of capacity for work in day/person, also causes damage to the economy, being associated with the costs of first aid, the transfer of the victim to easier work, lower labor productivity, cost of repair or adjustment of equipment.

Economic consequences associated with occupational diseases P_d, are made up of the same three components – the losses of the organization of O_d, conditional losses of C_d and social costs of S_d.

The conditional loss of production can be determined with sufficient accuracy by the average cost of production P produced per working hour and the loss of working time T in connection with the illness of the victim:

$$Cd= PT. \tag{14}$$

Other economic losses are defined similarly to the case of injuries, except for losses associated with damage to equipment and its repair.

The economic costs of P_d, connected with the presence of harmful production factors, additionally contain expenditures for special food, individual means of protection, the maintenance of sanatoria and dispensaries, the need to establish a shorter working day, additional vacations, additional intra-shift breaks, reduced length of service before retirement etc.

The information databases needed to analyze the relationship between environmental characteristics and the health of the population of the adjacent territory, including the sanitary protection zone, are divided into three types:

- database on hazardous and harmful environmental factors;
- database on the exposition of the population living near industrial zones;
- database on the health status of the population adjacent to the industrial zone, including the sanitary protection zone, including diseases caused by environmental pollution (environmentally caused diseases).

Ecologically caused diseases include diseases and pathological conditions that have developed among workers in a particular territory under the influence of harmful factors of the production environment in the form of a "nonspecific" and "specific" pathology.

Indicative ecologically caused diseases include somatic and other diseases which are common among the population adjacent to the industrial zone, including the sanitary protection zone, whose frequency over a certain period of time exceeds the recorded values over the previous 5 to 10 years of observations. The reason for their growth can be the action of known both local and regional unfavorable environmental factors.

Information databases on the health of the population may contain the following types of information:

- results of periodic medical examinations;
- available and regularly updated reporting data on disease and mortality levels of the population, as well as medical records from the case histories;
- registers of special data (congenital malformations, oncological diseases, occupational diseases, etc.);
- results of bio monitoring (determination of lead in blood and other exposures);
- account of doctors' notices about all cases of diseases;
- general data on the number of visits to doctors and ambulance calls.

In the study of a particular case of a disease, the incidence rate is defined as the incidence of cases per person (probability of disease):

$$P(G_{dis}) = \frac{N_{I\,dis}}{N_{pop}}, \quad (15)$$

where $N_{i\,dis}$ – incidence number of i disease per year in total, person; N_{pop} – total number of employees of the enterprise in the examined area, person.

In addition, a difficult task is to determine the quantitative characteristics of the probability of the factor and the reasons that the disease can be caused. First of all, it is necessary to take into account the factors that describe the degree of environmental pollution. The probability of a negative factor is the probability of a negative effect for a person and is defined as the probability of occurrence of this effect under specific conditions. Most of the methods previously offered by specialists use for calculating this probability the ratio of the level of contamination to the permissible value, which in our work is determined by the criterion of general pollution P_{total}.

The main principle on which calculation of the pollution factor assessment acting on the i type of the disease is based is the use in calculation of the value of the individually absorbed dose of a harmful substance that has got into the human body by airborne droplets, through drinking water or food, etc. for a certain time period.

Thus, when analyzing the effect of the level of environmental pollution on the incidence of the population of the examined territory, an important problem is the factor-based accounting of the pollution effect on the examined coefficient.

3. Results

The received data on high, medium, moderate and low pollution should be corrected taking into account the level of danger of each degree of pollution. For this, it is necessary to introduce calculation coefficients to obtain an average estimate of the contribution of the hazardous influence of each factor to the overall assessment of the environmental conditions of the examined territories.

The result of the logarithm of the hazard scale for each of the environmental components and bringing them all to the same basis (the initial hazard indicator) is to obtain a single-scale scale. This allows, taking into account the level of danger of each degree of pollution, to introduce the calculated hygienic hazard coefficient of the health status of the population of the territory adjoining the industrial zone, including the sanitary protection zone, and to bring all the indicators, expressed in different units of measure, to a single system of dimensionless indexes, that gives the possibility of their further summation and the obtaining of a total index of environmental pollution in the territories of industrial zones of large cities and adjacent elements of urban infrastructure [9-10].

Gradation levels of pollution of individual components of the environment, which are used to calculate the hazard coefficients, are scientifically valid from the point of view of the health impact of the population adjacent to the industrial zone, as defined by specialists on the basis of hygienic or epidemiological studies. Our approach to assessing the dangers of pollution levels proceeds from the mentioned above. Therefore, according to the degree of danger, these levels are quite comparable. Conducting instrumental measurements with subsequent calculations makes it possible to obtain maps of environmental assessment of the state of the environment in the area of industrial zones where the distribution of the values of total air pollution, water bodies, soil cover, and pollution with physical fields can be visually represented.

4. Discussion

The use of the offered criteria that assess the quality of the state of the environment and the effectiveness of measures to ensure environmental safety will make it possible to select the complex of ecologically effective and economical architectural and planning, organizational, technological and special engineering and environmental measures in urban planning as possible when designing new and operating existing industrial facilities. At the same time, the city itself or the urban agglomeration, as an environment, can significantly reduce the impact of pollutants formed during production processes.

5. Conclusions

Thus, the completed stage of research and the results obtained can be used as a basis for further development aimed at creating a methodological basis for the assessment, selection, calculation, design and implementation of a complex of ecologically efficient and economical architectural, planning, organizational, technological and special engineering and environmental measures for ensuring the ecological security of built-up areas in general.

References

[1] Adler F R and Tanner C J 2013 *Urban Environmental systems: Ecological Principles for the Built Environment* (London, Cambridge University Press)
[2] Forman R T 2014 *Urban Ecology: Science of Cities* (London, Cambridge University Press)
[3] Sheina S and Khamavova A 2016 *Procedia Engineering* 1960–5
[4] Bespalov V and Kotlyarova E 2015 *Inzhenernuy vestnik Dona* **4 (38)** 74
[5] Ilyichev V, Emelyanov S, Kolchunov V, Bakayeva N and Kobeleva S 2015 *Procedia Engineering* **117** 126–31

[6] Bespalov V and Kotlyarova E 2011 *Terra Economicus* **4-3 (9)** 121–3
[7] Watts G 2017 *Urban Forestry & Urban Greening* **26** 11–7
[8] Jayasooriya V M, Ng A W M, Muthukumaran S and Perera S 2017 *Urban Forestry & Urban Greening* **21** 34–47
[9] Caiado R G G, Freitas Dias R, Mattenvironment L V, Quelhas O L G and Filho W L 2017 *Journal of Cleaner Production* **165** 890–904
[10] Bespalov V and Kotlyarova E 2017 *Matec Web of Conferences* 05005

Published under license in *Journal Title* ***by IOP Publishing Ltd***

Cost aspects of membrane bioreactors for wastewater treatment

N Makisha[1], O Saveliev[1] and S Katella[2]

[1] Moscow State University of Civil Engineering, Moscow, Russia
[2] Moscow Technological University, Moscow, Russia

e-mail: ole6542@yandex.ru

Abstract. There is a tendency of more and more intensive annual growth of popularity of membrane process reactor technology. Membrane bioreactors (MBRs) have become serious competitors to other methods on the part of the quality of water cleaning. This is the new and perspective technology; however it is still rather expensive, although not so complicated to operate it. The choice of such a technology demands taking into account all its advantages and disadvantages. Nowadays customers are able to choose between different varieties of configurations of MBRs, smart choice of which may provide economic efficiency.

1. Introduction

One of the most advanced and dynamically developing technologies in the field of wastewater treatment is the technology of a membrane biological reactor (MBR). Membrane bioreactors combine the processes of microfiltration and ultrafiltration with traditional biological purification of active silt. As it follows from the very name of the technology, the main elements of MBR are membranes, which serve as a powerful barrier against viruses and bacteria, resulting in a high degree of purification.

At the moment two main types of MBRs are distinguished:

- Immersion membrane bioreactor – with this configuration, the membrane module is immersed in the aerotank and filtration is provided by the influence of a vacuum. In the aeration tank the mixture in the form of activated sludge and waste water is filtered through the surface of the membranes, after what the purified water is removed from the aeration tank, and the active sludge and the delayed substances are suspended. Immersion MBRs are compulsorily equipped with aeration systems.
- External membrane bioreactor – in this version of the technological design, the membrane module is placed outside the aeration tank and carries the functions of the secondary settler and post-treatment systems and does not participate in the biological treatment itself. A mixture of activated sludge and waste water is diverted to an external membrane module, which is represented by a set of tubular membranes through which this mixture passes.

2. Materials and methods

The membrane module, as well as the type of membrane, in its turn, is the main pricing factor of MBR. Currently, there are 2 types of membrane modules used in MBR: external (pressure) and submersible. Each of them from the economic and operational point of view has both its advantages and disadvantages. Let us consider which of the types of membrane modules can be preferred due to an economic point of view.

Content from this work may be used under the terms of the Creative Commons Attribution 3.0 licence. Any further distribution of this work must maintain attribution to the author(s) and the title of the work, journal citation and DOI.
Published under licence by IOP Publishing Ltd

IOP Conf. Series: Earth and Environmental Science **177** (2018) 012037 doi:10.1088/1755-1315/177/1/012037

According to its design type, the undisputable favorite is the immersible MBR module, since it has a simpler assembly configuration. This parameter is extremely important, as it implies a number of other significant advantages that greatly simplify, as well as reduce the cost of servicing the operated MBR. With a simpler type of construction, such a parameter as the possibility of modernizing the bioreactor is substantially increased, and at the same time the labor intensity demanded for the service of the aggregate is reduced. All these factors positively affect both the initial cost of ICBMs and the depreciation costs that are unavoidable during the operation of equipment.

Of course, the gain in simplicity of design entails certain costs. In this case, this is the specific productivity of the membrane bioreactor. For pressure MBRs this indicator is several times higher (3–4 times). Usually it amounts about 40–100 l/m^2 * h, while for immersive membrane modules of bioreactors it does not exceed 10–30 l/m^2 * h. Loss in performance can be a serious drawback if such parameter, as the volume of filtered water per unit of time is highly important. Moreover, it is critical in cases with large volumes of wastewater that require urgent filtration. However, in some cases, when the time for which the filtration is performed is not so important, it can be neglected.

Pressure and immersive MBR modules also differ in methods of reducing membrane contamination. In pressure MBR are used special technology Cross Flow (what means transverse flow), air flushing, backwashing and chemical cleaning, whereas in immersive modules air mixing, backwashing and chemical cleaning are practiced. However, we should mention that reverse flushing is not always possible due to design parameters, which makes the immersive MBR modules less universal.

It should also be added that by virtue of the simpler design of the immersive modules of membrane bioreactors, it is possible to achieve a higher packing density of membranes in comparison with pressure modules. In the case of pressure MBRs, it is rather low and amounts to only 50–200 m^2/m^3, while in submerged bioreactors this parameter is brought to 600 m^2/m^3, what is from 3 to 12 times higher. Unfortunately, such a huge gap in packing density of membranes partially causes the loss of immersive MBRs before the pressure ones in specific productivity.

On the other hand, simplicity of design and low specific productivity allow to reduce energy consumption, which plays an important role in the long-term commercial operation of membrane bioreactors. The consumption of power by pressure modules of MBRs exceeds 0.5 kW*h/m^3, while the operation of immersive bioreactors requires no more than 0.2–4.4 kW*h/m^3.

Based on the parameters discussed above, we can draw a preliminary conclusion about the economic expediency of each particular system. After comparing the characteristics of pressure and immersive modules of membrane bioreactors, we come to the conclusion that the installation and subsequent operation of immersive MBRs is much less costly both in terms of mounting and in terms of use. Of course, in cases requiring high specific power, it is necessary to use MBRs of pressure type. Particularly, for servicing large urban settlements and municipal centers.

In membrane modules are used different types of membranes. It can be hollow fibers, flat or tubular, each of which are installed depending on the quality of the contaminated water and the economic efficiency. These membranes are made of a polymeric material, but in individual cases, tubular sheath may consist of ceramics.

The hollow fiber membranes surpasses their competitors in terms of fiber packing density, which directly affects the quality of the water being purified. Their density differs between 300 m^2/m^3 and 600 m^2/m^3, which is much higher than the density of flat (50–150 m^2/m^3) and tubular (<300 m^2/m^3) membranes. It is also worth mentioning that the hollow fiber membranes are the leaders not only in this parameter, but also in terms of the material consumption, which will influence the increase or decrease in the cost of the construction.

Another important characteristic of membranes is their specific productivity. The more water with a suspension can pass through membrane fibers, while passing a benign cleaning, the more efficient the structure will work, and therefore, save resources. In this issue, flat grids will undoubtedly be dominant with their high specific productivity, leaving hollow fiber and tubular membranes far behind.

The membranes have tendency to become clogged by the formation of sediments on them. There is such conception as a "critical flow", what means the highest permissible flow, surplus of which leads to the unacceptably intensive growth of sediment, that becomes a serious obstacle for the normal functioning of the membrane module. It is obvious that all kinds of membranes are clogged with varying intensity. Flat membranes are inclined to becoming clogged faster than any others. This fact indicates that they will demand more frequent cleaning, what will significantly increase the cost of operation, so that the quality of the treated water and the efficiency of cleaning do not decrease. Tubular membranes will be clogged longer than their competitors due to their structure. The hollow fiber membranes in this category occupy an average position, what makes them the most balanced option.

There is an effective way to fight with sludge formation, that historically came from ultrafiltration plants for the purification of natural waters. It is flushing with a return stream of the filtrate, what requires additional energy costs. Usually modules with flat membranes do not allow backwashing, due to possible damage and, Consequently, its further replacement. But with the use of hollow fiber and tubular membranes, hydraulic flushing takes place. Because of it, the membranes will stay at service longer, before they will have to be removed and cleaned. The intervals between backwashes and their duration settle within 10–60 minutes and 15–300 seconds, respectively.

In the course of operation, sooner or later the question of replacing individual membranes due to damage and rupture of the used ones arises. In this issue the hollow fiber membranes are inferior to the tubular and flat membranes, which, if necessary, can be replaced by separate elements, and the structure will be brought back into operation, ensuring high-quality throughput. In the same case, it is absolutely unavoidable to completely change the membrane module, if the damage is critical indeed, but if not, then probably places of fractures will clog up with sludge and the module will continue to function. If we rely on the parameters of the membranes that we have listed above, we can conclude that preference will be given to the hollow fiber membranes, which are used in both pressure and submersible modules due to the advantages of its characteristics. In addition to most of its positive qualities and negative minima, in comparison with other options, hollow fiber membranes have the lowest cost.

While analyzing the available basic parts of the membrane biological reactor, we can conclude that we have 6 different modernizations. By combining the types of membrane modules with membrane types, we can achieve not only effective cleaning, but also economic benefits when taking into account all the weak and strong sides of these technologies.

During the operation of membrane biological reactors, a number of serious problems arise on the grounds of the construction of this technology. Excluding the subtleties of the functioning of the biological treatment itself, we can distinguish the following characteristic problems by priority:

- Contamination of membranes and mesh filters;
- Damage to membranes or severe contamination;
- Failure of communication lines of automation systems;
- Failure of membrane blowing systems;
- Failure of blowers and aerators;
- Contamination of nets or grids;
- Refusal of recycling;
- Failure of auxiliary membrane equipment.

In order to restore the permeability of membranes in the course of operation, treatment with solutions of reagents is carried out, usually with oxidants. In other words, chemical washing takes place. As reagents are generally applied solutions of sodium hypochlorite with a concentration of 0.2–1 % or citric acid with concentrations of 0.2–0.3 %. The frequency of this procedure makes up on average 1 time in a few months, depending on the amount of contamination and the rate of clogging of membranes. As a preventive measure, treatment with sodium hypochlorite solution can be carried out more often. For example, several times a month. The pressure modules are washed by circulating the reagent solution supplied by the pump from a separate tank, when the immersive modules are either moved to a separate special container or washed right at place. This procedure takes several hours by duration.

In a number of cases where it is not possible to achieve the necessary degree of membrane cleaning from clogging, it becomes necessary to extract membrane blocks for their further mechanical washing with water jets from deposits that have already accumulated on the membrane. Contamination of membranes and mesh filters are the consequences of poor pre-cleaning, what means there is an accumulation of hair, scraps of rags and other fibrous materials on the membrane fibers and in the cavities of the membrane blocks. Considering this fact, it is necessary to provide a quality preliminary cleaning in order to minimize the frequency of membrane contamination, what in other case would lead to a corresponding increase in the cost of their operation.

If the operation of aerators and blowers fails, the process of reducing the permeability of membranes occurs and the precipitate is formed more quickly. Such a problem will not affect the quality of purified water until a certain point, but this will affect energy consumption, because of the need to remove excess sediment. Repair of aerators and blowers is rather expensive and time-consuming.

If we look at the cost of operation in figures, here are the indicators we can observe, during the cost analysis: the capital costs for the construction of a wastewater treatment plant with MBRs range from 475703.4 thousand rubles to 79283.9 thousand rubles (6000–1000 Euros) for 1 m^3/ day depending on the system's productivity. The cost of the membrane plant itself with all the auxiliary equipment amount 30–60 %, which is an impressive part of the total sum. The cost of membrane blocks varies from 5946.29 thousand rubles to 11892.59 thousand rubles (75–150 euro/m^2) with an average specific productivity of 15–30 l/h per 1 m^2 of membrane area. The cost of treating domestic waste water on membrane modules ranges from 6.34 rubles to 11.89 rubles (0.08 to 0.15 Euros) per 1 m^3, and besides, lower values are obtained with the use of hollow fiber modules. Total operating costs amount 19.03–19.85 rubles (0.24–0.25 Euros) per 1 m^3.

If we compare the technological scheme with the membrane biological reactor with the traditional technological scheme of wastewater treatment, a number of serious differences will become noticeable, such as cleaning efficiency and economic benefits. Evidently, both of these parameters are inextricably connected with each other. The advantages of a scheme with an MBR over a traditional designed one are determined by such features as:

- Full retention of all suspended solids and microorganisms, and as a result, achievement of the maximum effect of purification from suspended substances, increase of the effect of purification by COD (chemical oxygen demand) and BOD (biochemical oxygen demand), disinfection of purified water without the use of reagents, low sensitivity to fluctuations in flow and quality of source water;
- Minimum residence time of water in the zone of separation of the solid phase;
- Complete retention of microorganisms in the reactor that significantly changes the conditions for auto-selection of microorganisms of activated sludge.

All this can allow to change the parameters of the reactor (aerotank) operation: at high hydraulic loads on the reactor can be increased the age of the activated sludge, including accumulating slowly growing kinds of microorganisms (nitrifying agents, microorganisms, oxidizing bioresistant compounds); we are able to prolong the time of suspension of suspended substances in the reactor until their complete biological destruction; we also have possibility to exclude the influence of sedimentation characteristics of activated sludge on the quality of purified water; and to increase the stability of the system to fluctuations in the concentrations of contaminants in the source water due to good adaptation of biocenoses.

However, if we would pay attention to the economy, we can face a number of problems, caused by the costs at all stages of using the technology with MBR, rather than the traditional technological scheme. Among them are high capital costs, and the specific cost of the membrane units themselves is practically independent from productivity. There is also the problem of inevitable contamination of membranes and consequent subsequent costs, waste for electricity and membrane replacement, a more complex process control and automation system, the difficulty in providing a sufficient aeration level at high active sludge concentrations specific for MBRs.

In order to graphically see the price difference between the technological scheme with MBR and the traditional technological scheme of wastewater treatment, their technical and economic indicators are demonstrated in Table 1.

Table 1. Technical and economic indicators are demonstrated

Variants of schemes	Costs of construction and assembly works (thousands of rubles)
Variant 1:	
Primary clarifiers, type design 902-2-363.83	23190.04
Aerotank-mixer, type design 902-2-211	33028.34
Secondary clarifiers, type design 902-2-347	15730.82
Filters for post-treatment, type design 902-4-11.84	27601.56
Total:	99550.77
Variant 2:	
Aerotank-mixer, type design 902-2-211	10307.34
Submerged Membrane Blocks	88043.48
Total:	98350.82

3. Results

This assessment of technical and economic characteristics of the possible scheme with MBRs and the traditional scheme of treatment facilities showed that, despite the high cost of MBRs, generally, the total cost of construction of such schemes does not differ from each other (Fig. 1). The savings in the second variant of construction is achieved due to the fact that there is no need to operate additional facilities that differ from one another in technical and technological features. It is also should be mentioned that, with the second option, fewer personnel are required, which contributes to a reduction in payroll payments and social deductions that are quite complicated to calculate. There is a decrease in costs at the stage of capital construction due to a reduction in the construction area, and under the traditional construction scheme, the cost of the land plot reaches 45–50 % of the cost of construction and installation works.

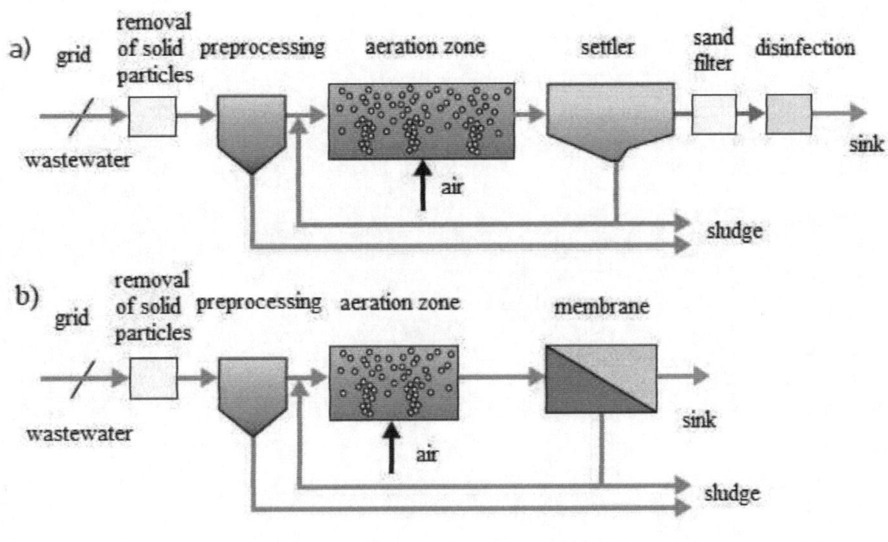

Figure 1. A conventional purification scheme (a) and a membrane bioreactor scheme (b)

4. Conclusions

A description of the types of membrane modules and the types of membranes themselves was given, which are the main elements of membrane biological reactors that require proper operation to maintain normal functioning of the system and save on repairment and replacement works of individual parts for prolonging the service life of the equipment.

The emphasis was on the cost of capital construction, the cost of consumables and other consequent elements, and an analysis of possible savings by selecting the necessary configuration of the system was conducted.

During the comparison of the technological scheme with MBRs and the traditional scheme of wastewater treatment the pros and cons of both systems were revealed. We can now conclude that subsequently in terms of economic efficiency both these schemes will be equal to each other and the choice should be made on the basis of the quality of treatment and the individual requirements imposed by treatment facilities.

Acknowledgement

The reported study was funded by The Head Regional Shared Research Facilities of the Moscow State University of Civil Engineering

References

[1] Kulakov A and Makisha N 2017 *MATEC Web of Conf.* **112** 10019
[2] Gogina E and Gulshin I 2016 *Procedia Eng.* **153** 189–94
[3] Ruzhitskaya O and Yantsen O 2016 *Int. J. of App. Eng. Res.* **11(5)** 3496–98
[4] Kuzina O 2016 *Matec web of conf.* **73** 07013
[5] Gulshin I 2017 *IOP Conf. Series: Earth and Env. Sci.* **90(1)** 012198
[6] Pervov A G 2015 *Desalin.* **368** 140–51
[7] Makisha N and Panteleeva I 2017 *MATEC Web of Conf.* **106** 70015
[8] Pervov A G and Andrianov A P 2015 *Petrol. Chem.* **55(5)** 373–88
[9] Volkov A, Kuzina O 2016 *Procedia Eng.***153** 838
[10] Galkina E 2017 *Sci. rev.* **21** 155
[11] Pervov A G, Andrianov A P, Gorbunova T P and Bagdasaryan A S 2015 *Petrol. Chem.* **55(10)** 879–86
[12] Galkina E 2017 *Sci. rev.* **8** 131

IOP Conf. Series: Earth and Environmental Science **177** (2018) 012038 doi:10.1088/1755-1315/177/1/012038

Possibilities of saline soil use for the development of Ural and Siberia environment

L N Skipin, O V Udartseva, V S Petuhova and E V Zhilyakov

Tyumen Industrial University, 38 Volodarsky street, Tyumen, 625000, Russia

E-mail: skipinln@tyuiu.ru

Abstract. The problem of rational use of saline soils, as applied to the specific region, becomes one of the most important in Russia as well as abroad. The article reflects potential possibility to include these soils in the development of urban environment. This will allow us to extend the degree of use of the demographic capacity, as applied to the Siberian and Ural cities. The areas of these lands in respect to this territory makes up more than 10 mln ha. To include these soils in the development of the urban environment of small and large cities on the level of zonal soils, it's required necessary chemical melioration.

1. Introduction

Its carrying out with the use of gypsum, phosphogypsum and other meliorants-coagulants allows fundamental increasing of solonetzes fertility with their following involvement in the gardeners' non-commercial partnerships, individual construction and as agricultural fields for the suburban base. This will allow reduction of plough lands involvement in the urban environment. It's important to note that the action of meliorants-coagulants lasts more than 40–50 years and their dosage is up to 20 tons per ha. The indicated regions possess mineral, production and scientific resources for the solution of such important problem in conditions of Siberia and Ural.

The problem of saline soils use is now one of the main problems in the land-use system. Saline soils include such soils which contain highly soluble salts in their content in volumes that are toxic for the plants. The source of salts for these soils is rock, volcanic eruption product, eolian transfer of salts, atmosphere precipitations, ground waters, irrigative ground-waters, vegetation (saltwort). The most widespread among the saline soils are solonetzes containing the increased amount exchangeable sodium in the illuvial horizon (more than 15 % of thecation exchange capacity) [1].

It's important to note that such soils can adjoin with their large missivesto the towns of Siberia and Ural and other regions or be inside them. Their biological productivity is very low 3–5 cwt/ha of dry mass while it's multiple times higher on the zonal soils. Such phenomenon decreases biogenic circulation of the substances in the vicinity as well asinside the urban environment. At the same time presence of such soils almost excludes their use for gardeners' non-commercial partnerships, individual construction, recreational zones and for other purposes [2].

Nowadays the solonetz soils can serve as an important reserve for the creation of demographic capacity of small as well as big cities by the availability of territory for the individual construction, satisfaction of citizens' needs on gardening plots and possibility of their use as agricultural fields for suburban base. The approximate index of the need of one thousand citizens on gardening plots is from 500 to 2000 ha [3].

Content from this work may be used under the terms of the Creative Commons Attribution 3.0 licence. Any further distribution of this work must maintain attribution to the author(s) and the title of the work, journal citation and DOI.
Published under licence by IOP Publishing Ltd

Solonetz soils have a series of negative hydro physical and chemical properties: high dispersion, swelling ability, weak water permeability, lack of structure, slow physical ripening of the soil, adhesiveness under humidification, low water mobility, decreased air exchange, strong compression of soil material in case of drying, solidity and larger resistance, alkaline reaction of the environment, toxicity of the salts, calcium deficit and sodium excess [1].

It suffices here to remark that the area of indicated soils in complex with zonal soils in Ural and Western Siberia is more than 10 mln ha (table 1) [4, 5].

Table 1. Structure of solonetz complexes in Ural and Western Siberia, thousands of ha

Region	Total	Solonetz availability, %		
		10–30	30–50	>50
Orenburg	1431	720	99	606
Chelyabinsk	1365	721	92	551
Kurgan	2203	636	190	1377
Sverdlovsk	631	225	50	302
Tyumen	389	176	30	182
Omsk	3364	485	678	1812
Novosibirsk	4317	2311	233	2258
Kemerovo	146	1826	44	21
Altai Territory	1798	1183	143	472
	15254	6108	1562	7581

The territory including Orenburg, Chelyabinsk, Kurgansk, Tuymen, Omsk, Novosibirsk regions and Altai territory are characterized by large heterogeneity of natural conditions. There are landscapes of the dry steppe with south and ordinary chernozems (Omsk, Chelyabinks, Orengburg, Novosibirsk regions and Altai territory), landscapes of forest-steppe with kolkis with ordinary solonetz-like chernozems (Chelyabinsk, Kurgansk, Omsk region) and landscapes of Northern forest-steppe with meadow-chernozem soils, lixiviated and gray forest soils (Tuymen, Kurgansk, Chelyabinsk, Omsk, Novosibirsk regions and Altai territory) [6].

With the ameliorative diagnostics of solonetz soils, a number of important characteristics are taken into account, in particular, soil complexes where among zonalones, the solonetzes occupy 10–30 %, 30–50 and more than 50 %. Fertile lands with less than 10 % of solonetzes are not included in the land reclamation fund. On complex soils with the availability of 10 to 30 to 50 % of solonetz, selective chemical melioration (if solonetzes are located in the shallow spots) is expedient, and over 50 % is continuous. Hydrodynamic regime determines the numerous ameliorative properties of the solonetzes. The type of this mode is established by the level of ground wateroccurrence, taking into account the content of soil-forming and basement rocks.

The expediency of melioration depends on the type of hydrological regime. The character of solonetz use is determined according to the regime. Meadow solonetzes can be developed only when the mineralization of groundwater doesn't exceed the critical values. Melioration of meadow-steppe and steppe solonetzes doesn't depend on the state of ground waters (table 2) [5].

Table 2. Critical mineralization of ground waters
depending on the depth of their occurrence and the rock composition

Depth of the groundwater occurrence, m			Mineralization, g/l
Heavy clays, light loams	Medium loams	Medium clay, heavy loam	
<1.0	<1.2	<1.5	<1.5
1.0–1.2	1.2–1.5	1.5–2.0	1.5–3.0
1.2–1.5	1.5–2.0	2.0–2.5	3.0–5.0
1.5–2.0	2.0–2.5	2.5–3.0	5.0–10.0
2.0–2.5	2.5–3.0	3.0–3.5	>10.0
>2.5	>3.0	>3.5	-

The character of the solonetzes salinization determines the features of their utilization methods. The most important salinity indicators are the depth of the salt horizon, the chemistry and the degree of salinity. The most laborious in the ameliorative plan is the development of solonetzes with a high level of location of the maximum salts and with a high content of soda.

The degree of solonetzicity of soils depends on the exchangeable sodium and plays a decisive role in the calculation of the dose of chemical ameliorant. According to the content of exchangeable sodium, solonetzes are divided into residual (less than 10 % of the capacity of absorption), low-sodium (10–20 %), medium sodium (20–40 %) and high sodium (more than 40 %). Methods of melioration of solonetzes must be strictly differentiated in connection with their qualitative and quantitative diversity [4].

With the melioration of solonetz soils (improvement if physical properties, creation of powerfulfertile soil level, replacement of sodium with calcium, removal of water soluble salts from the root habitable layer) to increase the efficiency of grown cultures it's recommended to apply two main methods: chemical and self-melioration. Chemical method of solonetz amelioration provides for the introduction of chemical ameliorants from the outside. This method is the only mean to increase the solonetz fertility with deep occurrence of carbons and gypsum (deeper than 40–50 cm). On the steppe solonetzes, the gypsuming is reasonable to combine with the watering. Under the high occurrence of carbons and gypsum it's used the method of self-amelioration at the expense of use of deep trench and step tillages [5].

2. Materials and Methods

The experiments carried out in Western Siberia and Ural showed high efficiency of chemical amelioration. The term of consequences lasts more than 40–50 years [1–3, 6–11].

Local deposits of gypsum are in the Altai Territory where there were revealed about 20 deposits. The most promising for use is the Jirin deposit (Lake Jira, Kuluddinsky district), the reserves of which are sufficient for complete reclamation of all solonetzes of the Altai Territory and neighboring regions of other regions. Its effectiveness has been tested and proven in the West Siberian region of Ural and the Republic of Kazakhstan [4].

Large reserves of natural gypsum are located in the Perm Krai (Kungur), near the city of Salikamsk there is a carnallite deposit, which has proven itself as a good coagulant.

However, it should be noted that the waste product of chemical industry – phosphogypsum, is very promising as the ameliorant. Besides gypsum, it contains 0.5–1.5 % of labile phosphorus (P_2O_5). Its huge reserves are in Sverdlovsk region of the town Revda.

A similar effect on the soil is provided by green vitriol – a waste of paint and varnish industry. It is used in the Chelyabinsk region and is a fine-grained powder, which has good flowability and weak caking ability, its dose compared with gypsum should be increased by 1.6 times.

It is important to note that theoretical and practical issues related to the development of solonetzes have been substantiated in the works of scientists in Siberia and other regions. However, taking into account the genetic characteristics of solonetzes in different zones and even within a certain region, it is necessary to continuously improve the appliedmethods of chemical and agrobiological melioration, as well as their integrated use in relation to specific agro-ameliorative conditions.

The goal of the studies is to reveal long-term influenceof ameliorantscontaining the calcium on the water-physical and physically-chemical properties of solonetzes and their efficiency as applied to Western-Seberian region.

Study objectives: to provide theoretical and practical reasoning of the dose of introduced ameliorant taking into account the outflow intensity of exchange reactions products; to follow the change of quantitative and qualitative composition of the salts in the soil under the influence of different dose gypsuming, to find the possibilities of displacement of exchangeable sodium from the soil adsorption complex during the long ameliorative period, to determine the response of cultures-phytoameliorants to the long-term consequence of chemical melioration. To determine the possibility of gypsum solonetz using for the extension of demographic capacity of small and large town of Ural and Siberia.

Research methods: experiments were conducted on hydromorphic chernozem-meadow saline sulphate-sodic medium-saline deep-carbonate deep-gaseous cortical high-sodium columnar solonetzes in the Tyumen region. Experiments on chemical reclamation began in 1972, plots area 400 m^2, tier is 4-fold. Exchangeable sodium was determined by the method of Gedroits. The analysis of aqueous extract was carried out according to the following procedures: alkalinity from the soluble carbonates – potentiometrically, chloride ions – complexometric, potassium and sodium in difference.

The studies have shown that during the observation period the level of subsaline groundwaters varied from 1.2 to 2.4 m, it was maximal after the total snow melting. Long fallow solonetz was used as control. Saline regime of the long fallow solonetz showed that the general content of salts in the meter-deep layer varied within the limits of 48–58 t/ha, their maximum amount took place in the dry years. Soda in amount of 28.2 t/ha prevailed in the composition of salts, chlorides made up to 7 t/ha and sulphatesmade up 5.3 t/ha. Predominance of soda in the solonetzes added them sulphate-sodium carbonate salinization that led to the strengthening of their alkalinity and toxicity. Introduction of phosphogypsum, in 1972 changed significantly the qualitative and quantitative content of water extract in the whole soil crossover (fig. 1).

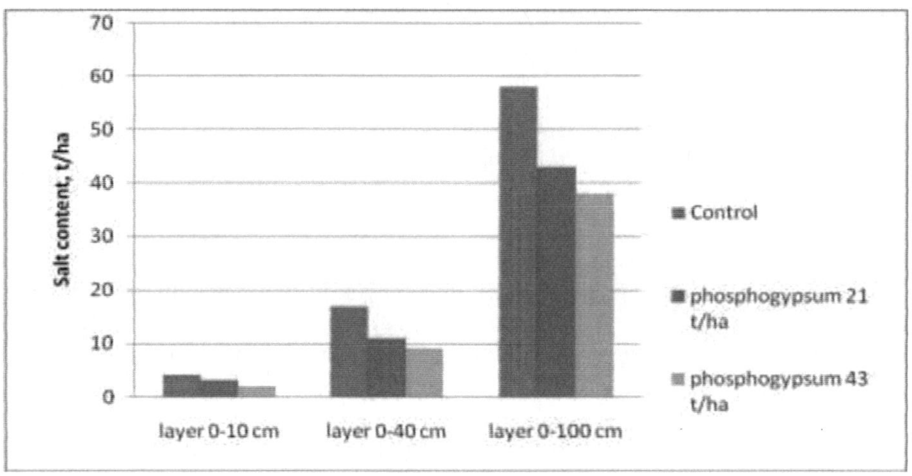

Figure1. Afteraction of phosphogypsum doses on the leaching out of salts from the crusted solonetz during 34 years

So, the amount of salts in the meter-deep layer decreased in comparison with control variant by 15–20 t/ha. At the same time significant decrease was noted at the expense of soda content decrease. Its amount in this layer dropped by 3.6–5.4 times. With the full dose of phosphogypsum leaching of soda from the soilcrossover took place more intensively. At the same time, content of exchange reactions products in a form of sulphates increased by 2 times. This phenomenon didn't lead to the intensification of salt regime stress, since the toxicity of sulphates is few times lower than the soda toxicity.

The data in the figure 2 shows that the introduction of ameliorants in the full and half doses led to the significant weakening of solonetzicity. Thus, initially the content of exchangeable sodium in crusted solonetz (0–40 cm layer) was on the level from medium-sodium to the high-sodium (28,8–53,4 % of the exchange capacity), then after the gypsuming in the declared doses the level of solonetzicity decreased to the residual and low-sodium, 2.2–19 %, respectfully. The thesis about the short duration or total absence of ameliorative process in conditions of close location of the groundwaters hasn't been confirmed. It's important to note that the phosphogypsum dose for 0–15 cm layer (21 t/ha) didn't yield to the complete predicted dose of the layer for 0–30 cm (43 t/ha).

The amount of exchangeable sodium maintained in 0–20 cm layer was on the same level as with half as well as full doses of ameliorant.

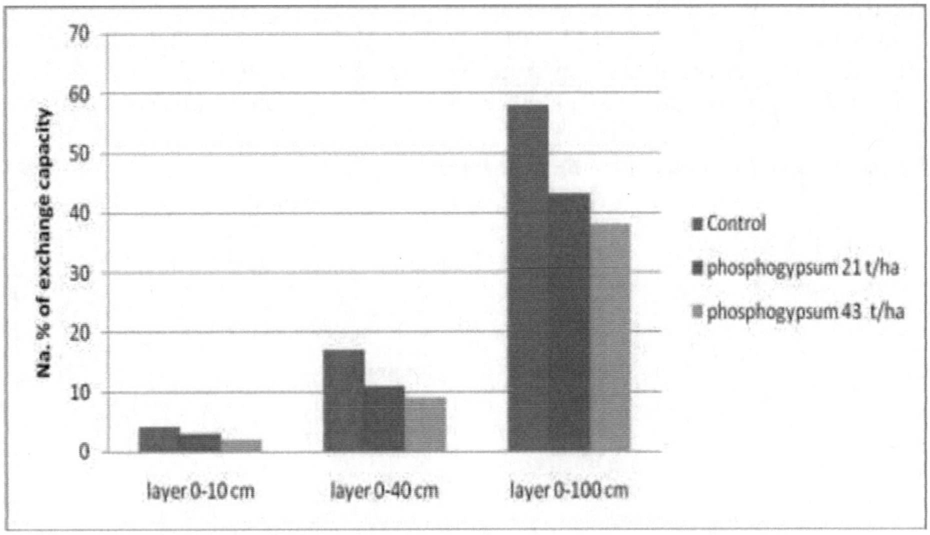

Figure 2. Afteraction of phosphogypsum doses (34 years) on the replacement of exchangeable sodium from SAC

The productivity of the dry mass of the grass stand of perennial grasses over the years of research at half the dose was 21.2 c/ha, with the total dose being 24.7 c/ha. In the fallow area (control), it did not exceed 3.2 c/ha.

3. Conclusions

The use of phosphogypsum waste in the crusted solonetzes allows the production of perennial grasses for more than 30 years, with a significant improvement in the basic chemical properties of these soils. For the reclamation of ameliorated solonetzes, it is sufficient to apply a half dose of phosphogypsum for the 0–30 cm layer calculated according to Gedroits.

Increase of the fertility of solonetz soils up to the level of zonal soils will allow them to be used for the increasing of the demographic capacity of small and large cities in Ural and Siberia. It will also allow exclusion of the reduction of forests near the urban environment for construction and gardeners' partnership. For the introduction of ameliorant and its embedding, special road service vehicles can be used. Projects for chemical reclamation are capable of forming regional centers of agrochemical service.

References

[1] Fedotkin V A 1993 *Solonetzes of Siberia and the Urals* (Novosibirsk)
[2] Khusainov A T, Skipin L N and Guzeeva S A 2012 *Possibilities for reclamation and development of solonetzes of Western Siberia* (Tyumen – Kokshetau)
[3] Sapega V A 2002 *Practical work on ecology* (Tyumen)
[4] 1976 *Melioration of solonetzes of the Southern Urals and Western Siberia* (recommendations)
[5] 1983 *Recommendations for amelioration of solonetz soils* (Moscow)
[6] Berezin L V 2006 *Melioration and use of Siberian solonetz* (Omsk)
[7] Yeremchenko O Z 1997 *Natural-anthropogenic changes in solonetz soils in the Southern Trans-Urals* (Perm: Publishing house of the Perm University)
[8] Oborin A I 1958 On melioration and development of solonetzes in conditions of irrigated agriculture of the chernozem strip of Western Siberia *Problems of melioration of solonetzes* (Moscow: Publishing House of the USSR Academy of Sciences) 239–78

[9] Parfenov A I et al 1986 Efficiency of chemical melioration of solonetzes in the forest-steppe zone of the Omsk Region *Features of melioration of solonetzes of saline soils of Western Siberia* (Omsk) 4–10
[10] Semendyaeva N V and Elizarov N V 2014 Dynamics of salt composition of Baraba solonetzes during the 27–32-year-old action of gypsum *Vestnik NSAU* **1 (30)** 41–5
[11] Skipin L N 2000 *Solonentzes of Siberia: ecological aspects of development* (Tyumen)

Published under licence in Journal Title ***by IOP Publishing Ltd.***

Implementation of Project Management in Eco-Industrial Parks Development in Russian Cities

M A Liubarskaia[1, 2], M V Tsurkan[3] and A M Vorotnikov[4, 5]

[1] Saint-Petersburg State University of Economics, 30/32 Naberezhnaya kanala Griboedova, Saint-Petersburg, 191023, Russia
[2] Baltic Academy of Tourism and Entrepreneurship, 13A Petrozavodskaya str., Saint-Petersburg, 197110, Russia
[3] Tver State University, 33 Zhelyabova str., Tver, 170100, Russia
[4] Russian Presidential Academy of National Economy and Public Administration under the President of the Russian Federation, 84 bldg 2 prospect Vernadskogo, Moscow, 119571, Russia
[5] Expert of PORA Center, 12 bldg. 1 2-nd Zvenigorodskaya str., Moscow, 123100, Russia

E-mail: lioubarskaya@mail.ru

Abstract. Recently solid waste management problems show growing trend in the most of Russian cities. This trend proves the relevance of the research on project approach to development of eco-industrial parks focused on diminishing waste flows going to landfills and increasing recycling and reusing of secondary materials. The purpose of the paper is to systematize the ideas on the possibilities of using project management methods for amplification of eco-industrial parks in the cities of Russia. The research clarifies the term "eco-industrial park" and concepts of its application to Russian practice. Different parameters, including composition of the waste flows and types of final products, were used for developed classification of eco-industrial parks. Based on the study of the experience of eco-industrial parks functioning in the cities of Russia, the authors analyzed the prospective and limitations for implementation of project approach to the creation of effective waste management schemes in these parks. The results of the research are of practical importance for public authorities, existing and potential participants of eco-industrial parks, specialists in the field of waste management, researchers in the field of environmental science and economics.
Keywords: project management, eco-industrial park, project, city, waste management

1. Introduction

Russian cities face many waste management problems, such as exhaustion of landfills' capacities, growing territories of unauthorized dumps, undeveloped system of separate collection of municipal solid waste, shortage of solid waste processing facilities, disunity of secondary resources market. 85 % of municipal solid waste is currently going to the facilities that do not provide any economic or ecologic security, only 10 % is going to the landfills fulfilling the requirements on environmental safety and as little as 5 % is going to waste processing plants [1].

In the last decade, most of the countries began to consider municipal and industrial solid waste as a valuable resource to be involved in the production process within eco-industrial parks. Therefore,

Content from this work may be used under the terms of the Creative Commons Attribution 3.0 licence. Any further distribution of this work must maintain attribution to the author(s) and the title of the work, journal citation and DOI.
Published under licence by IOP Publishing Ltd

from 100 % of municipal solid waste it is possible to get 5 % of secondary material resources, 25 % of certified fuel for cement plants or complexes for the production of electric and thermal energy, 40 % of compost, man-made soil and only 20 % is a residue that can be sent in compressed form to a landfill site [2].

In Russia, usage of industrial waste as secondary materials is not a common practice. So, cities with developed industries face even more environmental problems than others. The most burdensome city of the planet, according to UNESCO, is the Russian city of Karabash, a recipient of waste from copper production. A toxic chromium lake of seven million tons was artificially created in Pervouralsk. 70 thousand tons of chemicals require processing or disposal in Dzerzhinsk

All these problems need to be solved as soon as possible to stop the growth of their negative impact on the environment and people's health. In the world practice, eco-industrial parks are well-recognized as structures aimed to diminish environmental degradation, involve municipal and industrial waste in technological processes and increase energy efficiency and resource saving.

Many of these eco-industrial parks functioning based on the principles of project management, which allows optimizing the usage of financial resources for municipal and industrial waste utilization under time constraints. Russian cities have some, but very limited experience of eco-industrial parks operation. There are almost no scientific papers, books or manuals considering the possibilities of applying project management for the development of eco-industrial parks.

2. Clarification of the Term "Eco-industrial park" and Concepts of Its Application to Russian Practice

In the world practice, the term "eco-industrial park", or "eco-technical park", is often refer to a group of enterprises cooperating with each other and with the local community in order to efficiently exchange resources, information, materials, water, energy, infrastructure and natural habitat. It leads to economic benefits and improving the quality of the environment [3].

The goal of eco-industrial parks development is to achieve economic, environmental and social benefits while reducing the use of raw materials and energy resources, involving secondary resources in the technological cycle, reducing the level of environmental impact of existing industrial facilities and eliminating the accumulated environmental damage [4].

In the Russian practice, the normative definition of the concept of "eco-industrial park" is included in the Strategy for the Development of Industry for the Processing, Utilization and Disposal of Waste of Production and Consumption by 2030.

According to this Strategy, "eco-industrial park" is a complex of objects united by energy links, including buildings and facilities, technological and laboratory equipment used for the processing, utilization and neutralization of waste, ensuring continuous converting waste into industrial products, as well as the implementation of scientific and (or) educational activities in the field of waste management.

At the discussion stage, there is a draft on amending the Federal Law 488-FZ "On Industrial Policy in the Russian Federation". This Law define the "eco-industrial park" as a set of objects of industrial infrastructure and equipment located in the territory of one constituent entity of the Russian Federation or territories of several constituent entities of the Russian Federation, which are used by industrial enterprises for processing, utilization and decontamination of wastes and ensuring on their basis the production of industrial products.

By waste in both documents authors mean the waste of production and consumption, the term also used in the Federal Law 89-FZ "On waste of production and consumption".

By authors' opinion, for classifying an "industrial technopark" as "eco-industrial park" the following criterial could be used:

1) availability of exchange options for material and energy resources both inside the park and with remote partners;

2) organization of the cascade system of water supply and integrated wastewater treatment of residents (tenants) of the park (and in some cases also integrated with municipal sewage of nearby settlements);
3) organization of a common system for managing the material and energy resources (through the identification of a focal point and (or) "anchor resident");
4) use of the general infrastructure of the park, including administrative, transport, waste management, energy and water supply systems.

The coordination center of the "eco-industrial park" is the administrative body, responsible for organizing network for exchange of material and energy resources within the enterprises inside the park and beyond. The anchor resident of the "eco-industrial park" is a key organization, which provides most of the material and energy flows for industrial symbiosis. It could be company from energy sector, large communal or industrial enterprise.

From authors' point of view, the key difference between the "industrial technopark" and "eco-industrial park" is the functioning of the "eco-industrial park" on the basis of the principles of industrial symbiosis, which implies:
1) reuse of by-products and (or) waste, that means exchange of product-specific materials between two or more companies to replace the use of commercial products or raw materials;
2) sharing of utilities (infrastructure facilities), that means joint use and management of such resources as energy, water, electricity and heat, as well as joint operation of water treatment facilities and gas cleaning equipment [5];
3) joint using of services, that means provision the general needs of companies for the implementation of auxiliary activities such as fire safety, transportation and others.

In addition, there is an approach that allows the authors to identify the concept of "eco-industrial park" with the notion of "industrial-ecosystem", that presents the "model of the ecologically equipped industrial area" [6]. According to this, the creation of an eco-industrial park is the development and transformation of an existing industrial system into an "industrial-ecosystem" that means the integration of the industrial system and ecosystem based on the ecofriendly waste management mechanism.

Industrial system (enterprise, industrial city) is a spatially defined set of industrial entities and living organisms, united by a single environment of existence.

Ecosystem is a spatially defined set of living organisms and their habitat, united by material and energy exchange and information interaction.

Ecofriendly waste management mechanism can be defined as management of material and energy flows through information interaction between ecosystem components and the environment, ensuring self-organization and sustainable development of the ecosystem [7].

The authors' approach, which is also used by domestic practitioners, unites all the above definitions and allows us to propose the following interpretation of the concept of "eco-industrial park" as a complex of innovative solutions, equipment, facilities unified with ultimate goal to replace in technological processes up to 100 % of raw materials and energy with secondary ones.

3. Classification of Existing and Planned Eco-industrial Parks in the Cities of Russia

For systematization of eco-industrial parks, which already exist or just planned to be created in Russian cities, it's rational to propose the following parameters within the current legal framework:
1) orientation on the type of waste;
2) sources of generation of the flows of material and energy resources;
3) availability of a key organization that generates most of the exchange flows of material and energy resources;
4) level of localization.

Depending on the orientation on the type of waste, eco-industrial parks could be:
1) oriented on utilization of the components of industrial solid waste;
2) oriented on utilization of the components of municipal solid waste;

3) oriented on joint utilization and neutralization of municipal and industrial solid waste (with possibility to involve the components of medical and biological waste).

Depending on the sources of flows of material and energy resources, eco-industrial parks could be:
1) with industrial enterprise (enterprises) as a source of material and energy resources;
2) with households as sources of material and energy resources;
3) with mixed sources of material and energy resources .

Depending on the availability of a key organization generating the most of the exchange flows of material and energy resources, eco-industrial parks could be:
1) park without an anchor resident (main participant);
2) park with anchor resident (main participant).

Depending on the level of localization, eco-industrial parks could be:
1) Federal eco-industrial parks, supervised by Union of waste processing companies; final products are made in the park from secondary materials supplied from regional eco-industrial parks (total capacity 1.5 million tons);
2) Regional eco-industrial parks, supervised by Regional Waste Management Operator; activities are aimed at minimization of waste landfilling by selection and utilization of valuable components, neutralization of organic, medical and biological waste, and disposal only certain types of waste (the proportion of decontamination and disposal is 60-80 %);
3) Corporate or municipal eco-industrial parks, primarily aimed at addressing the relevant problems of the urban community.

4. The Concept of Project Management Implementation for Eco-industrial Parks Development in the Cities of Russia

The implementation of project management for the development of eco-industrial parks in the cities of Russia assumes the formation of projects as a way to solve the problems facing the management of the eco-industrial parks. In this context, the project is a set of interrelated activities aimed at recycling or processing of certain type of waste under time and resource constraints to achieve a measurable result obtained by the project team (temporary organizational structure).

Each project is considered as an object of management, which requires planning, organization and control of the labor, financial and material resources, to achieve the project objectives in the most effective way.

Application of project management for the development of eco-industrial parks shows the most effective results as a method of organizing the operation of the park since each project is an innovative business unit. The project team achieves the results through the application of modern methods, tools and technologies. Thus, the implementation of each project is a transition to a new qualitative level.

The project team could be formed from the number of residents of eco-industrial park with possibility for inviting the external experts. The resident (tenant) of the eco-industrial parks is an enterprise that participates in industrial symbiosis and (or) uses the integrated services of the eco-industrial parks (administrative, infrastructural), an industrial enterprise that accepts water, energy and material resources as input, and produces products, water, energy and waste [8].

Therefore, residents (tenants), as well as representatives of local (regional) authorities, expert communities, industry associations, unions, research institutes and other interested parties are the possible team members. If there is an anchor resident in the eco-technical park, he can also be a part of the project team. The head of the project team could be the manager of any of the above structures. That is determined individually for each project.

The implementation of the project management for the development of eco-industrial parks assumes the presence of a project office in its structure, which is responsible for methodological and organizational support, planning and monitoring the project portfolio, implementation and development of the information system for project planning and monitoring, and the formation of consolidated project reporting. Building a well-structured framework for information exchange is an "urgent task in order to achieve further process simulation and optimization in the eco-industrial parks" [9]. Some experts talk about big prospective for implementation of a "cyber-infrastructure system based on a virtual, hierarchical representation of an eco-industrial park, by employing ideas and technologies from Industry 4.0" [9].

A conceptual scheme for the implementation of project management for the development of eco-industrial parks is presented in Figure 1.

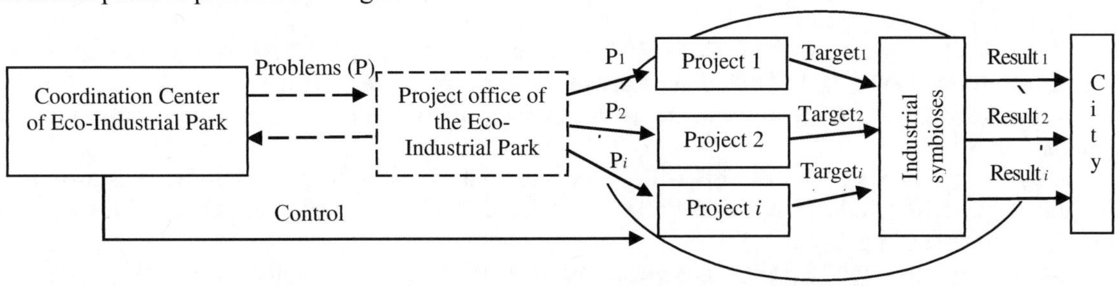

Figure 1. The Concept of Project Management Implementation for Eco-industrial Parks Development in the Cities of Russia
Source: compiled by authors

Projects aimed at the solution of similar or interrelated tasks can be combined into a program, a set of projects and programs to form a portfolio of eco-industrial parks.

The introduction of project management for the development of eco-industrial parks in the cities of Russia requires the participants to have relevant competencies.

Project competencies are a type of professional competencies that form the knowledge and skills of employees of the structures included in the eco-industrial parks to implement project management in accordance with their role in the project.

The study of foreign experience in the operation of eco-industrial parks (in the cities of Denmark, Finland, Great Britain) showed that the introduction of the project approach allows achieving the following effects:

1. Increase in productivity of secondary raw materials, received within the framework of specific projects and used for energy supply of cities, on average by 17 %;
2. An increase in the number of educational projects in the field of energy efficiency and energy conservation, waste management in city schools, is average of 62 %;
3. Reduction of the cost of secondary raw materials due to optimization of labor resources, on average by 5 %;
4. Attracting bank funds in the framework of project lending, which is the most attractive for investors (the best example is the eco-industrial park Kalundborg, Denmark);
5. Increase in the number of budget subsidies, as they are allocated for specific projects.

All these effects can be obtained in Russia. In addition, the authors believe that the introduction of a project approach to the activities of eco-industrial parks can contribute to the systematization of waste management in cities required for a particular project.

5. The Project Management Implementation Practice at the Eco-industrial Parks in the Cities of Russia

Currently, eco-industrial parks are established only in several cities of Russia, among them Tambov, Novokuznetsk, Nizhny Novgorod, Kazan, Naberezhnye Chelny, Kurgan.

On the territory of the eco-industrial park in the city of Tambov, there is a technological complex for the processing of municipal solid waste and other types of waste based on the technology of thermolysis. This technological complex allows accomplishing the following tasks:

1) deep degree of utilization (up to 90 %);
2) insignificant (in comparison with the economic and ecological effect) tariff growth for the population of the city where the eco-industrial park is located;

3) extension of the service life of the existing landfills (up to 3–4 times);
4) reduction the size of the sites planned for organizing new landfills.

Technology of thermolysis allows to receive in Eco-industrial Park Tambov several types of by-products:
1) synthetic black oil fraction – mass output of 8 %, is used as a component of liquid heating fuel for municipal needs of the city;
2) artificial coal – mass output of about 30 %, demanded as a fuel in ferrous metallurgy, power generation, production of automobile tires, and as RDF for cement plants;
3) fuel synthesis gas – mass output up to 20 %, is used for own needs of eco-industrial park as a secondary fuel.

The use for the needs of the city of artificial oil and coal instead of raw ones allows to stay almost with the same tariff (slight increase from 35.42 rubles a month per person to 43.75 rubles in 2018 using the claimed technology).

Eco-industrial park with specific processing of industrial waste is created in Novokuznetsk, the structure of waste generated in the city is shown in Figure 2.

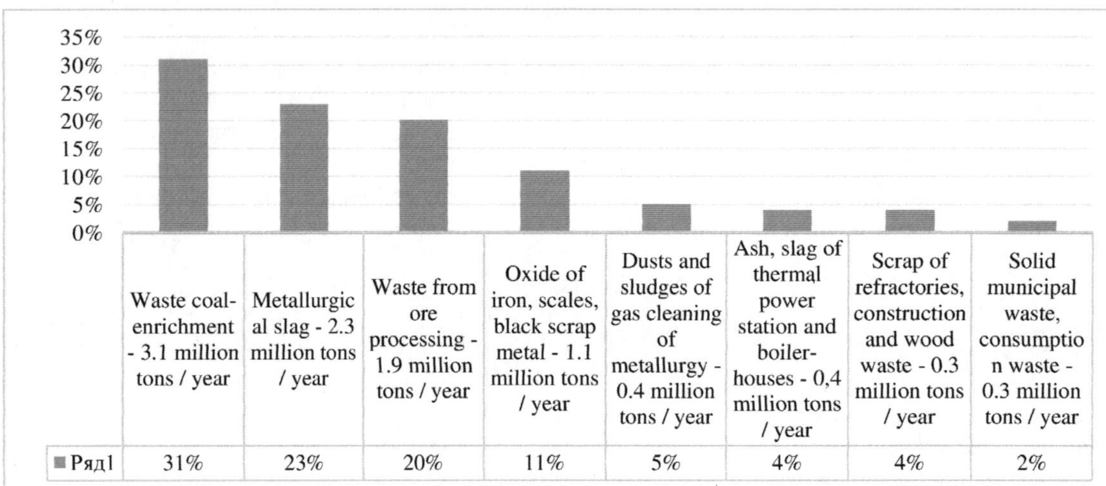

Figure 2. Structure of waste generated in Novokuznetsk
Source: compiled by authors

The reasons for creating of eco-industrial park in the city of Novokuznetsk are the following:
1) availability of facilities with intensive waste generation, such as mining and metallurgy, thermal power engineering (the metallurgical plant of JSC "EVRAZ ZSMK", the OAF JSC "EVRAZ Ruda", the aluminum plant of OAO "RUSAL Novokuznetsk", the ferroalloy plant of OAO "Kuznetsk Ferroalloys", 4 coal preparation plants, 3 thermal power plants and over 100 coal-fired boiler houses;
2) significant resource base of accumulated industrial waste, up to 500 million tons;
3) operating waste processing plants, development of processing of accumulated industrial waste;
4) self-organization of waste-processing enterprises in the Kuzbass Association of Waste Recyclers, that is real support for the waste processing enterprises;
5) waste recycling is one of the strategic goals for the further development of the urban economy, it is included in the Strategy of Social and Economic Development of the Novokuznetsk Municipal District by 2035;
6) availability of a developed industrial and transport infrastructure, production sites with communication support in the city;
7) modern landfill for municipal solid waste with a mechanical sorting capacity of 200 thousand tons per year;

8) availability of metallurgical enterprises as potential consumers of industrial waste and by-products as materials;
9) machine-building base for the production of equipment for the waste-processing industry;
10) scientific school developing new technologies in the field of waste processing, and so on [6].

The organizational structure of the Eco-industrial Park "Novokuznetsk" and interaction with external stakeholders are presented in Figure 3.

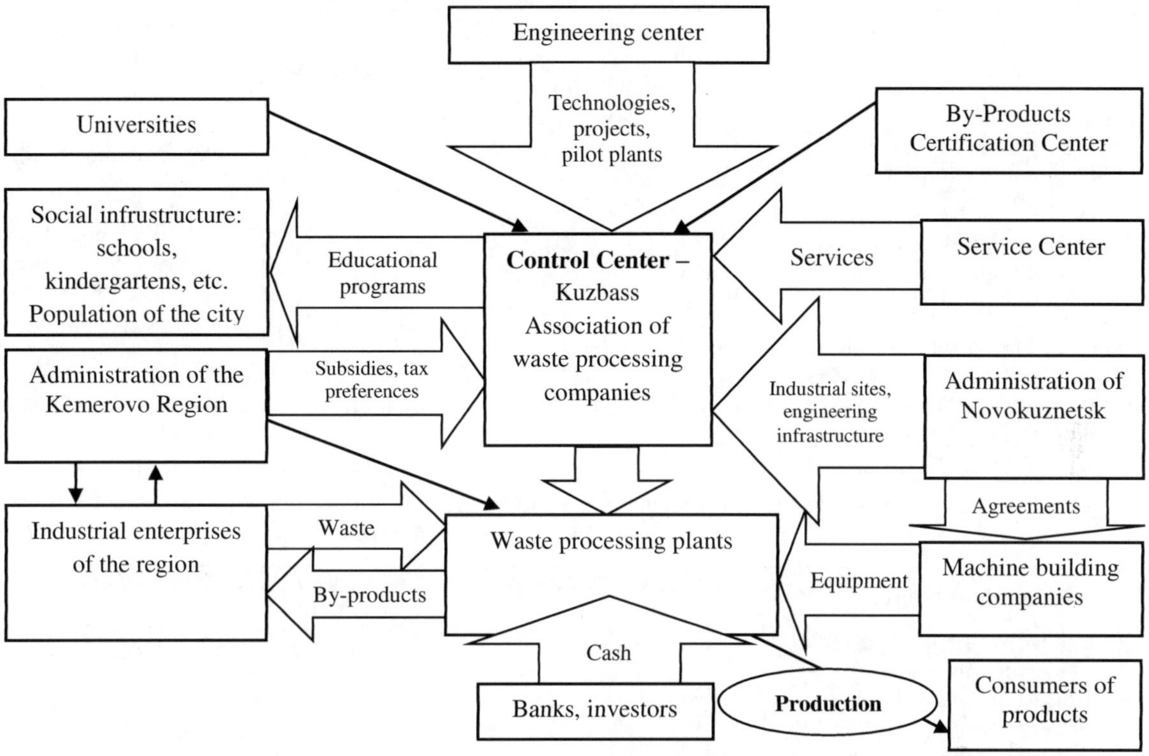

Figure 3. Organizational structure of Eco-industrial Park "Novokuznetsk"
Source: compiled by authors

Administration of the Kemerovo Region is involved in organization of waste supply process for the eco-technical park. Universities in the region is involved in personnel training and consulting the project team. The certification center conducts tests, certification at the request of the eco-industrial park.

In the Eco-industrial Park "Novokuznetsk" experience of project management was obtained in the framework of the projects presented in Table 1.

In the eco-industrial park of Nizhny Novgorod the importance of the project approach is evaluated, which confirms the implementation of the projects presented in Table 2.

The regional eco-industrial park in the city of Volzhsky is under the opening process. The start of operation is planned for the first half of 2018. This park also implements project management principles, and at this stage, already united:
1) the largest waste processing complex in the Southern Federal District with a capacity of up to 450 thousand tons waste per year;
2) modern landfill for municipal solid waste with a capacity of up to 2.5 million cubic meters per year;
3) facility for utilization of construction and demolishing waste with a capacity of more than 100 thousand tons per year;

4) facility for neutralization of medical, biological and other types of waste with the capacity of more than 12 thousand tons per year.

Table 1. Projects, completed in the framework of Eco-industrial Park "Novokuznetsk"

Name of the Project	Capacity, thousand tons per year	Investments, millions of RUR	By-Product
Recycling of waste refractories	waste sorting: more than 130 waste recycling: 180 production of by-products: 120	More than **30**	–
Recycling of waste of coke production located in the "resin" lake	25	**60** (2006–2015) **97** (2016–2018)	Sleeper impregnated oil
Processing of steelmaking slags of metallurgical enterprises	More than 100	More than **160**	Man-made iron concentrate, crushed stone
Creation of waste processing complex for coke and chemical and coal-enrichment industries	162	More than **600**	Fuel, fuel additives and coal sorbents
Processing of ash and slag wastes (aluminosilicate gold mine)	15	**48**	Facade and fireproof protective paint

Source: compiled by authors

Table 2. Projects planned for implementation in Nizhny Novgorod eco-industrial park

Name of the Project	Recycled waste, thousand tons per year	Investments, millions of RUR	Completion of the project, year
Utilization of waste from galvanic and metallurgical industries	300	**800**	2021
Utilization of industrial rubber waste	1	**55**	2019
Recycling of polymer waste	2,1	**77**	2020

Source: compiled by authors

At the stage of primary operation or creation at present there are 10 similar eco-industrial parks: 6 in the Vologda region; 2 in the Rostov region; one of each in the Astrakhan Region and the Republic of Kalmykia.

6. Conclusion

Eco-industrial parks play an important role in circular economy, which is recognized as the most recourse and energy efficient type of economy [10]. Currently, the development of eco-industrial park in Russia is at the stage of formation, the following results can be attributed to the expected results of their functioning in the cities:
1) Improvement of the environmental situation;
2) Increase in the share of recycled waste and demanded by-products;
3) Containment of tariffs for waste processing and energy generation for city residents;
4) Growth of the number of jobs for qualified specialists;
5) Increase in the revenue of the city budget.

To increase the manageability and efficiency of eco-industrial parks, it is necessary to include project management principles in their operation process. The positive experience is already obtained in the eco-industrial parks of Novokuznetsk and Tambov. At the same time, the expansion of project management to eco-industrial parks may face a number of difficulties, such as the lack of project competencies among potential project participants and primarily employees of resident organizations of eco-technical parks.

References

[1] Vergun P 2018 Development of systems for a closed cycle of waste management *J. Mun. Solid Waste* **2** 52–3
[2] Taddeo R 2016 Local industrial systems towards the eco-industrial parks: the model of the ecologically equipped industrial area *J. of Cleaner Production* **131** 189–97
[3] 2018 Eco-industrial parks of Russia *J. Mun. Solid Waste* **1** 55
[4] Mariev V, Smirnova T and Kiseleva S 2017 Eco-industrial parks as a basis for an integrated waste management system and secondary resources (world experience) *Mat. Of Round Table Ecological-oriented risk management and ensuring the safety of socio-economic and socio-political systems and natural-technogenic complexes (State University of Management)* 102–10
[5] Ramos M, Boix M, Aussel D, Montastruc L and Domenech S 2016 Water integration in eco-industrial parks using a multy-leader-follower approach *J. Comp. & Chem. Eng.* **87** 190–207
[6] Mariev V and Smirnova T 2017 Factors of success of eco-technical parks in the world *J. Mun. Solid Waste* **2** 14–7
[7] Mariev V and Smirnova T 2017 Formation of the system of eco-technical parks in the conditions of the Russian Federation *J. Mun. Solid Waste* **3** 21–3
[8] Zhou L, Pang M, Sikorski J, Garud S, Kleinelangshorst M, Karimi I and Kraft M 2017 System development for eco-industrial parks using ontological innovation *J. Energy Procedia* **105** 2239–44
[9] Zhou L, Zhang C, Karimi I and Kraft M 2017 J-Park Simulator, an intelligent system for information management of eco-industrial parks *J. Energy Procedia* **142** 2953–8
[10] Gomez A, Gonzalez F and Barcena M 2017 Smart eco-industrial parks: A circular economy implementation based on industrial metabolism *Resources, Conservation and Recycling* (In press, corrected proof, Available online 12 September 2017)

Published under licence in *Journal Title* by IOP Publishing Ltd

IOP Conf. Series: Earth and Environmental Science **177** (2018) 012040 doi:10.1088/1755-1315/177/1/012040

Planning and management of urban environment using the models of degradation theory

A A Varlamov[1], V I Rimshin[2] and S Y Tverskoi[3]

[1] Magnitogorsk state technical University G. I. Nosov, 27 Lenin Ave, Magnitogorsk, 455000, Russia
[2] Scientific-research Institute of building physics (NIISF RAASN), 21 Locomotive travel, Moscow, 127238, Russia
[3] Design Institute "Magnitogorskgrazhdanproekt", 79 K. Marksa Ave, Magnitogorsk, 455028, Russia

e-mail: Mgrp@mgn.ru

Abstract. Urban systems are the most complex multi-level objects. The construction and design of such systems is preceded by their analysis and deep study. The article considers the issue of long-term planning of urban systems. The planning is based on the forecast of evaluation of buildings, structures and other urban systems durability. The longevity forecast is proposed to be implemented on the basis of models of degradation theory. The theory of degradation is developed as a general energy method for estimating the longevity of objects. It is based on the law of conservation of energy. To analyze the longevity of urban environment, it is proposed to adopt the simplest model of theory of degradation. The proposed work provides an explanation for the description of proposed model. The urban environment is divided into a number of simpler systems. The more the degree of system separation into simpler ones, the more accurately one can describe the behavior of the entire system in time. Each simple system is described by the simplest model of degradation theory. The general behavior of the system is understood as a simple sum of individual degradation models. A feature of the proposed theory is that the simplest model can describe the entire system or the sum of its individual parts at once. The model makes it possible to consider the issue of equalizing the longevity of individual systems. The analysis of durability of a multicompartment building is taken as an example. In terms of analyzing the given example, individual stages of systems operation, that are still subject to discussion, are identified.

1. Introduction

Urban environment is a complex multi-level object. Such an object includes multilevel development, ecological systems, energy systems, transport and intellectual systems. All this structure is in constant movement in time and space. Each system and all systems together after their occurrence begin to decay, degrade and gradually die off. Some systems are systematically restored or destroyed. Some structures can collapse from unforeseen influences. New systems, structures and their connections are emerging. It is necessary to manage and plan the ongoing processes. To do this, it is necessary to be able to assess the longevity of an individual building, a separate system, and also the whole city, urban systems.

Content from this work may be used under the terms of the Creative Commons Attribution 3.0 licence. Any further distribution of this work must maintain attribution to the author(s) and the title of the work, journal citation and DOI.
Published under licence by IOP Publishing Ltd

Each planning and management should develop through modeling and design. Therefore, in recent years, the designing is understood not as a static process of assessing strength and deformability, but also as a process of durability assessing. The term "durability" is now often replaced by the term "service life". The planning process includes the renovation process, which can also be both natural and systematic. When renovating, not only the built-up area itself is affected, but all adjacent territories in greater or lesser extent (also depending on distance) [1].

2. Models of durability of buildings and structures

A lot of scientific works are devoted to questions of practical and philosophical longevity. In work [2] modern approaches to the evaluation of technical durability are described in sufficient details. In particular, it is proposed to introduce into the design standards a calculation based on a new limit state – durability.

Models of durability as a whole can be divided into general and differential (simplified and detailed) [2, 3]. A general degradation model is proposed to be written in the following way [4].

$$\mu(D) = D(x_1; x_2; \ldots x_n; t). \quad (1)$$

$M(D)$ – degradation average; $x_1; x_2; \ldots x_n$ – models of object and environment parameters, t – age.

Such a record makes it possible to use a mathematical model for a multidimensional system both in the form of a general model and in the form of a cumulative integrated model.

The models of each object parameter are usually considered separately. Thus, one of the most well-known models for evaluating the durability of a reinforced concrete structure operating in an aggressive environment is the model proposed by Tuuti [5]. The crack growth model, for example, can be used depending on the influence of various factors and finally determines the longevity as the element-wise effect of different factors on the growth of the critical crack [6].

Several possible forms of models can be considered for one parameter, the compatibility criterion of which is the substitution of the coefficients, which subsequently lead to similar results [7].

A distinction is made between planning using normalized characteristics or by results of field surveys. The object is divided into parts and then, summing the longevity of individual parts, the overall longevity of the object is determined [8]. Physical wear according to BCH [9] is determined by summing the average-weighted wear of individual sections of a structure or building at a replacement or estimated cost. Figure 1 shows some known models of physical deterioration of buildings and structures. Well-known graphs and dependencies are obtained empirically and are based on observation experiments and estimating the costs of routine repairs. The capital repair expenditures can extend the service life of structures. Figure 1b shows the option of reducing the physical wear of a building with the subsequent extension of its service life. It can be seen from the figures that the physical wear of different objects is described by approximately the same physical laws. However, some discrepancies in the behavior of objects are revealed, especially at the beginning of their life cycles. The graphs show a significant zone of stable operation. In general, the shape of the graphs reflects the approach to finding the criterion of durability and a purely practical approach to repair and restoration of structures.

3. Approaches of degradation theory to the assessment of physical wear of urban environment

The theory of degradation is based on a general energy approach to the interaction of environmental objects [11]. Theory presupposes that each object has its own form of energy. The form of energy is made up of individual elementary forms. Based on the assumption of the initial form of energy, a model is constructed for the interaction of an object with the environment. The final shape of the object energy is established after comparing the obtained model with the physical object of construction. The basis of the model is the energy potential of the object. Potentials of energy are of three types: constant, ascending and descending. The ascending potentials, built in the direction of their complication, look like this:

$$B_0^* = B_0(L-t)/t\,;\; B_1^* = B_1(L^2-t^2)/2Lt\,;\; B_2^* = B_2(L^3-t^3)/2L^2t \ldots. \quad (2)$$

In the given expression B_i^* – is a simple potential of an object; B_i – acceleration of the object's energy in the moment of death; L – life duration of the object; t –current time.

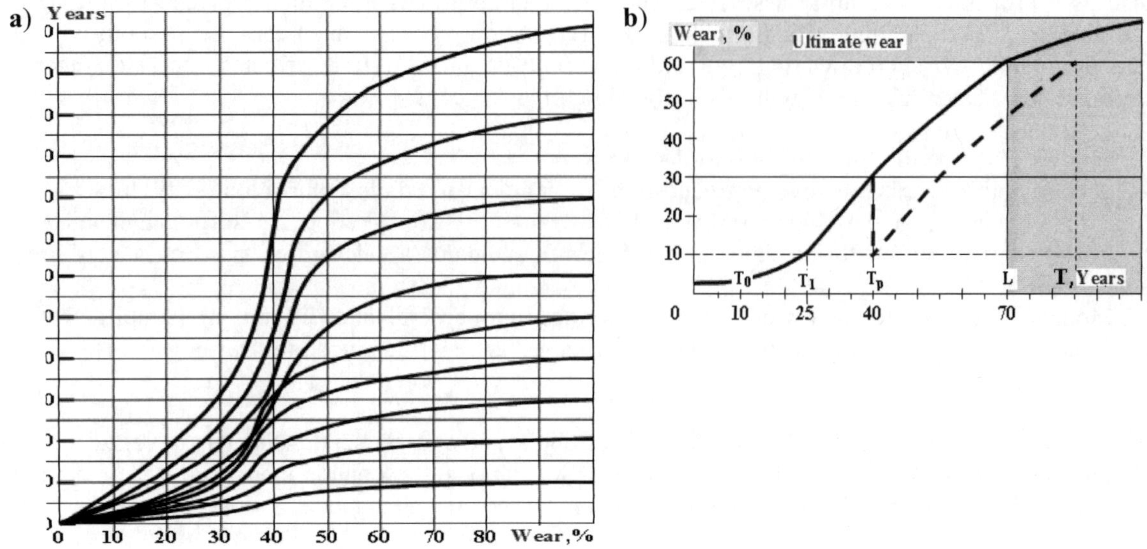

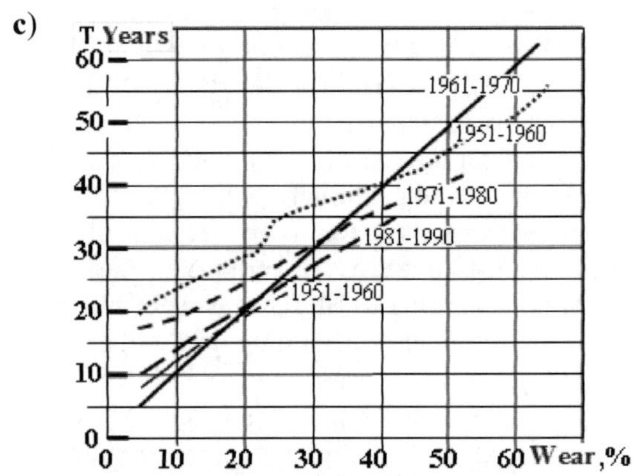

a) assessment of reinforced concrete bridge physical wear (the dotted line shows the restoration of wear) [8]
b) evaluation of deterioration of residential buildings in Moscow [10]
c) wear of reinforced concrete three-layer panels in accordance with BCH [9]

Figure 1. Models of physical deterioration of buildings and structures

The total potential of an object is obtained by summing the potentials of energies. A feature of simple potentials is that for simple potentials a time potential is allocated that does not depend on the energy of the object. In this case it is possible to study the behavior of objects in time, regardless of their cost and volume.

Analysis of the behavior of potentials has also revealed some specific features of energy behavior over time. One of the main features is the existence of the initial zone of "elastic work" of the material (designation a). The zone separates the operation of the object without the occurrence of additional defects. "Elastic zone" characterizes the initial defects of the object. The smaller the "elastic zone" the better the object. In complex objects, such as buildings and structures, the initial defective zone is greater than in individual components of the object.

The simplest form of potential (the first dependence in (2)) is the average of all the ascending and descending forms of potential. Therefore, it can be used to predict the behavior of an object [11]. The actual wear of an object in comparison with the simplest form will be somewhat more or less intensive, but the tendency of its behavior in time will remain.

As already noted in the simple form, the potential does not depend on the value of energy. Therefore, here we see the interrelation of all objects in time, since in the first simple form the behavior of all objects is described by a single curve. For a compound object, the difference will be only in the multipliers used for each component of the object.

4. Approach to the analysis of physical wear of an object

A simple stable interaction potential is taken as the basis of the analysis [12, 13]. It is the basis to construct the dependence of work of object A in time:

$$A = B^{*}(L\,t \cdot \ln t/a - 0{,}5(t^2 - a^2) - (L - a)(t - a)) \quad (3)$$

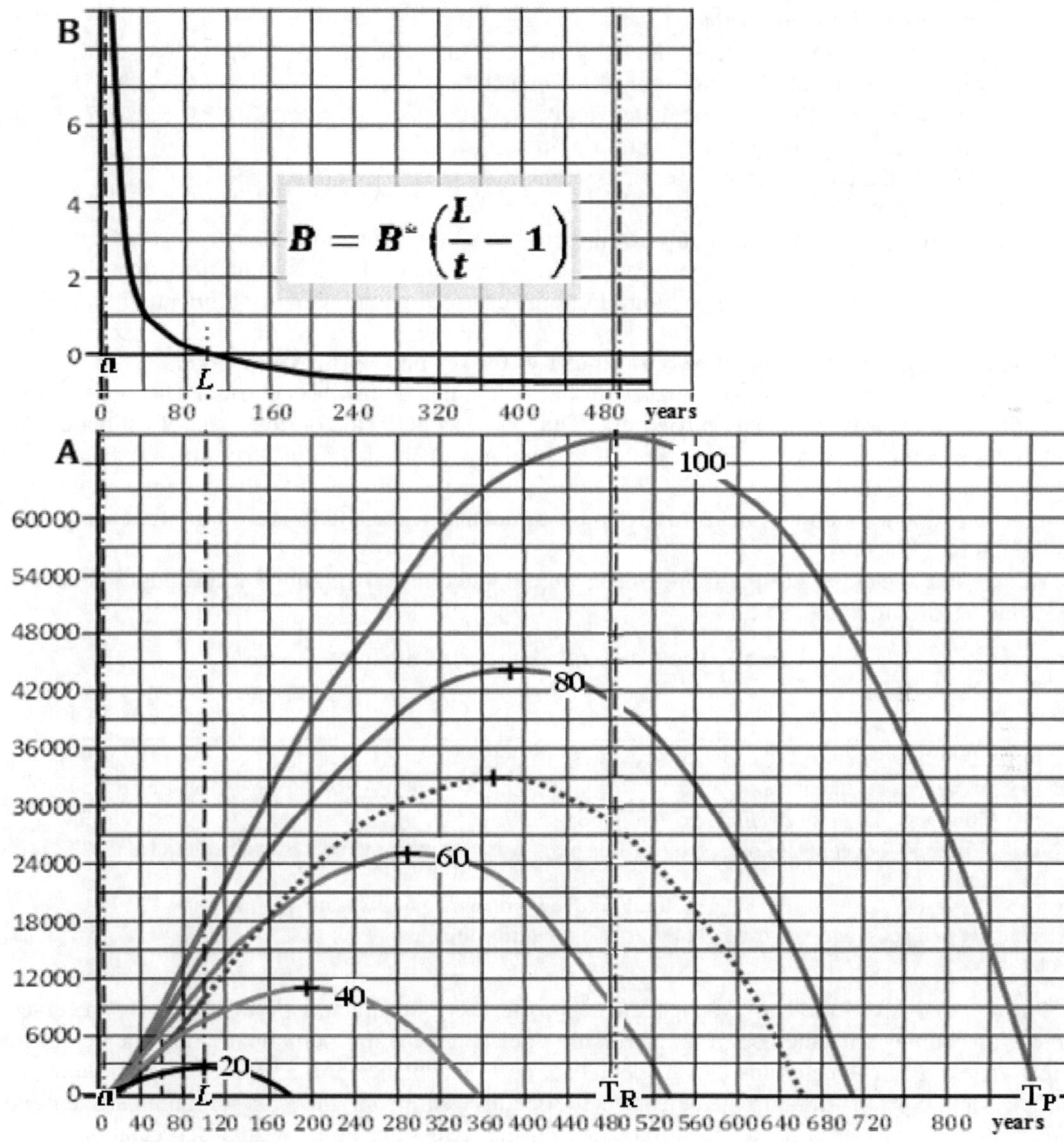

Figure 2. Graphs of potential B and work A of structure

For clarity, Figure 2 shows graphs of the value of B for a lifetime of L = 100 years and A for a lifetime of L = 100, 80, 60, 40 and 20 years. The graphs are constructed with the duration of defect-free operation of an object for 4 years. That means that all defects occurring in the object within 4 years will have less influence on the object than existing defects. The energy characteristic B * is taken equal to one. Therefore, the graphs shown characterize the behavior of the object only in time.

IOP Conf. Series: Earth and Environmental Science **177** (2018) 012040 doi:10.1088/1755-1315/177/1/012040

Fig. 2 shows the work schedule of the object with a lifetime of L = 100 and a duration of operation without defects = 10 years. That means that the number of initial defects in the object is large (compared to a = 4) and is compensated only after 10 years of operation. The decrease in the life span of the object due to initial defects was 30 %. This explains the significant reduction in cost and duration of the object operation due to poor-quality of construction. In a complex facility, which is a building, the initial increased number of defects is more likely than in a simple product. Therefore, the reduction of initial defects is so important in the construction of buildings.

The graphs highlight the indicative time values. They are all interconnected This is the time corresponding to the following values:

L – the lifetime of an object, determined by the comfort and safety of operation, in the present understanding means 100 % physical wear of the object;

a – the time when the effect of initial defects of the object on its durability exceeds the influence of current defects arising during the operation of the object;

Tr – the time when the object fails to resist external influences;

Tp – the time of complete destruction of the object.

The indicative points are shown only for the lifetime of the object of 100 years.

As we can see from the figure, the service life of a product 0.6 ***L*** is much less than the duration of its destruction. That means the need for additional costs for demolishing or renovating the object.

5. Example of physical wear analysis of an object basing on the theory of degradation

As a subject of research, a 16-floor apartment house in Chelyabinsk was chosen. The analysis of cost formation of its construction was carried out. The data was collected on the basis of estimates for the construction with the prices as of 2000, provided by OJSC "Design Institute for Planning and Development of Cities and Settlements" Magnitogorskgrazhdanproekt". To evaluate the physical wear in value terms, a decision was made to group the structures according to their minimum useful life for residential buildings [9].

The structural elements of the frame.house were divided into five groups according to the period of operation according to the table 1.

Table 1. The structural elements of the frame house

Name	Cost th. Rub.
Group 1 (100 years) – reinforced concrete structures	10560.451
Group 2 (80 years) – external walls	5856.531
Group 3 (60 years) – airbricks, staircases, ventilation shafts	889.183
Group 4 (40 years) – openings, floors, interior finishing, partition walls, windows, stained-glass windows, doors	15602
Group 5 (20 years) – exterior finishing, automation, low-voltage devices, heating, ventilation, water supply and sewerage, electric lighting, roofing, interior finishing	12178
Total	45085

For the analysis in the above dependence (3), the energy characteristic of object B * receives the dimension "ruble" with the use of a conversion coefficient for the numerical axis. The actual initial defectiveness **a** can be determined by a special act upon acceptance of the object into operation.

The total cost of structural elements will be different from the cost of an apartment house construction. That has to do with the fact that during assessment of physical depreciation, only the direct costs of structure erection are taken into account, that is, the cost of materials without taking into account labor costs and other additional works. The data in the table are used to construct the graphs shown in Figure 3. The graphs are plotted on the basis of graphs A according to Figure 2, where the energy characteristic of object B * is selected according to the table in accordance with the scale of prices.

6. Conclusion

It is clear from the schedule that after forty years of operation the wear of a building starts to increase rapidly. Slowdown is projected at 80 years of operation. At this level a building's durability is predicted [14, 15]. The maximum wear of a building is projected after 120 years. The greatest impact on the wear of buildings is provided by systems with a short service life and relatively cheap. Therefore, the restoration of small systems of group 5 (shown in dotted lines in the figure) after forty years of operation led to an increase in the service life of the entire building for 30 years. The recovery of systems of Group 6 resulted in an additional increase in the service life by 5 years. Changing of individual components makes it possible to optimize the building operation.

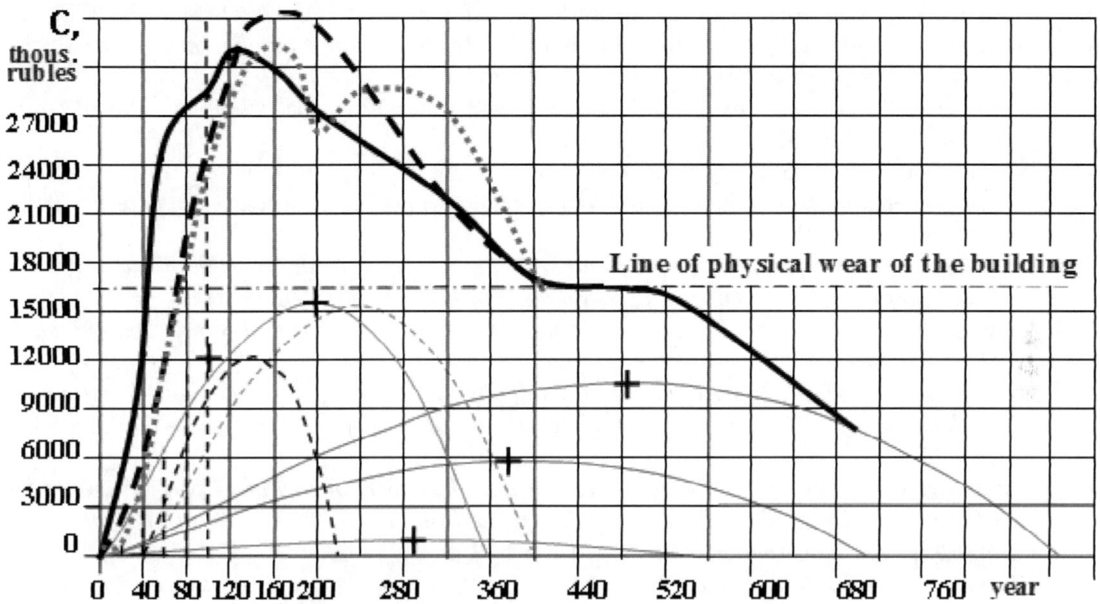

Figure 3. Operation schedule of a complex object

The cumulative schedule shows the physical wear of a complex object.

Within the existing approach, when assessing the physical wear of a building, a linear graph of physical wear can be taken into account.

The upper diagram in Figure 2 can be taken as a simple generalized depreciation diagram of an object.

The considerable value of a falling line of physical wear shows that the existing structures of the building in question (heavy reinforced concrete) significantly increase the cost of construction and unreasonably increase the service life of the building. In fact, some reinforced concrete structures will have even longer service life.

The existing approach to assessing the physical wear of buildings has a more economic approach than a technical one.

In fact, this article demonstrates the simplest approach to assessing the physical wear of an object. Only the diagram of object operation was considered (as is customary with the existing approach). When assessing the wear, it is necessary to consider the whole complex of diagrams, starting with the energy distribution diagram in time, the distribution diagram of potential, the power and operation distribution diagrams of the object.

At first, it is supposed to analyze the degradation behavior of the environment using models of degradation theory. With the accumulation of experience, the model of degradation theory can be used in planning and design of urban systems.

References

[1] Rimshin V I and Kimjaeva T V 2017 *Zhilishchnoe Stroitel'stvo* **12** 8–13
[2] Karpenko N I, Karpenko S N, Yarmakovskiy V N and Erofeev V T 2015 On modern methods of ensuring the durability of reinforced concrete structures *Academia. Architecture and construction* **1** 93–102
[3] Puhonto L M 2004 *Durability of reinforced concrete structures of engineering structures* (Moscow: Publishing House ACB)
[4] Report of RILEM Technical Committee 130-csl *Durabiliti desing of concrete structures* Edited by A. Sarja and E. Vesicary (E & SPON)
[5] Tuuti K 1982 Corrosion of steel in concrete *Swedish Cement and Concrete Institute* (CIB) **4** 82
[6] Piradov K A, Dudchenko K S and Marchenko S M 2005 *Concrete and reinforced concrete – paths of development Plenary reports* (Moscow) **4** 396–99
[7] Andrade K 2005 *Concrete and reinforced concrete – paths of development Plenary reports* (Moscow) **1** 321–30
[8] Shesterikov V I 2005 Forecasting the service life of reinforced concrete structures of road bridges when planning their repair *Concrete and reinforced concrete – paths of development* (Moscow) **5** 78–91
[9] 1990 *Regulations on the organization and carrying out reconstruction, repair and maintenance of buildings, objects of utility and social and cultural appointment* BCH 53–86
[10] Kievskiy I L and Leonov V V 2017 Prediction of physical wear of buildings *Zhilishchnoe Stroitel'stvo* **7** p 17–20
[11] Varlamov A A 2015 *Highly efficient technologies as an integral part of the development of modem society Judgment about the development of technology* (Odessa: KUPRIENKO SV) 106–25
[12] Varlamov A A 2016 *Evaluation of the repair and durability of buildings and structures Safety of the building fund of Russia. Problems and solutions: materials of the international academic readings* (Kursk: Kursk State. un-t.) 153–57
[13] Varlamov A A 2014 *Investigation of changes in elastic deformations of Architecture Building. Education* **2** (4) 73–7
[14] Varlamov A A 2017 Carryng out of researches at an estimation of dependences of the modulus of elasticity from structure of concrete *Architecture. Building. Education* **1** (9) 19–26
[15] Varlamov A A 2017 Evaluation of physical wear and durability of buildings *Bulletin of construction technique* **11** 24–5

IOP Conf. Series: Earth and Environmental Science **177** (2018) 012041 doi:10.1088/1755-1315/177/1/012041

System approach to evaluation of the building heating efficiency using geothermal heat pump in the climatic conditions of Western Ural

Ya I Vaysman[1], A A Surkov[1], Yu I Surkova[1] and A V Kychkin[2]

[1] Institute for the Environmental Protection, Perm National Research Polytechnic University, 29 Komsomolsky prospect, Perm, 614990, Russia
[2] Institute for the Automation Microprocessors, Perm National Research Polytechnic University, 29 Komsomolsky prospect, Perm, 614990, Russia

e-mail: alex.a.surkov@gmail.com

Abstract. The current paper is devoted to the estimation the effectiveness of the application of the geothermal heat pump application in the climatic conditions of Western Ural. The system approach for assessing the efficiency of the heating system of the research unit (model of a residential house) on the basis of the heat pump is proposed. This approach involves multi-criteria analysis of the facility operation based on the data of field experiments and simulation results, as well as evaluating the economic efficiency of the heat pump as compared with traditional heat sources.

1. Introduction

According to the current economic estimates, it is considered that in countries with a cold climate, to which Russia belongs, it is appropriate to consider the use of geothermal heat pumps (GHP) using ground as a source of low-potential heat. Such pumps are well proven for year-round use. Heating of the building with GHP involves the use of thermal energy stored in the ground or in water with a small consumption of electrical energy. Since the soil temperature is virtually constant throughout the year, the Heat Transfer Coefficient Conversion (HTCC) throughout the year is a constant value ranging from 3 to 4.5.

In Perm National Research Polytechnic University (PNRPU) in cooperation with the Ministry of Industry of Perm Krai, a project of the creation of the energy-efficient autonomous research module (EEARM) with a GHP and a building dispatching system was developed and implemented. In this building, the high-tech development of the PNRPU in the field of energy and resource saving is concentrated, which makes it possible to use this facility to develop full-scale experiments of the renewable energy sources use, including GHP, in the climatic conditions of the Western Urals (Figure 1) [1].

Content from this work may be used under the terms of the Creative Commons Attribution 3.0 licence. Any further distribution of this work must maintain attribution to the author(s) and the title of the work, journal citation and DOI.
Published under licence by IOP Publishing Ltd

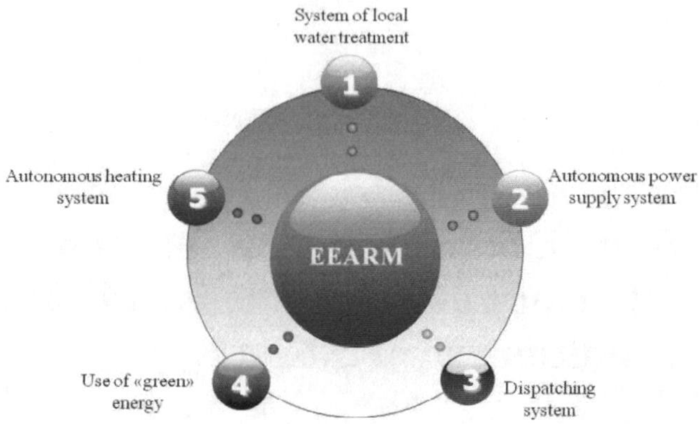

Figure 1. External appearance and engineering systems of the research module

The structural design of the research module is a frame with the use of glued wooden structures (GWS) as supporting columns, beams and floor panels of the full factory assembly. External walls are frame, filled with mineral wool Rockwool Light Batts (Figure 1). Windows are panoramic, wood-aluminum European standard (Euro-windows of the 4th generation) with tempered, low-emission glass. Roof is pitched and is made from profiled sheet. The total area of the building is 200 m^2. Taking into account the heat losses, which amounted to 15 kW, the building on energy efficiency corresponds to class B (high).

Calculated outdoor temperatures for the design of the heating system were adopted in accordance with SNiP 23-01-2003 "Construction climatology": for the cold period – To = -35 ºC; for the warm period – To = 21.8 ºC; duration of the heating period is 229 days.

Functionally, the module is divided into several units: meeting room, showroom, lecture room, boiler room, server room.

A system approach to assessing the efficiency of heating EEARM with the use of a geothermal heat pump (GHP) provides for a multicriteria analysis of the building functioning from different points of view, combined with an integral indicator of energy efficiency, based on the results of field experiments and simulation results [2].

2. Formalization of the task of energy-efficient management of the heat supply system for EEARM

Let's describe the models of heat supply control for each *i-th* building premise at given environmental parameters – E_i and room parameters I_i with the following vector:

$$M = \left\{ f_{ij} \left(E_i(t), I_i(t) \right), i = \overline{1,m} \right\} , \quad (1)$$

where f_{ij} – function that determines the dependence of the control parameters for a given *j-th* temperature in the *i-th* room; *m* – number of heated premises in the building.

We shall establish the limitations on the thermal energy supply at each instant of time on the interval *T*:

$$\forall t \in T : Q_i \in \left[\underline{Q}_i, \overline{Q}_i \right], i = \overline{1,m} \quad (2)$$

and the total limit on the amount of energy used:

$$\sum_{i=1}^{m} \int_{t \in T_i} Q_i(t) dt \leq Q_0 , \quad (3)$$

where $\underline{Q}$, $\overline{Q}$ - lower and upper limits of the supplied heat energy to the room; Q_0 – maximum volume of thermal energy of all sources in the T interval of time, $T_i \subseteq T$ – time interval of heating of the *i-th* room.

Let`s write down the control parameters that are required to solve the problem of energy-efficient heating control of the *i-th* room at the *j-th* indoor air temperature, as vector P:

$$P = \left\{ P_{ij}, i = \overline{1,n} \right\}, \tag{4}$$

where n – number of values of the air temperature in the room.

The target value of the indoor temperature at the time interval under consideration can be written as:

$$Temp = \left\{ Temp_j, i = \overline{1,n} \right\} \tag{5}$$

The results of the control functions implementation for models M (1) using S scenarios for given constraints (2) and (3) will be described by the following tuple: $K =< M, S, Q >; i = \overline{1,m}; j = \overline{1,n}$. In this case, the scenarios describe the algorithm for solving the problem of supplying the coolant to the building premises at the target air temperature values (5).

At the same time, the optimal result of the EEARM heat supply system management will be achieved with minimal energy costs:

$$K^* = \sum_{i=1}^{m} \int_{t \in T_i} Q_i \left(t; P_{ij} \right) dt \to \min(Q_i) \tag{6}$$

The optimal distribution of heat energy through the premises is achieved due to the established control parameters P_{ij}^* described by the most rational scenario S*.The implementation of this scenario can be achieved with the help of heat management controllers using expert knowledge that describes the nonlinear dependencies of the target temperature values and the amount of thermal energy used for the specified building parameters [3, 4].

3. Management of the heat supply system of the EEARM and monitoring of its operation

To ensure the objective function (6) with the help of the built-in automation system, scenarios for controlling the building heating according to the coolant temperature and the indoor temperature were realized.

The “indoor temperature” control script provides minimum heat energy consumption in relation to other management scenarios, including “season management”, “day-time management”, “day-of-week management”, “coolant temperature control”. In this case, to ensure a constant set temperature within the control room, the heat transfer temperature of the system is controlled by a heat pump. When the set temperature in the room is reached, the heating is reduced and the set air temperature is maintained. This controls the temperature of the coolant at a constant pump output. Such a system effectively operates at different ambient temperatures outside the heated room. However, with good thermal insulation of the room, there will be inertia, since the sensor inside the room will later react to the temperature change from the outside.

The relative savings in the consumption of heat energy for heating the premises during the heating period can be calculated by the following expression:

$$\Delta Q = \frac{|Q_1 - Q_2|}{Q_1} \cdot 100\% \quad , \qquad (7)$$

where $Q_1 = L \cdot c \cdot (t_0 - t_н)$ – heat consumption, W; L – volume flow of coolant, m^3/h; c – specific mass heat capacity of coolant, J/(kg°C); $t_п$ – cooled coolant temperature, °C; t_0 – temperature of the heat medium supplied to the heat pump, °C.

To assess the operation of the heat supply system of the research module with the embedded algorithm, data were collected from the coolant temperature sensors in the inlet ($t_п$) and output (t_0) piping of the heat pump and the environment (t_{oc}).

From 01.03.2017 to 13.03.2017 and from 19.03.2017 to 02.05.2017 the monitoring of the implementation of the scenario "control over coolant temperature" was carried out. The results are presented in Figure 2 and 3. He volumetric flow of the heat carrier is taken as a constant value. The data visualization is performed in the OpenJEVis – a specialized software product for processing and visual analysis of energy data [5, 6].

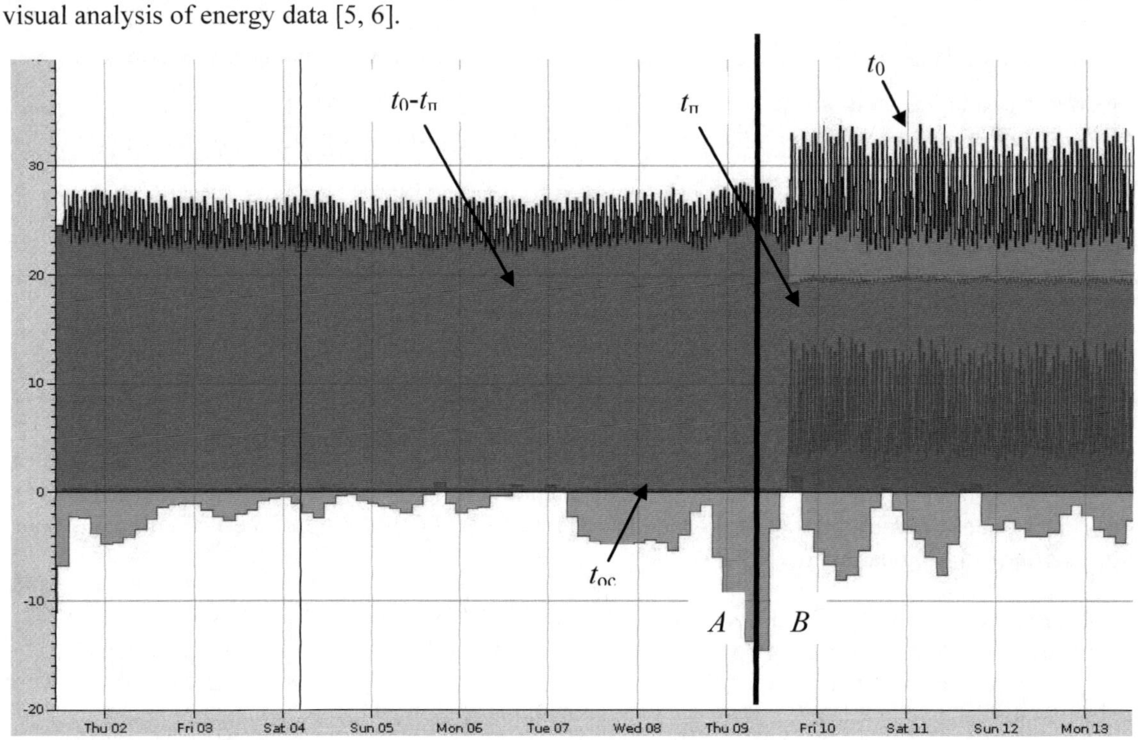

Figure 2. Monitoring results of coolant temperature and ambient temperature, interval A – control scenario disabled, interval B – scenario "control by coolant temperature"

Figure 2 shows that the temperature change (t_0-$t_П$) when the script does not occur, i.e. thermal energy is continuously fed into the room with a constant value. Besides, according to Figure 3, it is evident that the temperature change (t_0-$t_П$) occurs at the time when t_{oc} is below zero, which corresponds to the control scenario. If, in the first scenario, the temperature difference was approximately 20 °C at an ambient air temperature of 7 °C, then with the activation of the second scenario the difference is only 3 °C at the same external temperature.

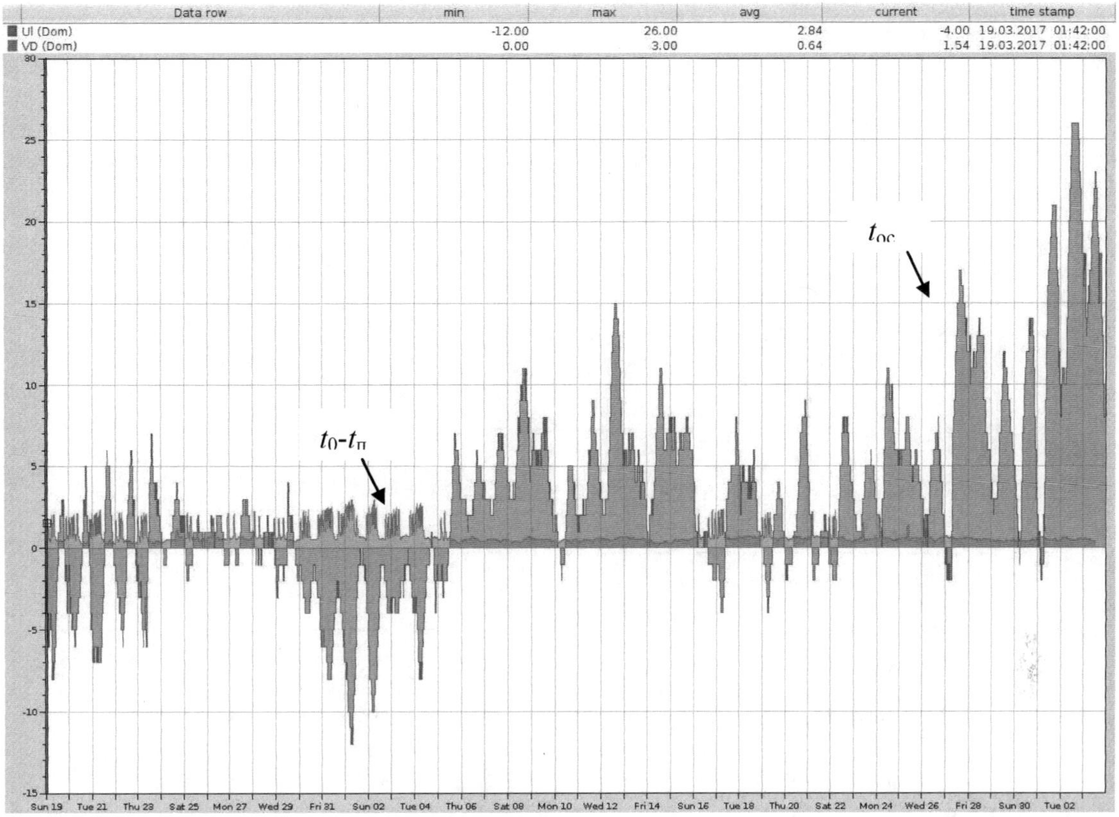

Figure 3. Results of monitoring the temperatures of the coolant and the environment, the control scenario "on the temperature inside the premises"

Taking into account the constant flow of the heat carrier and using expression (7), it is easy to estimate that the use of the second scenario gives an almost 7-fold effect on saving thermal energy under the given conditions of experimental studies. In particular, for the sampling of the statistical data presented in Figure 2, using the expression (7), the relative savings of the heat consumption during the operation of the "coolant temperature control" scenario reaches 23 %.

4. Calculation of the effectiveness of the use of GHP from the economic point of view

To calculate the economic efficiency of the heat pump in the long term, an analysis of the growth in tariffs for electricity and gas supply for the period 2010–2017 was made. The average value of relative growth rates for electricity was – 7.17 % for the gas – 8.35 %. These values were used for further calculations as indicators of the average annual growth in the cost of tariffs for electricity and gas.

Calculation of heating costs was carried out for two scenarios: first – 100 % heating of the building's areas, the second – reducing the temperature in the unattended building premises to 9 °C, the total area of which reaches 50 % of the total area.

As a result of the analysis of the data obtained, it can be concluded that, from the point of view of economic efficiency, the use of an electric boiler for a heat supply system is inexpedient; the costs for a heating period will be several times higher than using a heat pump or a gas boiler. If there is an opportunity to connect to gas supply, the heat supply system based on a gas boiler is the most cost-effective source of heat supply. At the same time, a heating system with a heat pump produces a good economic effect, is safer compared to gas, and environmentally friendly.

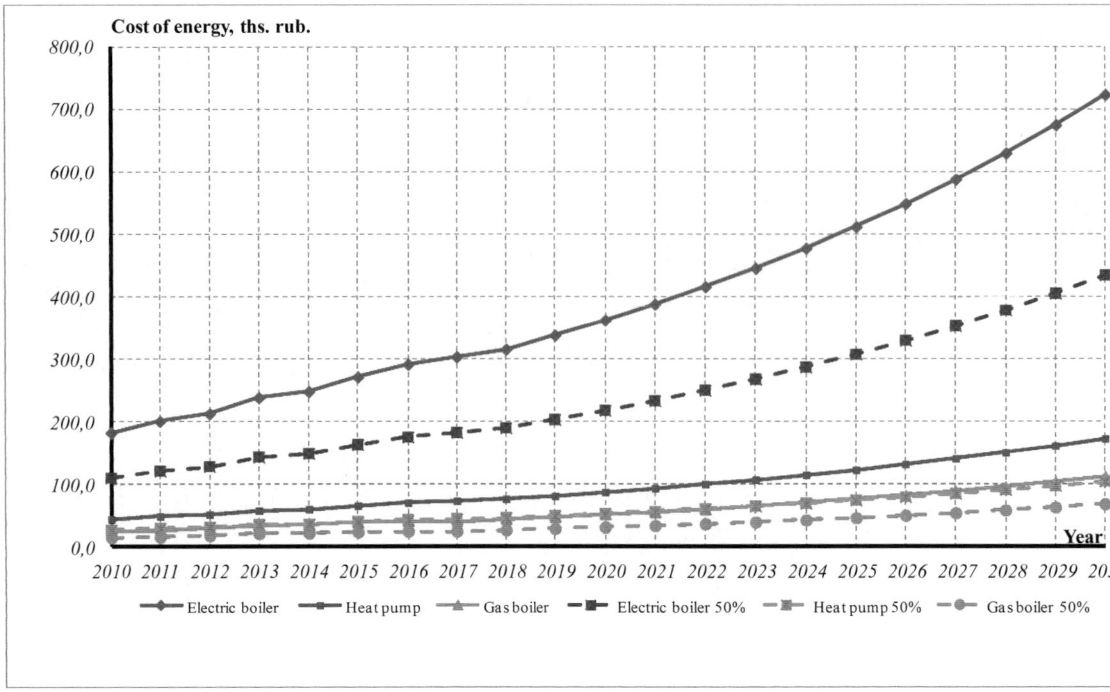

Figure 4. Dynamics of heating costs increase for EEARM in 2010–2018, and the forecast for 2019–2030

5. Conclusion

Based on the results obtained, the following conclusions can be drawn:

1. The implementation of the scenario "control by coolant temperature" gives a relative saving of heat consumption up to 23 %.

2. The use of the building dispatching system in the management of the heating system allows efficient distribution of heat energy through the premises for various purposes with the provision of reduced internal temperatures in 50 % of the premises, which allows to reduce the integral energy consumption of the building by at least 30 %.

3. With the further growth of tariffs for energy carriers, in particular gas, the use of a heat pump can become an alternative to traditional energy sources, in particular, the cost of energy supply of 50% of the building volume with a heat pump is comparable to the cost of full-scale heating using a gas boiler. In this case, the heat pump is a safer source of heat.

Acknowledgements

This work was supported by PNRPU grant, agreement No. 2016/PI-2 "Development of a methodology for monitoring and utilization of the heat flows as low-grade resource of the enterprise".

References

[1] Lyakhomskii A V, Perfil'eva E N, Kychkin A V and Genrikh N A 2015 A software-hardware system of remote monitoring and analysis of the energy data *Russian Electrical Engineering* **86** (6) 314–19

[2] Faizrakhmanov R A, Frank T, Kychkin A V and Fedorov A B 2011 Sustainable energy consumption control using the My-JEVis energy management data system *Russian Electrical Engineering* **82** (11) 607–11

[3] Kychkin A V 2016 Synthesizing a System for Remote Energy Monitoring in Manufacturing, *Metallurgist* **59** (9–10) 752–60

[4] Kychkin A V and Mikriukov G P 2016 Applied data analysis in energy monitoring system *Regional Energy Problems* **2** (31) 84–92

[5] Noskov V and Kychkin A 2017 Data Analysis and Visualization for Industrial Enterprise Water Supply System *Proceedings of the 5th International Conference on Applied Innovations in IT* 113–18

[6] Vaysman Ya I, Surkov A A, Surkova Yu I and Kychkin A V 2017 Simulation Model for Assessing the Efficiency of a Combined Power Installation Based on a Geothermal Heat Pump and a Vacuum Solar Collector *IOP Conference Series: Earth and Environmental Science:* 2 *International Conference on Sustainable* Cities **72** 012018

Published under licence in *Journal Title* by IOP Publishing Ltd.